Fixed Point Theory and Its Related Topics II

Fixed Point Theory and Its Related Topics II

Editor

Hsien-Chung Wu

MDPI • Basel • Beijing • Wuhan • Barcelona • Belgrade • Manchester • Tokyo • Cluj • Tianjin

Editor
Hsien-Chung Wu
National Kaohsiung Normal
University
Taiwan

Editorial Office
MDPI
St. Alban-Anlage 66
4052 Basel, Switzerland

This is a reprint of articles from the Special Issue published online in the open access journal *Axioms* (ISSN 2075-1680) (available at: https://www.mdpi.com/journal/axioms/special_issues/fixed_point_theory_relatedII).

For citation purposes, cite each article independently as indicated on the article page online and as indicated below:

LastName, A.A.; LastName, B.B.; LastName, C.C. Article Title. *Journal Name* **Year**, *Volume Number*, Page Range.

ISBN 978-3-0365-2173-2 (Hbk)
ISBN 978-3-0365-2174-9 (PDF)

© 2021 by the authors. Articles in this book are Open Access and distributed under the Creative Commons Attribution (CC BY) license, which allows users to download, copy and build upon published articles, as long as the author and publisher are properly credited, which ensures maximum dissemination and a wider impact of our publications.

The book as a whole is distributed by MDPI under the terms and conditions of the Creative Commons license CC BY-NC-ND.

Contents

About the Editor vii

Preface to "Fixed Point Theory and Its Related Topics II" ix

Yaé Ulrich Gaba, Hassen Aydi and Nabil Mlaiki
(ρ, η, μ)-Interpolative Kannan Contractions I
Reprinted from: *Axioms* **2021**, *10*, 212, doi:10.3390/axioms10030212 1

Helga Fetter Nathansky and Jeimer Villada Bedoya
Cascading Operators in CAT(0) Spaces
Reprinted from: *Axioms* **2021**, *10*, 20, doi:10.3390/axioms10010020 9

Hsien-Chung Wu
Using the Supremum Form of Auxiliary Functions to Study the Common Coupled Coincidence Points in Fuzzy Semi-Metric Spaces
Reprinted from: *Axioms* **2021**, *10*, 5, doi:10.3390/axioms10010005 19

Ehsan Lotfali Ghasab, Hamid Majani, Manuel De la Sen and Ghasem Soleimani Rad
e-Distance in Menger PGM Spaces with an Application
Reprinted from: *Axioms* **2021**, *10*, 3, doi:10.3390/axioms10010003 51

Youssef Errai, El Miloudi Marhrani and Mohamed Aamri
Fixed Points of *g*-Interpolative Ćirić–Reich–Rus-Type Contractions in *b*-Metric Spaces
Reprinted from: *Axioms* **2020**, *9*, 132, doi:10.3390/axioms9040132 59

Nopparat Wairojjana, Nuttapol Pakkaranang, Habib ur Rehman, Nattawut Pholasa and Tiwabhorn Khanpanuk
Strong Convergence of Extragradient-Type Method to Solve Pseudomonotone Variational Inequalities Problems
Reprinted from: *Axioms* **2020**, *9*, 115, doi:10.3390/axioms9040115 73

Chinda Chaichuay and Atid Kangtunyakarn
The Split Various Variational Inequalities Problems for Three Hilbert Spaces
Reprinted from: *Axioms* **2020**, *9*, 103, doi:10.3390/axioms9030103 89

Nopparat Wairojjana, Habib ur Rehman Manuel De la Sen and NuttapolPakkaranang
A General Inertial Projection-Type Algorithm for Solving Equilibrium Problem in Hilbert Spaces with Applications in Fixed-Point Problems
Reprinted from: *Axioms* **2020**, *9*, 101, doi:10.3390/axioms9030101 105

Godwin Amechi Okeke, Mujahid Abbas and Manuel de la Sen
Inertial Subgradient Extragradient Methods for Solving Variational Inequality Problems and Fixed Point Problems
Reprinted from: *Axioms* **2020**, *9*, 51, doi:10.3390/axioms9020051 129

Mujahid Abbas, Fatemeh Lael and Naeem Saleem
Fuzzy b-Metric Spaces: Fixed Point Results for ψ-Contraction Correspondences and Their Application
Reprinted from: *Axioms* **2020**, *9*, 36, doi:0.3390/axioms9020036 153

Hassen Aydi, Zoran D. Mitrović, Stojan Radenović and Manuel de la Sen
On a Common Jungck Type Fixed Point Result in Extended Rectangular b-Metric Spaces
Reprinted from: *Axioms* **2020**, 9, 4, doi:10.3390/axioms9010004 . . . . **165**

About the Editor

Hsien-Chung Wu is a Professor of the Department of Mathematics, National Kaohsiung Normal University University, Taiwan. He is the sole author of more than 120 scientific papers published in international journals. He is an area editor of the *International Journal of Uncertainty, Fuzziness and Knowledge-Based Systems* and an associate editor of *Fuzzy Optimization and Decision Making*. His current research includes the nonlinear analysis in mathematics and the applications of fuzzy sets theory in operations research.

Preface to "Fixed Point Theory and Its Related Topics II"

This book contains the successful submissions to a Special Issue of *Axioms* on the subject area of "Fixed Point Theory and Related Topics". Fixed point theory arose from the Banach contraction principle and has been studied for a long time. Its application mostly relies on the existence of solutions to mathematical problems that are formulated from economics and engineering. Fixed points of functions depend heavily on the considered spaces that are defined using the intuitive axioms. Different spaces will result in different types of fixed point theorems

Hsien-Chung Wu
Editor

(ρ, η, μ)-Interpolative Kannan Contractions I

Yaé Ulrich Gaba [1,2,3,4,*], Hassen Aydi [4,5,6,*] and Nabil Mlaiki [7]

1. Institut de Mathématiques et de Sciences Physiques (IMSP/UAC), Laboratoire de Topologie Fondamentale, Computationnelle et leurs Applications (Lab-ToFoCApp), Porto-Novo BP 613, Benin
2. African Center for Advanced Studies (ACAS), P.O. Box 4477, Yaounde 7535, Cameroon
3. Quantum Leap Africa (QLA), AIMS Rwanda Centre, Remera Sector, Kigali KN 3, Rwanda
4. Department of Mathematics and Applied Mathematics, Sefako Makgatho Health Sciences University, P.O. Box 60, Ga-Rankuwa 0208, South Africa
5. Institut Supérieur d'Informatique et des Techniques de Communication, Université de Sousse, H. Sousse 4000, Tunisia
6. China Medical University Hospital, China Medical University, Taichung 40402, Taiwan
7. Department of Mathematics and General Sciences, Prince Sultan University, P.O. Box 66833, Riyadh 11586, Saudi Arabia; nmlaiki@psu.edu.sa
* Correspondence: yaeulrich.gaba@gmail.com (Y.U.G.); hassen.aydi@isima.rnu.tn (H.A.)

Abstract: We point out a vital error in the paper of Gaba et al. (2019), showing that a (ρ,η,μ) interpolative Kannan contraction in a complete metric space need not have a fixed point. Then we give an appropriate restriction on a (ρ,η,μ)-interpolative Kannan contraction that guarantees the existence of a fixed point and provide an equivalent formulation. Moreover, we show that this formulation can be extended to the interpolative Reich-Rus-Ćirić type contraction.

Keywords: (ρ,η,μ)-interpolative Kannan contraction; fixed point; metric space.

MSC: Primary 47H05; Secondary 47H09, 47H10

1. Introduction and Preliminaries

A mapping T on a metric space (X, d) is called Kannan if there exists $\lambda \in [0, \frac{1}{2})$ such that
$$d(Tx, Ty) \leq \lambda [d(x, Tx) + d(y, Ty)], \quad \text{for all } x, y \in X.$$

Kannan [1] proved that if X is complete, then a Kannan mapping admits a fixed point. Please note that this well-known Kannan contraction that does not require a continuous mapping. Recently, Karapinar [2] proposed a new Kannan-type contractive mapping via the notion of interpolation and proved a fixed point theorem over metric space. The interpolative method has been used by several researchers to obtain generalizations of other forms of contractions [3–5]. This notion of interpolative contractions gives directions to investigate whether existing contraction inequalities can be redefined in this way or not. The purpose of this paper is to revisit the approach to attain a more general and less restrictive formultion of Karapinar's result [2]. Some examples are given to illustrate the new approach.

Throughout this manuscript, we denote an interpolative Kannan contraction by IKC and a (ρ, η, μ)-interpolative Kannan contraction by (ρ, η, μ)-IKC. The main result of Karapinar [2] is as follows:

Theorem 1. ([2] Theorem 2.2)
Let (X, d) be a complete metric space and $T : X \to X$ be an interpolative Kannan type contraction, i.e., a self-map such that there are $\rho \in [0, 1)$ and $\eta \in (0, 1)$ so that

$$d(Ta, Tb) \leq \rho d(a, Ta)^\eta d(b, Tb)^{1-\eta} \qquad (1)$$

for all $a, b \in X$ with $a \neq Ta$. Then T has a unique fixed point in X.

This theorem has been generalized in 2019 by Gaba et al. [6], where they initiated the notion of (ρ, η, μ)-IKCs. In [6], the authors have defined (ρ, η, μ)-IKC and proved a fixed point theorem for such mappings. The definition of that mapping is given as follows:

Definition 1. *(See [6]) Let (X, d) a metric space and $T : X \to X$ be a self-map. We shall call T a (ρ, η, μ)-IKC or GIK (Gaba Interpolative Kannan) contraction, if there exist $0 \leq \rho < 1$ and $0 < \eta, \mu < 1$ with $\eta + \mu < 1$ such that*

$$d(Ta, Tb) \leq \rho\, d(a, Ta)^\eta d(b, Tb)^\mu \qquad (2)$$

whenever $a \neq Ta$ and $b \neq Tb$.

Theorem 2. *(See [6]) Let (X, d) be a complete metric space and $T : X \to X$ be a (ρ, η, μ)-IKC. Then T has a fixed point in X.*

The interpolative strategy has been successfully applied to a variant types of contractions (see [7,8]). One of our goals in this paper is to show that Theorem 2 has a gap by giving an illustrated example. We will also give its proof correctly.

2. An Error in the Fixed Point Theorem for GIK Contractions

Theorem 2 is not true in general. The next example proves our assertion.

Example 1. *Let $X = \left\{\frac{1}{4}, \frac{1}{6}\right\}$ be endowed with the usual metric and $T : X \to X$ be given as*

$$T\left(\frac{1}{4}\right) = \frac{1}{6};\ T\left(\frac{1}{6}\right) = \frac{1}{4}.$$

We have:

$$0.0833 = \left|\frac{1}{4} - \frac{1}{6}\right| \leq \frac{3}{5} \cdot \left|\frac{1}{4} - \frac{1}{6}\right|^{(1/3)} \cdot \left|\frac{1}{6} - \frac{1}{4}\right|^{(1/3)} = 0.1144.$$

Hence, T is a GIK contraction with $\rho = \frac{3}{5}$ and $\eta = \mu = \frac{1}{3}$. Here, X is complete, but T has no fixed point in X.

In the proof of Theorem 2 proposed by Gaba et al. in [6], the vital error emanated from the fact that the inequality, for the real numbers a, η, μ such that $0 < \eta \leq \mu$:

$$a^\eta \leq a^\mu$$

holds if and only if $a \geq 1, \eta \leq \mu$.

3. Revisiting the GIK Contraction Fixed Point Theorem

We provide an alternative formulation to the existence of (ρ, η, μ)-IKCs.

Theorem 3. *(GIK fixed point revisited) Let (X, d) be a complete metric space such that $d(a, b) \geq 1$ for $a \neq b$ and $T : X \to X$ be a GIK contraction. Then T has a fixed point in X.*

Proof. Following the steps of the proof of [2] (Theorem 2.2), we build the sequence $(a_n)_{n \geq 1}$ of iterates $a_n = T^n a_0$, where $a_0 \in X$ is an arbitrary starting point. Without loss of generality, making the hypothesis that $a_{n+1} \neq a_n$ for each nonnegative integer n, we observe that

$$d(a_n, a_{n+1}) = d(Ta_{n-1}, Ta_n) \leq \rho\, d(a_{n-1}, a_n)^\eta\, d(a_n, a_{n+1})^\mu,$$

i.e.,
$$d(a_n, a_{n+1})^{1-\mu} \leq \rho\, d(a_{n-1}, a_n)^{\eta} \leq \rho\, d(a_{n-1}, a_n)^{1-\mu}$$

since $\eta < 1 - \mu$ and $d(a_n, a_{n+1}) \geq 1$.

Similar to the proof of [2] (Theorem 2.2), the usual strategy ensures that there is a unique fixed point $a^* \in X$. □

Example 2. *(See [6] Example 1.) Take $X = \{a, b, z, w\}$. We equip with metric:*

	a	b	z	w
a	0	5/2	2	5/2
b	5/2	0	3/2	1
z	4	3/2	0	3/2
w	5/2	1	3/2	0

Consider on X the self-map T given as $Ta = a$, $Tb = w$, $Tz = a$ and $Tw = b$. We observed that the inequality:

$$d(Ta, Tb) \leq \rho d(a, Ta)^{\eta} d(b, Tb)^{\mu}$$

is satisfied for:

$$\eta = \frac{1}{8},\ \mu = \frac{3}{4},\ \rho = \frac{8}{9} \leq \frac{9}{10};$$

$$\eta = \frac{1}{9},\ \mu = \frac{3}{4},\ \rho = \frac{8}{9} \leq \frac{9}{10};$$

$$\eta = \frac{1}{8},\ \mu = \frac{4}{5},\ \rho = \frac{8}{9} \leq \frac{9}{10}.$$

In all above cases, $\eta + \mu < 1$, i.e., $\mu < 1 - \eta$ and the hypotheses of Theorem 3 are satisfied. Moreover, the map clearly possesses a unique fixed point.

On the other hand, when a metric d is such that $d(a, b) \geq 1$ whenever $a \neq y$, the inequality

$$d(Ta, Tb) \leq \rho d(a, Ta)^{\eta} d(b, Tb)^{1-\eta}$$

could just be replaced by the existence of two reals η, μ so that $\eta + \mu < 1$,

$$d(Ta, Tb) \leq \rho d(a, Ta)^{\eta} d(b, Tb)^{\mu}.$$

4. Equivalent GIK Formulations

Let (X, d) be a metric space. Denote by $\Gamma(GIK)$ the set of all GIK contractions on X. For a mapping $T : X \to X$, T is an s-GIK contraction if there are $0 \leq \rho < 1, 0 < \eta, \mu < 1$ with $\eta + \mu < 1$ so that

$$d(Ta, Tb) \leq \rho\, [d(a, Ta) d(b, Tb)]^{\frac{\eta+\mu}{2}}$$

whenever $a \neq Ta, b \neq Tb$.

Let us denote by $\tilde{\Gamma}(GIK)$ the set of all s-GIK contractions on X.

Theorem 4. *In a metric space (X, d), such that $d(a, b) \geq 1$ for $a \neq b$, we have the equality*

$$\Gamma(GIK) = \tilde{\Gamma}(GIK).$$

Proof. Clearly,

$$\tilde{\Gamma}(GIK) \subset \Gamma(GIK)$$

since for any s-GIK contraction T, one has

$$d(Ta, Tb) \leq \rho \left[d(a, Ta)d(b, Tb)\right]^{\frac{\eta+\mu}{2}} \iff d(Ta, Tb) \leq \rho \, d(a, Ta)^{\frac{\eta}{2}} d(b, Tb)^{\frac{\mu}{2}}$$

and

$$\frac{\eta}{2} + \frac{\mu}{2} = \frac{\eta+\mu}{2} < \eta + \mu < 1.$$

Now, let $T \in \Gamma(GIK)$, so there are $0 \leq \rho < 1$ and $0 < \eta, \mu < 1$ with $\eta + \mu < 1$ so that

$$d(Ta, Tb) \leq \rho d(a, Ta)^\eta d(b, Tb)^\mu \tag{3}$$

whenever $a \neq Ta, b \neq Tb$.

Additionally, due to symmetry,

$$d(Ta, Tb) = d(Tb, Ta) \leq \rho d(b, Tb)^\eta d(a, Ta)^\mu. \tag{4}$$

Multiplying the inequalities (3) and (4), it follows that

$$d(Ta, Tb) \leq \rho \left[d(a, Ta)d(b, Tb)\right]^{\frac{\eta+\mu}{2}}. \tag{5}$$

□

So far, in our discussions regarding GIK contractions, we overlooked the case where $\eta + \mu = 1$. This case is actually central in the present investigation. Indeed, in the definition of a (ρ, η, μ)-IKC, if we allow the sum $\eta + \mu$ to attain 1, one can see that the IKC in the sense of Karapinar [2] is a particular case of a GIK. In particular, we have:

Definition 2. *Let (X, d) a metric space and $T : X \to X$ be a self-map. T is called an extended (ρ, η, μ)-IKC or extended GIK contraction, if there are $0 \leq \rho < 1$ and $0 < \eta, \mu < 1$ with $\eta + \mu \leq 1$ so that*

$$d(Ta, Tb) \leq \rho \, d(a, Ta)^\eta d(b, Tb)^\mu \tag{6}$$

whenever $a \neq Ta$ and $b \neq Tb$.

For a metric space (X, d), let's denote by $e\text{-}\Gamma(GIK)$ the set of all extended GIK contractions on X. Moreover, if $\Gamma(IK)$ denotes the set of all interpolative Kannan type contractions, it is clear that:

Corollary 1. *In a metric space (X, d), such that $d(a, b) \geq 1$ for $a \neq y$, we have*

$$\Gamma(IK) \subset e\text{-}\Gamma(GIK).$$

For a mapping $T : X \to X$, T is an s-GIK contraction if there are $0 \leq \rho < 1$ and $0 < \eta, \mu < 1$ with $\eta + \mu < 1$ so that

$$d(Ta, Tb) \leq \rho \left[d(a, Ta)d(b, Tb)\right]^{\frac{\eta+\mu}{2}}$$

whenever $a \neq Ta$ and $b \neq Tb$.

Furthermore, if we plug $\eta + \mu = 1$ in (5), we achieve

$$d(Ta, Tb) \leq \rho \left[d(a, Ta)d(b, Tb)\right]^{\frac{1}{2}}, \tag{7}$$

which naturally leads to

Corollary 2.

$$T : X \to X \in \Gamma(IK)$$

$T : X \to X$ and there is $0 \leq \rho < 1$ so that

$$d(Ta, Tb) \leq \rho \left[d(a, Ta)d(b, Tb)\right]^{\frac{1}{2}}$$

whenever $a \neq Ta$ and $b \neq Tb$.

5. GI-RRC Contractions

As an extension of interpolative Kannan-type contractive mappings, Karapinar et al. introduced Interpolative Reich-Rus-Ćirić type contractions (see [9]). The definition is given below:

Definition 3. ([9]) In a metric space (X, m), a mapping $T : X \to X$ is called an interpolative Reich-Rus-Ćirić type contraction if it satisfies

$$m(Ta, Tb) \leq \rho [d(a, y)]^{\mu} [m(a, Ta)]^{\eta} [m(b, Tb)]^{1-\eta-\mu}$$

for all $a, b \in X \setminus \mathrm{Fix}(T) = \{\sigma \in X : T\sigma = \sigma\}$ for some $\rho \in [0, 1)$ and for $\eta, \mu \in (0, 1)$.

Theorem 5. ([9]) Let (X, d) be a complete metric space and $T : X \to X$ be an interpolative Reich-Rus-Ćirić type contraction mapping. Then T has a fixed point in X.

In the present paper, we introduce the concept of $(\rho, \eta, \mu, \gamma)$-interpolative Reich-Rus-Ćirić type contractions, which we also call GI-RRC contractions.

Definition 4. Let (X, d) a metric space and $T : X \to X$ be a self-map. T is named a (ρ, η, μ)-Reich-Rus-Ćirić contraction or GI-RRC (Gaba Interpolative Reich-Rus-Ćirić) contraction, if there exist $0 \leq \rho < 1, 0 < \eta, \mu, \mu < 1$ with $\eta + \mu + \gamma < 1$ such that

$$d(Ta, Tb) \leq \rho [d(a, y)]^{\eta} [d(a, Ta)]^{\mu} [d(b, Tb)]^{\gamma} \tag{8}$$

for all $a, b \in X \setminus \mathrm{Fix}(T)$.

Let us denote by $\Gamma(GI - RRC)$ the set of all GI-RRC contractions on X. A mapping $T : X \to X$, T is an s-GI-RRC contraction if there exist $0 \leq \rho < 1, 0 < \eta, \mu, \gamma < 1$ with $\eta + \mu + \gamma < 1$ such that

$$d(Ta, Tb) \leq \rho \, d(a, b)^{\eta} [d(a, Ta)d(b, Tb)]^{\frac{\mu+\gamma}{2}}$$

whenever $a \neq Ta$ and $b \neq Tb$.

Let us denote by $\tilde{\Gamma}(GI - RRC)$ the set of all s-GI-RRC contractions on X.

Theorem 6. In a metric space (X, d), such that $d(a, b) \geq 1$ for $a \neq b$, we have the equality

$$\Gamma(GI - RRC) = \tilde{\Gamma}(GI - RRC).$$

Proof. Clearly,
$$\tilde{\Gamma}(GI - RRC) \subset \Gamma(GI - RRC)$$

since for any s-GI-RRC contraction T, one has

$$d(Ta, Tb) \leq \rho \, d(a, b)^{\eta} [d(a, Ta)d(b, Tb)]^{\frac{\mu+\gamma}{2}} \iff d(Ta, Tb) \leq \rho d(a, b)^{\eta} \, d(a, Ta)^{\frac{\eta}{2}} \, d(b, Tb)^{\frac{\mu}{2}}$$

and

$$\frac{\mu}{2} + \frac{\gamma}{2} = \frac{\mu + \gamma}{2} < \mu + \gamma < 1.$$

Now, let $T \in \Gamma(GI - RRC)$, so there are $0 \leq \rho < 1, 0 < \eta, \mu, \gamma < 1$ with $\eta + \mu + \gamma < 1$ so that
$$d(Ta, Tb) \leq \rho d(a,b)^\eta d(a, Ta)^\mu d(b, Tb)^\gamma \tag{9}$$
whenever $a \neq Ta, b \neq Tb$.

Additionally, due to symmetry,
$$d(Ta, Tb) = d(Tb, Ta) \leq \rho d(b,a)^\eta d(b, Tb)^\mu d(a, Ta)^\gamma. \tag{10}$$

Multiplying the inequalities (9) and (10), it follows that
$$d(Ta, Tb) \leq \rho \, d(a,b)^\eta [d(a, Ta)d(b, Tb)]^{\frac{\mu+\gamma}{2}}. \tag{11}$$

□

To include the interpolative Reich-Rus-Ćirić type contraction in our study, we allow $\eta + \mu + \gamma = 1$ in the following definition:

Definition 5. *Let (X, d) be a metric space and $T : X \to X$ be a self-map. T is named an extended (ρ, η, μ)-Reich-Rus-Ćirić contraction or extended GI-RRC contraction, if there exist $0 \leq \rho < 1, 0 < \eta, \mu, \mu < 1$ with $\eta + \mu + \gamma \leq 1$ such that*
$$d(Ta, Tb) \leq \rho [d(a,y)]^\eta [d(a, Ta)]^\mu [d(b, Tb)]^\gamma \tag{12}$$
for all $a, b \in X \setminus Fix(T)$.

For a metric space (X, d), let's denote by $e\text{-}\Gamma(GI - RRC)$ the set of all extended GI-RRC contractions on X. Moreover, if $\Gamma(GI - RRC)$ denotes the set of all interpolative Kannan type contractions, then it is clear that:

Corollary 3. *In a metric space (X, d) so that $d(a,b) \geq 1$ for $a \neq b$, we have*
$$\Gamma(GI - RRC) \subset e\text{-}\Gamma(GI - RRC).$$

For a mapping $T : X \to X$, T is an s-GI-RRC contraction if $0 \leq \rho < 1, 0 < \eta, \mu, \gamma < 1$ with $\eta + \mu + \gamma < 1$ such that
$$d(Ta, Tb) \leq \rho \, d(a,b)^\eta [d(a, Ta)d(b, Tb)]^{\frac{\mu+\gamma}{2}}$$
whenever $a \neq Ta, b \neq Tb$.

Furthermore, if we plug $\eta + \mu + \gamma = 1$ in (11), we achieve
$$d(Ta, Tb) \leq \rho \, [d(a, Ta)d(b, Tb)]^{\frac{1}{2}}, \tag{13}$$
which naturally leads to:

Corollary 4.
$$T : X \to X \in \Gamma(GI - RRC)$$

$T : X \to X$ *and there exists* $0 \leq \rho < 1$ *such that*
$$d(Ta, Tb) \leq \rho d(a,b)^\eta [d(a, Ta)d(b, Tb)]^{\frac{1-\eta}{2}}$$
whenever $a \neq Ta, b \neq Tb$.

6. Conclusions

In this paper, we provided conditions under which a (ρ, η, μ)-IKC on a complete metric space can lead a fixed point. Moreover, we show how this new formulation can be extended to the interpolative Reich-Rus-Ćirić type contraction. The authors' plan is, in another manuscript (part 2 of the present manuscript), to enlarge the scope of this new formulation to the frame of different type of interpolative contractions.

Author Contributions: All authors contributed equally and significantly in writing this article. All authors have read and agreed to the published version of the manuscript.

Funding: This work does not receive any external funding.

Institutional Review Board Statement: Not applicable.

Informed Consent Statement: Not applicable.

Data Availability Statement: No data were used for the present study.

Acknowledgments: The first author would like to acknowledge that his contribution to this work was carried out with the aid of a grant from the Carnegie Corporation provided through the African Institute for Mathematical Sciences. The last author would like to thank Prince Sultan University for funding this work through research group Nonlinear Analysis Methods in Applied Mathematics (NAMAM) group number RG-DES-2017-01-17.

Conflicts of Interest: The authors declare that they have no competing interest concerning the publication of this article.

References

1. Kannan, R. Some results on fixed points. *Bull. Calcutta Math. Soc.* **1968**, *60*, 71–76.
2. Karapinar, E. Revisiting the Kannan type contractions via interpolation. *Adv. Theory Nonlinear Anal. Appl.* **2018**, *2*, 85–87. [CrossRef]
3. Aydi, H.; Chen, C.M.; Karapinar, E. Interpolative Ciric-Reich-Rus type contractions via the Branciari distance. *Mathematics* **2019**, *7*, 84. [CrossRef]
4. Errai, Y.; Marhrani, E.; Aamri, M. Some remarks on fixed point theorems for interpolative Kannan contraction. *J. Funct. Spaces* **2020**, *2020*, 2075920.
5. Karapinar, E.; Alqahtani, O.; Aydi, H. On interpolative Hardy-Rogers type contractions. *Symmetry* **2019**, *11*, 8. [CrossRef]
6. Gaba, Y.U.; Karapinar, E. A new approach to the interpolative contractions. *Axioms* **2019**, *8*, 110. [CrossRef]
7. Aydi, H.; Karapinar, E.; de Hierro Francisco, R.L. ω-Interpolative Ciric-Reich-Rus type contractions. *Mathematics* **2019**, *7*, 57. [CrossRef]
8. Debnath, P.; de La Sen, M. Fixed-points of interpolative Ćirić-Reich-Rus-type contractions in metric spaces. *Symmetry* **2020**, *12*, 12. [CrossRef]
9. Karapinar, E.; Agarwal, R.P.; Aydi, H. Interpolative Reich-Rus-Ćirić type contractions on partial metric spaces. *Mathematics* **2018**, *6*, 256. [CrossRef]

Article

Cascading Operators in CAT(0) Spaces

Helga Fetter Nathansky [1,*,†] and Jeimer Villada Bedoya [2,†]

1. Centro de Investigación en Matemáticas (CIMAT), 36240 Guanajuato, Mexico
2. Institute of Mathematics, Marie Curie Skłodowska University, 20-031 Lublin, Poland; villadabedoyaj@office.umcs.pl
* Correspondence: fetter@cimat.mx
† These authors contributed equally to this work.

Abstract: In this work, we introduce the notion of cascading non-expansive mappings in the setting of CAT(0) spaces. This family of mappings properly contains the non-expansive maps, but it differs from other generalizations of this class of maps. Considering the concept of ∆-convergence in metric spaces, we prove a principle of demiclosedness for this type of mappings and a ∆-convergence theorem for a Mann iteration process defined using cascading operators.

Keywords: cascading non-expansive mappings; CAT(0) space; fixed point property; Mann iteration

1. Introduction

In [1], Lennard et al. introduced a class of nonlinear operators in Banach spaces called *cascading non-expansive mappings* which generalizes the non-expansive mappings. These mappings arise naturally in the setting of Banach spaces which contain an isomorphic copy of ℓ_1 or c_0 and some important concepts like reflexivity [1] and weak compactness [2] have been characterized in terms of the fixed point property for this family of operators.

Although these mappings were introduced in the framework of Banach spaces, its definition depends fundamentally on three properties of the space, the metric, the completeness, and the concept of convexity, so it makes sense to define cascading operators in metric spaces where there is a notion of convexity.

A nonlinear setting which is natural to extend the concept of cascading operator is that of uniquely geodesic metric spaces, since in these spaces the notion of geodesic allows a definition of convex sets. In relation to fixed point problems, it has been specially fruitful to consider the subclass of CAT(0) spaces, which possesses a metric structure that is similar to the one in Hilbert spaces.

In Section 3, we introduce the cascading operators in the setting of CAT(0) spaces and, following the reasoning by Lennard et al. in [1], we distinguish this family from other collections of operators, which encompass the most common generalizations of *asymptotically non-expansive mappings*, studied in metric spaces.

In Section 4, mainly inspired by the results by Dhompongsa et al. in [3] about asymptotically non-expansive mappings and Khamsi et al. in [4] concerning *asymptotically pointwise non-expansive mappings*, we establish a *demiclosedness principle* for cascading non-expansive mappings in CAT(0) spaces and derive some fixed point results for this family of operators. We also prove a ∆-*convergence theorem* for a Mann iteration process ([5]) using a cascading operator.

2. Preliminaries

A *geodesic* joining two points x, y in a metric space (X, d) is a mapping $\gamma : [0, 1] \to X$ such that $\gamma(0) = x$, $\gamma(1) = y$ and, for any $t, t' \in [0, 1]$, we have that:

$$d(\gamma(t), \gamma(t')) = |t - t'| d(x, y).$$

A metric space (X,d) is *geodesic* if every two points in X are joined by a geodesic. (X,d) is said to be *uniquely geodesic*, if, for every $x,y \in X$, there is exactly one geodesic joining x and y for each $x,y \in X$, which we denote by $[x,y]$. The point $\gamma(t)$ in $[x,y]$ is also denoted by $(1-t)x \oplus ty$.

As a subclass of the uniquely geodesic spaces, we have the CAT(0) spaces which usually are considered as the nonlinear analogue of Hilbert spaces. These spaces were introduced by Aleksandrov in [6].

Definition 1. *Let (X,d) be a uniquely geodesic metric space, $x,y,z \in X$. We say that (X,d) is CAT(0) if*

$$d\left(\left(\frac{x}{2} \oplus \frac{y}{2}\right), z\right)^2 \leq \frac{1}{2}d(x,z)^2 + \frac{1}{2}d(y,z)^2 - \frac{1}{4}d(x,y)^2.$$

The inequality from above is known as the CN inequality of Bruhat and Tits ([7]).

The following result is very useful in order to perform calculations in CAT(0) spaces.

Proposition 1. *([7] Proposition 2.2) Let X be a CAT(0) space. Then, for all $x,y,w \in X$ and $t \in [0,1]$,*

$$d((1-t)x \oplus ty, w) \leq (1-t)d(x,w) + td(y,w).$$

Definition 2. *A subset C of a uniquely geodesic space is convex if, for any $x,y \in C$, we have that $[x,y] \subset C$. If $K \subset X$, we define*

$$\overline{conv}(K) = \bigcap \{D \subset X : D \supset K, D \text{ is closed and convex}\}.$$

A class of mappings widely studied (see [8–10] among others) in the setting of metric fixed point theory is the class of asymptotically non-expansive mappings.

Definition 3. *Let (X,d) be a metric space. A mapping $T : X \to X$ is said to be asymptotically non-expansive if there exists a sequence of positive numbers (k_n), with $\lim_{n \to \infty} k_n = 1$, such that, for all $n \in \mathbb{N}$ y $x,y \in X$,*

$$d(T^n x, T^n y) \leq k_n d(x,y). \tag{1}$$

These functions were defined first in the context of normed spaces by Kirk in [8] and properly extend the collection of non-expansive mappings, that is, those functions $T : X \to X$, such that $d(Tx, Ty) \leq d(x,y)$.

Finally, given a metric space X, a nonempty set $C \subset X$ and a mapping $T : C \to C$,

$$\text{Fix}(T) = \{x \in C : Tx = x\}.$$

3. Cascading Non-Expansive Mappings

In this section, we introduce the notion of cascading non-expansive mappings in the setting of complete CAT(0) spaces and compare it to other types of functions that include the most common generalizations of asymptotically non-expansive mappings studied both in metric spaces and Banach spaces.

Definition 4. *Let (X,d) be a complete CAT(0) space and $C \subset X$ a closed convex set.*
Define $C_0 = C$, $C_1 = \overline{conv}(T(C))$,..., $C_n = \overline{conv}(T(C_{n-1}))$. If there exists $\{k_n\} \subset [1, \infty)$ with $k_n \to 1$ as $n \to \infty$ such that, for all $x,y \in C_n$

$$d(Tx, Ty) \leq k_{n+1} d(x,y),$$

we say that T is a cascading non-expansive mapping.

Next, we recall the notions of totally asymptotically non-expansive mapping, asymptotically pointwise non-expansive mapping and mapping of an asymptotically non-expansive type.

Definition 5. *[11] Let (X,d) be a metric space. A mapping $T : X \to X$ is called totally asymptotically non-expansive if there are nonnegative real sequences $(k_n^{(1)})$ and $(k_n^{(2)})$ with $k_n^{(1)}, k_n^{(2)} \to 0$, as $n \to \infty$ and a strictly increasing and continuous function $\psi : \mathbb{R}^+ \to \mathbb{R}^+$ with $\psi(0) = 0$ such that:*

$$d(T^n x, T^n y) \leq d(x,y) + k_n^{(1)} \psi(d(x,y)) + k_n^{(2)} \quad n \in \mathbb{N}, x, y \in X.$$

Remark 1. *This definition unifies several generalizations of the asymptotically non-expansive mappings.*

If $\psi(t) = t$, we get the nearly asymptotically non-expansive mappings ([12]).
If $\psi(t) = t$ and for all $n \in \mathbb{N}$, $k_n^{(1)} = 0$ and

$$k_n^{(2)} = \max\left(0, \sup_{x,y \in X} d(T^n x, T^n y) - d(x,y)\right),$$

we recover the asymptotically non-expansive mappings in the intermediate sense ([13]).
If $\psi(t) = t$ and for all $n \in \mathbb{N}$, $k_n^{(2)} = 0$, we have the asymptotically non-expansive mappings.

Definition 6. *[14] Let (X,d) be a metric space. A mapping $T : X \to X$ is called asymptotically pointwise non-expansive if there exists a sequence of mappings $\alpha_n : X \to [0, \infty)$ such that, for every $x \in X$, $\limsup_n \alpha_n(x) \leq 1$, it is verified that:*

$$d(T^n x, T^n y) \leq \alpha_n(x) d(x,y), \quad n \in \mathbb{N}, x, y \in X.$$

Definition 7. *[15] Let (X,d) be a metric space. A mapping $T : X \to X$ is said to be an asymptotically non-expansive type if, for every $x \in X$,*

$$\limsup_n (\sup\{d(T^n x, T^n y) - d(x,y)\} : y \in X\}) \leq 0.$$

In [1], some examples are given that prove that the collections of cascading non-expansive mappings and asymptotically non-expansive mappings differ, in the sense that, in general, neither collection is contained in the other. Taking as reference these examples, we distinguish the collection of cascading non-expansive mappings from the respective classes of functions given in Definitions 5, 6, and 7 in the setting of CAT(0) spaces. We recall that a linear space X is CAT(0) if and only if X is pre-Hilbert ([7]) Proposition 1.14 p. 167.

Example 1. *[1] (Example 2.5) Let $X = (\mathbb{R}, |\cdot|)$ and $K = \left[0, 1/\sqrt{2}\right]$. Let $\mathbb{Q}$ denote the set of rational numbers and $\mathbb{I} = \mathbb{R}/\mathbb{Q}$ be the set of irrational numbers. Define $U : K \to K$ such that:*

$$Ux = \begin{cases} \min\left(\sqrt{2}x, \frac{1}{\sqrt{2}}\right), & \text{if } x \in \mathbb{Q} \cap K, \\ Ux = 0, & \text{if } x \in \mathbb{I}/K. \end{cases}$$

If, for all $n \in \mathbb{N}$, $K_n = \overline{\text{conv}}(U(K_{n-1}))$, where $K_0 = K$, then $0, \frac{1}{2} \in K_n$, but

$$\left|U(0) - U\left(\frac{1}{2}\right)\right| = \left|0 - \frac{1}{\sqrt{2}}\right| = \sqrt{2}\left|0 - \frac{1}{2}\right|.$$

From this, we conclude that U is not a cascading non-expansive mapping. Observe that, for all $n \geq 2$ and $x \in K$, $T^n x = 0$ and hence T is totally asymptotically non-expansive, asymptotic pointwise non-expansive and of an asymptotically non-expansive type.

The example shows that the family of asymptotically non-expansive maps is not contained in the class cascading operators in $CAT(0)$ spaces.

Example 2. *In l_2, we consider the following norm:*

$$\|x\| = \left(\sum_{i=1}^{\infty} |\gamma_i x(i)|^2\right)^{1/2}$$

*where $(\gamma_i) \subset (0,1)$ is a sequence such that $\gamma_i \to 1$, $\frac{\gamma_{i+1}}{\gamma_i}$ is decreasing. It is straightforward to see that $X = (l_2, \|\cdot\|)$ is a Hilbert space and hence $\mathrm{CAT}(0)$ ([7] Proposition 1.14 p. 167).
Let $C = C_0 = \{x \in B_{l_2} : x(i) \geq 0, \quad i \in \mathbb{N}\}$ and, for $n \geq 1$,*

$$C_n = \{x \in C : x(i) = 0, \quad 1 \leq i \leq n\}.$$

Let $T : C \to C$ be such that

$$(Tx)(j) = \begin{cases} 0, & \text{if } j = 1 \\ x(j-1), & \text{if } j > 1. \end{cases}$$

It is easy to check that, for every $n \in \mathbb{N}$, $C_n = \overline{\mathrm{conv}}(T(C_{n-1}))$.
Let us see that T is a cascading non-expansive mapping. If $x, y \in C_n$, then

$$\begin{aligned}
\|Tx - Ty\| &= \left(\sum_{j=1}^{\infty} (\gamma_{n+j+1})^2 |x(n+j) - y(n+j)|^2\right)^{1/2} \\
&= \left(\sum_{j=1}^{\infty} \left(\frac{\gamma_{n+j+1}}{\gamma_{n+j}}\right)^2 (\gamma_{n+j})^2 |x(n+j) - y(n+j)|^2\right)^{1/2} \\
&\leq \frac{\gamma_{n+2}}{\gamma_{n+1}} \|x - y\|.
\end{aligned}$$

If $k_n = \frac{\gamma_{n+2}}{\gamma_{n+1}}$, then $k_n \to 1$; therefore, T is a cascading non-expansive mapping. However,

$$\|T^n e_1 - T^n e_2\| = \left(\gamma_n^2 + \gamma_{n+1}^2\right)^{1/2} = \left(\frac{\gamma_n^2 + \gamma_{n+1}^2}{\gamma_1^2 + \gamma_2^2}\right)^{1/2} \|e_1 - e_2\|$$

and, as $\left(\frac{\gamma_n^2 + \gamma_{n+1}^2}{\gamma_1^2 + \gamma_2^2}\right)^{1/2} \to \left(\frac{2}{\gamma_1^2 + \gamma_2^2}\right)^{1/2} > 1$, we deduce that T is neither totally asymptotically non-expansive, pointwise asymptotically non-expansive nor of an asymptotically non-expansive type.
This example shows that the family of cascading operators is not contained in the family of asymptotically non-expansive maps in $\mathrm{CAT}(0)$ spaces.

Example 3. *Consider the following equivalence relation over the set $\mathbb{N} \times [0,1]$:*

$$(n,0)\mathcal{R}(m,0) \quad \text{and} \quad (n,t)\mathcal{R}(n,t) \quad n,m \in \mathbb{N}, t \in [0,1].$$

Now, we define a metric on $X = (\mathbb{N} \times [0,1])/\mathcal{R}$ as:

$$d((n,t),(n,s)) = |t-s| \quad \text{and if } n \neq m, \quad d((n,t),(m,s)) = t+s.$$

It is easy to see that (X,d) is a complete $\mathbb{R}$-tree ([7] p. 167), and, consequently, it is a complete $\mathrm{CAT}(0)$ space.
Let $C_0 = X$ and, for every $n \in \mathbb{N}$, $C_n = ((\mathbb{N} - \{1, \ldots, n\}) \times [0,1])/\mathcal{R}$.

Let $(\gamma_n) \subset [1, \infty)$ be such that γ_n is a strictly decreasing sequence and $1 < \prod_{n=1}^{\infty} \gamma_n < \infty$. Define $T : C \to C$ such that, if $x = (m, t)$, then

$$Tx = (m + 1, \min(1, \gamma_m t)).$$

Observe that T is well defined. Let us see that T is a cascading non-expansive mapping. It can easily be checked that, for every $n \in \mathbb{N}$, $C_n = \overline{\text{conv}}(T(C_{n-1}))$.

Let $x, y \in C_n$ and consider the following cases:

Case 1. $x = (m, t)$, $y = (m, s)$ $(m \geq n + 1)$

$$d(Tx, Ty) = |\min(1, \gamma_m t) - \min(1, \gamma_m s)| \leq \gamma_m |t - s| \leq \gamma_{n+1} d(x, y).$$

Case 2. $x = (m, t)$, $y = (p, s)$ with $n + 1 \leq m < p$.

$$d(Tx, Ty) = \min(1, \gamma_m t) + \min(1, \gamma_p s) \leq \gamma_m t + \gamma_p s \leq \gamma_{n+1} d(x, y).$$

As $\gamma_n \to 1$ when $n \to \infty$, it follows that T is cascading non-expansive. However, T is not asymptotically non-expansive because, if $x = (1, t)$, $y = (1, s)$ with $s < t < \frac{1}{\prod_{n=1}^{\infty} \gamma_n}$, then:

$$T^n x = (1 + n, \min(1, \gamma_n \ldots \gamma_1 t))$$
$$= (1 + n, \gamma_n \ldots \gamma_1 t)$$

due to $t < \frac{1}{\prod_{n=1}^{\infty} \gamma_n}$.

Analogously, $T^n y = (1 + n, \gamma_n \ldots \gamma_1 s)$. Then,

$$d(T^n x, T^n y) = \gamma_n \ldots \gamma_1 d(x, y).$$

However, $\gamma_n \cdots \gamma_1 \to \prod_{n=1}^{\infty} \gamma_n > 1$ when $n \to \infty$. Consequently,

$$\limsup_n d(T^n x, T^n y) > d(x, y)$$

and T does not belong to the families given in Definitions 5–7.

Remark 2. In [16], the author studied several generalizations of the asymptotically pointwise non-expansive mappings in the context of complete CAT(0) spaces. However, throughout similar examples to those given above, it can be proved that the collection of cascading non-expansive mappings differs from such generalizations.

4. Fixed Point Results for Cascading Operators

Cascading non-expansive operators constitute a new object of study in the framework of CAT(0) spaces and, in general, they do not contain and are not contained in the collection of asymptotically non-expansive mappings as it was illustrated in Section 3. Thus, the theorems in this section are new and do not follow from the results related to asymptotically non-expansive maps.

Let (X, d) be a complete CAT(0) space, (x_n) be a bounded sequence in X and $x \in X$. Let $r(x, (x_n)) = \limsup_{n \to \infty} d(x, x_n)$. The asymptotic radius of (x_n) is given by

$$r((x_n)) = \inf \{r(x, (x_n)) : x \in X\}$$

and the asymptotic center of $r((x_n))$ is the set

$$A((x_n)) = \{x \in X : r(x, (x_n)) = r((x_n))\}.$$

It is a well known fact that, in complete CAT(0) spaces, $A((x_n))$ is a singleton ([17] Proposition 3.2).

The following notions of convergence were introduced by Lim and Kakavandi, respectively, in the setting of metric spaces. These notions resemble the weak convergence defined in Banach spaces and in fact they coincide with the weak convergence in Hilbert spaces ([18], p. 3452).

Definition 8. *Let (X, d) be a complete CAT(0) space and (x_n) be a bounded sequence in X.*

(i) ([19] p. 180) We say that (x_n) Δ-converges to $x \in X$ if $A((x_{n_k})) = \{x\}$ for every subsequence (x_{n_k}) of (x_n).

(ii) ([18]) We say that (x_n) weakly converges to $x \in X$ if

$$\lim_{n \to \infty} \left(d^2(x_n, x) - d^2(x_n, y) + d^2(x, y) \right) = 0, \quad y \in X.$$

Remark 3. *From Example 4.7 in [20], if follows that these notions of convergence are different.*

Let $(X, \|\cdot\|)$ be a Banach space, $C \subset X$ a nonempty closed convex set, and $T : C \to X$ be a mapping. If $I : X \to X$ denotes the identity map, it is said that $I - T$ is demiclosed at zero, if, for any sequence $(x_n) \subset C$ such that x_n weakly converges to x and $\|(I - T)(x_n)\| \to 0$, we have that $Tx = x$.

One of the fundamental results in metric fixed point theory for non-expansive mappings is the demiclosedness principle of Browder [21], which establishes that, if X is an uniformly convex Banach space, $C \subset X$ is a closed convex set and $T : C \to X$ is a non-expansive mapping, then $I - T$ is demiclosed.

Several works ([4,12,17,22,23] among others) have been devoted to prove demiclosedness principles both in Banach and metric spaces for mappings which generalize the non-expansive ones. The following theorem could be interpreted as a demiclosedness principle for cascading non-expansive mappings with respect to the convergence given in (i) in Definition 8.

Theorem 1. *Let (X, d) be a complete CAT(0) space and C a closed convex subset of X. Let $T : C \to C$ be a cascading non-expansive mapping and (k_n) be given as in Definition 4 with $\prod_{j=1}^{\infty} k_j < \infty$. If $(x_n) \subset C$ Δ-converges to w and $\lim_{n \to \infty} d(x_n, Tx_n) = 0$, then $Tw = w$.*

Proof. Since $d(x_n, Tx_n) \to 0$, there exists a subsequence (x_{n_l}) such that $d(x_{n_l}, T^l x_{n_l}) \to 0$ whenever $l \to \infty$. Let us see that, if $y_l = T^l x_{n_l}$, then y_l Δ-converges to w. Indeed, as (x_n) Δ-converges to w, for any $x \in X$:

$$\limsup_{j \to \infty} d(y_{l_j}, w) = \limsup_{j \to \infty} d(x_{n_{l_j}}, w) \leq \limsup_{j \to \infty} d(x_{n_{l_j}}, x)$$
$$= \limsup_{j \to \infty} d(y_{l_j}, x).$$

From this, we conclude that $w \in A(\{y_{l_j}\})$, but, since Proposition 3.2 in [17] implies that $A(\{y_{l_j}\})$ is a singleton, it follows that $A(\{y_{l_j}\}) = \{w\}$ and therefore (y_l) Δ-converges to w. By Proposition 3.2 in [24], $w \in D = \bigcap_{n=1}^{\infty} C_n$. Observe that

$$d(Ty_l, y_l) = d\left(T^{l+1} x_{n_l}, T^l x_{n_l}\right) \leq \left(\prod_{i=1}^{l} k_i\right) d(Tx_{n_l}, x_{n_l}) \to 0$$

when $l \to \infty$.

Hence, for $z \in X$, $r(z, (Ty_l)) = r(z, (y_l))$.

In particular,

$$r(Tw,(y_l)) = \limsup_{l \to \infty} d(Tw, Ty_l) \leq \limsup_{l \to \infty} k_l d(w,(y_l))$$
$$= r(w,(y_l)),$$

but, since (y_l) Δ-converges to w, $r(w,(y_l)) \leq r(Tw,(y_l))$.

Consequently, $r(w,(y_l)) = r(Tw,(y_l))$ and, since $A((y_l))$ is a singleton, we get that $w = Tw$. □

Theorem 1 also holds when we consider the notion of convergence given in ii) in Definition 8.

Corollary 1. *Let (X,d) be a complete $CAT(0)$ space and C a closed convex subset of X. Let $T : C \to C$ be a cascading non-expansive mapping and (k_n) be given as in Definition 4 with $\prod_{j=1}^{\infty} k_j < \infty$. If $(x_n) \subset C$ is such that $\lim_{n \to \infty} d(x_n, Tx_n) = 0$ and (x_n) weakly converges to w, then $Tw = w$.*

Proof. If (x_n) weakly converges to w, Proposition 2.5 in [20] implies that (x_n) Δ-converges to w, and the conclusion follows from Theorem 1. □

By considering the hypothesis of boundedness over C, we have that:

Corollary 2. *Let (X,d) be a complete $CAT(0)$ space and C a closed convex subset of X. Let $T : C \to C$ be a cascading non-expansive mapping and (k_n) be given as in Definition 4 with $\prod_{j=1}^{\infty} k_j < \infty$. If C is bounded, the set of fixed points of T, denoted by $\mathrm{Fix}(T)$, is a nonempty closed convex set.*

Proof. Let $x_0 \in C$ and $x_n = T^n x_0$. From [24] (p. 3690), (x_n) has a subsequence (x_{n_j}) which Δ-converges to w and by Proposition 3.2 in [24]

$$w \in \cap_{j=1}^{\infty} \overline{\mathrm{conv}}\{x_{n_j}, x_{n_{j+1}}, \ldots\} \subset \cap_{n=1}^{\infty} C_n.$$

Consequently, $\cap_{n=1}^{\infty} C_n$ is a nonempty set, and, since $T : \cap_{n=1}^{\infty} C_n \to \cap_{n=1}^{\infty} C_n$ is non-expansive, the conclusions follows from Theorem 5.1 in [4]. □

Lemma 1. *Let (X,d) be a complete $CAT(0)$ space and C a closed convex bounded subset of X. Let $T : C \to C$ be a cascading non-expansive mapping and (k_n) be as in Definition 4 with $\prod_{j=1}^{\infty} k_j < \infty$. Consider the following variant of the Mann iteration process ([5]):*

$$x_{n+1} = (1 - \alpha_n) T x_n \oplus \alpha_n T^2 x_n \tag{2}$$

where x_0 is any element in C and there exist $\beta_1, \beta_2 > 0$, such that, for all $n \in \mathbb{N}$, $0 < \beta_1 \leq \alpha_n \leq 1 - \beta_2 < 1$. It holds that:

1. *If $w \in \mathrm{Fix}(T)$, then $\lim_{n \to \infty} d(x_n, w)$ exists.*
2. *$\lim_{n \to \infty} d(x_n, Tx_n) = 0$.*

Proof. Remember that, by Corollary 2, $\mathrm{Fix}(T)$ is a nonempty set. Let $w \in \mathrm{Fix}(T)$.

1. By Proposition 1,

$$\begin{aligned} d(x_{n+1}, w) &= d((1 - \alpha_n) T x_n \oplus \alpha_n T^2 x_n, w) \leq (1 - \alpha_n) d(T x_n, w) \\ &\quad + \alpha_n d(T^2 x_n, T^2 w) \\ &\leq (1 - \alpha_n) k_n d(x_n, w) + k_{n+1} k_n \alpha_n d(x_n, w) \\ &= (1 + \alpha_n k_n (k_{n+1} - 1)) d(x_n, w) \end{aligned}$$

and, from Lemma 1.2 in [25], we get that $\lim_{n\to\infty} d(x_n, w)$ exists.

2. Let $r = \lim_{n\to\infty} d(x_n, w)$. Since $x_n \in C_n$ and $w \in \text{Fix}(T)$,

$$\limsup_{n\to\infty} d(Tx_n, w) = \limsup_{n\to\infty} d(Tx_n, Tw) \leq \limsup_{n\to\infty} k_{n+1} d(x_n, w) = r.$$

Similarly, $\limsup_{n\to\infty} d(w, T^2 x_n) \leq r$, so, by Lemma 4.5 in [17], we have that $\lim_{n\to\infty} d(Tx_n, T^2 x_n) = 0$.

On the other hand,

$$d(x_{n+1}, Tx_{n+1}) \leq d\left(x_{n+1}, T^2 x_n\right) + d\left(T^2 x_n, Tx_{n+1}\right)$$
$$\leq (1 - \alpha_n) d(Tx_n, T^2 x_n) + k_n \alpha_n d(Tx_n, T^2 x_n)$$
$$\leq (1 - \alpha_n + k_n \alpha_n) d\left(Tx_n, T^2 x_n\right) \to 0.$$

□

The following example shows a simple application of Theorem 1 and generalizes Example 3.

Example 4. *Let X be the space described in Example 3. For simplicity, we write (m, t) to represent the class $[(m, t)]$ if $t > 0$ and define $w_0 = [(m, 0)]$. Let $T : X \to X$ be a cascading operator for which the sequence (k_n) is given as in Definition 4, $\prod_{n=1}^{\infty} k_n < \infty$, and there exists $(m_0, t_0) \in X$, such that for all $N \in \mathbb{N}$,*

$$\{T^n (m_0, t_0) : n \in \mathbb{N}\} \cap \{(m, t) : m \geq N, 0 < t \leq 1\} \neq \emptyset. \quad (3)$$

(Example 3 shows that such T exists) Then, $w_0 = (m, 0) \in \text{Fix}(T)$.

Let (m_0, t_0) be the point for T satisfying (3). If $x_0 = (m_0, t_0)$ and $x_{n+1} = \frac{1}{2} Tx_n \oplus \frac{1}{2} T^2 x_n$, Lemma 1 implies that $d(Tx_n, x_n) \to 0$. Define $\pi_1 : X \to \mathbb{N}$ and $\pi_2 : X \to [0, 1]$ as

$$\pi_1((m, t)) = \begin{cases} m, & \text{if } t > 0 \\ 1, & \text{if } t = 0. \end{cases}$$

and $\pi_2((m, t)) = t$. From condition (3), by passing to a subsequence if necessary, we may assume that, for all $j \in \mathbb{N}$, $\pi_1\left(x_{n_j}\right) \neq \pi_1\left(Tx_{n_j}\right)$. Thus, $d\left(x_{n_j}, Tx_{n_j}\right) = \pi_2 x_{n_j} + \pi_2 Tx_{n_j}$ and $\lim_{j\to\infty} d\left(x_{n_j}, w_0\right) = 0$. Since convergence in metric implies Δ-convergence, x_{n_j} Δ-converges to w_0 and, from Theorem 1, it follows that $Tw_0 = w_0$.

Lemma 2. *Let (X, d) be a complete CAT(0) space and $T : C \to C$ be a cascading non-expansive mapping with (k_n) as in Definition 4 and $\prod_{n=1}^{\infty} k_n < \infty$. Let (x_n) be a sequence in C such that $\lim_{n\to\infty} d(x_n, Tx_n) = 0$ and $(d(x_n, w))$ converges for all $w \in \text{Fix}(T)$. Then, (x_n) Δ-converges to a fixed point of T.*

Proof. It is similar to the proof of Lemma 2.10 in [3]. □

Finally, from Lemmas 1 and 2, we conclude that the Mann iteration process defined in Equation (2), Δ-converges to a fixed point of T.

Theorem 2. *Suppose that C is a closed convex bounded subset of (X, d) and let $T : C \to C$ be a cascading non-expansive mapping with (k_n) as in Definition 4 and $\prod_{n=1}^{\infty} k_n < \infty$. Let x_0 be any initial point in C and (x_n) the sequence defined in Equation (2). Then, (x_n) Δ-converges to a fixed point of T.*

Proof. By Lemma 1, $d(x_n, Tx_n) \to 0$ when $n \to \infty$ and, for any $w \in \text{Fix}(T)$, $\lim_{n \to \infty} d(x_n, w)$ exists. Therefore, Lemma 2 implies that (x_n) Δ-converges to a fixed point w_0 of T. □

The theorems introduced in Section 4 are inspired by some well known results previously studied in CAT(0) spaces for asymptotically non-expansive maps. It would be interesting to find general fixed point theorems which include both families of maps and to determine conditions under which the two families coincide.

Author Contributions: Both authors contributed equally in the development of this work. Both authors have read and agreed to the published version of the manuscript.

Funding: This research was partially supported by Marie Curie Skłodowska University, 20-031 Lublin, Poland.

Institutional Review Board Statement: Not applicable.

Informed Consent Statement: Informed consent was obtained from all subjects involved in the study.

Data Availability Statement: Data sharing not applicable No new data were created or analyzed in this study. Data sharing is not applicable to this article.

Acknowledgments: We would like to thank the referees for their helpful comments.

Conflicts of Interest: The authors declare no conflict of interest.

References

1. Lennard, C.; Nezir, V. Reflexivity is equivalent to the perturbed fixed point property for cascading non-expansive maps in Banach lattices. *Nonlinear Anal. Theory Methods Appl.* **2014**, *95*, 414–420. [CrossRef]
2. Domínguez, T.; Japón, M. A fixed-point characterization of weak compactness in Banach spaces with unconditional Schauder basis. *J. Math. Anal. Appl.* **2017**, *454*, 246–264. [CrossRef]
3. Dhompongsa, S.; Panyanak, B. On Δ-convergence theorems in CAT(0) spaces. *Comput. Math. Appl.* **2008**, *56*, 2572–2579. [CrossRef]
4. Hussain, N.; Khamsi, M. On asymptotic pointwise contractions in metric spaces. *Nonlinear Anal. Theory Methods Appl.* **2009**, *71*, 4423–4429. [CrossRef]
5. Mann, W. Mean value methods in iteration. *Proc. Am. Math.* **1953**, *4*, 506–510. [CrossRef]
6. Alexandrov, A.D. A theorem on triangles in a metric space and some of its applications. *Trudy Mat. Inst. Steklov.* **1951**, *38*, 5–23.
7. Bridson, M.; Haefliger, A. *Metric Spaces of Non-Positive Curvature*; Springer Science & Business Media: Berlin, Germany, 2013.
8. Kirk, W.; Yañez, C.; Shin, S. Asymptotically non-expansive mappings. *Nonlinear Anal. Theory Methods Appl.* **1998**, *33*, 1–12. [CrossRef]
9. Khamsi, M. On asymptotically non-expansive mappings in hyperconvex metric spaces. *Proc. Am. Math.* **2004**, *132*, 365–373. [CrossRef]
10. Kohlenbach, U.; Leustean, L. Asymptotically non-expansive mappings in uniformly convex hyperbolic spaces. *arXiv* **2007**, arXiv:0707.1626.
11. Alber, Y.; Chidume, C.; Zegeye, H. Approximating fixed points of total asymptotically non-expansive mappings. *Fixed Point Theory Appl.* **2006**, *2006*, 10673. [CrossRef]
12. Karapinar, E.; Salahifard, H.; Vaezpour, S. Demiclosedness principle for total asymptotically non-expansive mappings in CAT(0) spaces. *J. Appl. Math.* **2014**, *2014*, 268780. [CrossRef]
13. Bruck, R.; Kuczumow, T.; Reich, S. Convergence of iterates of asymptotically non-expansive mappings in Banach spaces with the uniform Opial property. *Colloq. Math.* **1993**, *65*, 169–179. [CrossRef]
14. Kirk, W.; Xu, H. Asymptotic pointwise contractions. *Nonlinear Anal. Theory Methods Appl.* **2008**, *69*, 4706–4712. [CrossRef]
15. Kirk, W. Fixed point theorems for non-Lipschitzian mappings of asymptotically non-expansive type. *Isr. J. Math.* **1974**, *17*, 339–346. [CrossRef]
16. Nicolae, A. Generalized asymptotic pointwise contractions and non-expansive mappings involving orbits. *Fixed Point Theory Appl.* **2009**, *2010*, 458265.
17. Nanjaras, B.; Panyanak, B. Demiclosed principle for asymptotically non-expansive mappings in CAT(0) spaces. *Fixed Point Theory Appl.* **2010**, *2010*, 268780. [CrossRef]
18. Kakavandi, B. Weak topologies in complete CAT(0) metric spaces. *Proc. Am. Math.* **2013**, *141*, 1029–1039. [CrossRef]
19. Lim, T.C. Remarks on some fixed point theorems. *Proc. Am. Math.* **1976**, *60*, 179–182. [CrossRef]
20. Kakavandi, B.; Amini, M. Duality and subdifferential for convex functions on complete CAT(0) metric spaces. *Nonlinear Anal. Theory Methods Appl.* **2010**, *73*, 3450–3455. [CrossRef]
21. Browder, F. Semicontractive and semiaccretive nonlinear mappings in Banach spaces. *Bull. Am. Math. Soc.* **1968**, *74*, 660–665. [CrossRef]

22. Lin, P.-K.; Tan, K.-K.; Xu, H.-K. Demiclosedness principle and asymptotic behavior for asymptotically non-expansive mappings. *Nonlinear Anal. Theory Methods Appl.* **1995**, *24*, 929–946. [CrossRef]
23. Garcia-Falset, J.; Sims, B.; Smyth, M. The demiclosedness principle for mappings of asymptotically non-expansive type. *Houston J. Math.* **1996**, *22*, 101–108.
24. Kirk, W.A.; Panyanak, B. A concept of convergence in geodesic spaces. *Nonlinear Anal. Theory Methods Appl.* **2008**, *68*, 3689–3696. [CrossRef]
25. Zhou, H. Nonexpansive mappings and iterative methods in uniformly convex Banach spaces. *Georgian Math. J.* **2002**, *9*, 591–600.

Article

Using the Supremum Form of Auxiliary Functions to Study the Common Coupled Coincidence Points in Fuzzy Semi-Metric Spaces

Hsien-Chung Wu

Department of Mathematics, National Kaohsiung Normal University, Kaohsiung 802, Taiwan; hcwu@nknucc.nknu.edu.tw

Abstract: This paper investigates the common coupled coincidence points and common coupled fixed points in fuzzy semi-metric spaces. The symmetric condition is not necessarily satisfied in fuzzy semi-metric space. Therefore, four kinds of triangle inequalities are taken into account in order to study the Cauchy sequences. Inspired by the intuitive observations, the concepts of rational condition and distance condition are proposed for the purpose of simplifying the discussions.

Keywords: common coupled coincidence points; common coupled fixed points; distance condition; fuzzy semi-metric space; rational condition

MSC: 54E35; 54H25

1. Introduction

The common coupled coincidence points and common coupled fixed points in conventional metric spaces and probabilistic metric spaces have been studied for a long time in which the symmetric condition is satisfied. In this paper, we shall consider the fuzzy semi-metric space in which the symmetric condition is not satisfied. In this case, the role of triangle inequality should be re-interpreted. Therefore, four kinds of triangle inequalities are considered, which can also refer to Wu [1].

Schweizer and Sklar [2–4] introduced probabilistic metric space, in which the (conventional) metric space is associated with the probability distribution functions. For more details on the theory of probabilistic metric space, we can refer to Hadžić and Pap [5] and Chang et al. [6]. An interesting special kind of probabilistic metric space is the so-called Menger space. Kramosil and Michalek [7] proposed the fuzzy metric space based on the idea of Menger space. The definition of fuzzy metric space is presented below. Let X be a nonempty universal set associated with a t-norm $*$. Given a mapping M from $X \times X \times [0, \infty)$ into $[0, 1]$, the 3-tuple $(X, M, *)$ is called a fuzzy metric space when the following conditions are satisfied:

- for any $x, y \in X$, $M(x, y, t) = 1$ for all $t > 0$ if and only if $x = y$;
- $M(x, y, 0) = 0$ for all $x, y \in X$;
- $M(x, y, t) = M(y, x, t)$ for all $x, y \in X$ and $t \geq 0$; and,
- $M(x, y, t) * M(y, z, s) \leq M(x, z, t + s)$ for all $x, y, z \in X$ and $s, t \geq 0$.

The mapping M in the fuzzy metric space $(X, M, *)$ can be treated as a membership function of a fuzzy subset of the product space $X \times X \times [0, \infty)$. According to the first and second conditions of fuzzy metric space, the function value $M(x, y, t)$ means that the membership degree of the distance that is less than or equal to t between x and y.

In this paper, we are going to consider the semi-metric space that is completely different from the fuzzy metric space. The so-called fuzzy semi-metric space does not assume the symmetric condition $M(x, y, t) = M(y, x, t)$. Without this condition, the concept of triangle inequalities should be carefully treated. In this paper, there are four kinds of different triangle inequalities considered. It will be realized that, when the symmetric condition is satisfied, these four different kinds of triangle inequalities will be equivalent to the classical one. Being inspired by the intuitive observations, the concepts of rational condition and distance condition are proposed for the purpose of simplifying the discussions regarding the common coupled coincidence points and common coupled fixed points in a fuzzy semi-metric space.

Rakić et al. [8,9] studied the fixed points in b-fuzzy metric spaces. Mecheraoui et al. [10] obtained the sufficient condition for a G-Cauchy sequence to be an M-Cauchy sequence in fuzzy metric space. On the other hand, Gu and Shatanawi [11] used the concept of w-compatible mappings for studying the common coupled fixed points of two hybrid pairs of mappings in partial metric spaces. Petruel [12,13] studied the fixed point for graphic contractions and fixed point for multi-valued locally contractive operators. Hu et al. [14], Mohiuddine and Alotaibi [15], Qiu and Hong [16], and the references therein studied the common coupled coincidence points and common coupled fixed points in fuzzy metric spaces. Wu [17] also studied the common coincidence points in fuzzy semi-metric spaces. In this paper, the common coupled coincidence points and common coupled fixed points in fuzzy semi-metric spaces will be studied by considering four kinds of triangle inequalities. Although the common coupled fixed points are the common coupled coincidence points, the sufficient conditions will be completely different when considering the uniqueness.

This paper is organized, as follows. In Section 2, the concept of fuzzy semi-metric spaces will be introduced. Because the symmetric condition is not satisfied, four different kinds of triangle inequalities will be taken into account to study the common coupled fixed points. In Section 3, in order to study the Cauchy sequence in fuzzy semi-metric space, the auxiliary functions that are based on the supremun are proposed. In Section 4, while using the auxiliary functions proposed in Section 3, the desired property regarding the Cauchy sequence in fuzzy semi-metric space will be presented. In Section 5, many kinds of common coupled coincidence points in fuzzy semi-metric spaces will be investigated by considering the four different kinds of triangle inequalities. Finally, in Section 6, the common coupled fixed points shown in fuzzy semi-metric spaces will also be studied based on the four different kinds of triangle inequalities.

2. Fuzzy Semi-Metric Spaces

The concept of fuzzy semi-metric space is based on the concept of t-norm (triangular norm), which will be introduced below. Let $* : [0,1] \times [0,1] \to [0,1]$ be a function that is defined on the product set $[0,1] \times [0,1]$. We say that $*$ is a t-norm when the following conditions are satisfied:

- $a * 1 = a$.
- $a * b = b * a$.
- $b < c$ implies $a * b \leq a * c$.
- $(a * b) * c = a * (b * c)$.

The following properties regarding t-norm will be used in the further study.

Proposition 1. *We have the following properties.*

(i) *Suppose that the t-norm $*$ is left-continuous at 1 with respect to the first or second component. For any $a, b \in (0,1)$ with $a > b$, there exists $r \in (0,1)$ that satisfies $a * r \geq b$.*

(ii) *Suppose that the t-norm $*$ is left-continuous at 1 with respect to the first or second component. For any $a \in (0,1)$ and any $p \in \mathbb{N}$, there exists $r \in (0,1)$ satisfying $\overbrace{r * r * \cdots * r}^{p \text{ times}} > a$.*

(iii) *Given any fixed $a, b \in [0,1]$, suppose that the t-norm $*$ is continuous at a and b with respect the first or second component, and that $\{a_n\}_{n=1}^{\infty}$ and $\{b_n\}_{n=1}^{\infty}$ are two sequences in $[0,1]$ satisfying $a_n \to a$ and $b_n \to b$ as $n \to \infty$. Subsequently, we have $a_n * b_n \to a * b$ as $n \to \infty$.*

(iv) *Given any fixed $a, b \in [0,1]$, suppose that the t-norm $*$ is left-continuous at a and b with respect to the first or second component, and that $\{a_n\}_{n=1}^{\infty}$ and $\{b_n\}_{n=1}^{\infty}$ are two sequences in*

[0,1] satisfying $a_n \to a-$ and $b_n \to b-$ as $n \to \infty$. Afterwards, we have $a_n * b_n \to a * b$ as $n \to \infty$.

(v) Given any fixed $a, b \in [0, 1)$, suppose that the t-norm $*$ is right-continuous at a and b with respect to the first or second component, and that $\{a_n\}_{n=1}^{\infty}$ and $\{b_n\}_{n=1}^{\infty}$ are two sequences in [0,1] satisfying $a_n \to a+$ and $b_n \to b+$ as $n \to \infty$. Subsequently, we have $a_n * b_n \to a * b$ as $n \to \infty$.

Wu [1,17,18] proposed the concept of fuzzy semi-metric space. The formal definition is given below.

Definition 1. *Let X be a nonempty set and let M be a mapping from $X \times X \times [0, \infty)$ into $[0, 1]$. We say that (X, M) is fuzzy semi-metric space when the following conditions are satisfied:*

- *for any $x, y \in X$, $M(x, y, t) = 1$ for all $t \geq 0$ if and only if $x = y$;*
- *$M(x, y, 0) = 0$ for all $x, y \in X$ with $x \neq y$;*

The mapping M is said to satisfy the symmetric condition when $M(x, y, t) = M(y, x, t)$ for any $x, y \in X$ and $t \geq 0$.

Definition 2. *Let (X, M) be a fuzzy semi-metric space. We say that M satisfies the distance condition when, for any $x, y \in X$ with $x \neq y$, there exists $t_0 > 0$, such that $M(x, y, t_0) \neq 0$.*

Because the symmetric condition is not necessarily be satisfied in fuzzy semi-metric space (X, M), by referring to Wu [1,17,18], four kinds of triangle inequalities are proposed below.

Definition 3. *Let X be a nonempty set, let $*$ be a t-norm, and let M be a mapping that is defined on $X \times X \times [0, \infty)$ into $[0, 1]$.*

- *We say that M satisfies the $\bowtie$-triangle inequality when the following inequality is satisfied:*

$$M(x, y, t) * M(y, z, s) \leq M(x, z, t + s) \text{ for all } x, y, z \in X \text{ and } s, t > 0.$$

- *We say that M satisfies the $\triangleright$-triangle inequality when the following inequality is satisfied:*

$$M(x, y, t) * M(z, y, s) \leq M(x, z, t + s) \text{ for all } x, y, z \in X \text{ and } s, t > 0.$$

- *We say that M satisfies the $\triangleleft$-triangle inequality when the following inequality is satisfied:*

$$M(y, x, t) * M(y, z, s) \leq M(x, z, t + s) \text{ for all } x, y, z \in X \text{ and } s, t > 0.$$

- *We say that M satisfies the $\diamond$-triangle inequality when the following inequality is satisfied:*

$$M(y, x, t) * M(z, y, s) \leq M(x, z, t + s) \text{ for all } x, y, z \in X \text{ and } s, t > 0.$$

Remark 1. *Suppose that the mapping M satisfies the $\bowtie$-triangle inequality. Subsequently, we have*

$$M(a, b, t_1) * M(b, c, t_2) * M(c, d, t_3) \leq M(a, c, t_1 + t_2) * M(c, d, t_3) \leq M(a, d, t_1 + t_2 + t_3)$$

and

$$M(b, a, t_1) * M(c, b, t_2) = M(c, b, t_2) * M(b, a, t_1) \leq M(c, a, t_1 + t_2),$$

which implies

$$M(b, a, t_1) * M(c, b, t_2) * M(d, c, t_3) \leq M(d, a, t_1 + t_2 + t_3).$$

In general, we have

$$M(x_1, x_2, t_1) * M(x_2, x_3, t_2) * \cdots * M(x_p, x_{p+1}, t_p) \leq M(x_1, x_{p+1}, t_1 + t_2 + \cdots + t_p)$$

and

$$M(x_2, x_1, t_1) * M(x_3, x_2, t_2) * \cdots * M(x_{p+1}, x_p, t_{p+1}) \leq M(x_{p+1}, x_1, t_1 + t_2 + \cdots + t_p).$$

For the case of satisfying the ▷-triangle inequality, ◁-triangle inequality and ⋄-triangle inequality, we can refer to Wu [17].

Proposition 2 (Wu [1]). *Let (X, M) be a fuzzy semi-metric space. Then we have the following properties.*

(i) Suppose that the mapping M satisfies the ⋈-triangle inequality. Subsequently, M is non-decreasing in the sense of $M(x, y, t_1) \geq M(x, y, t_2)$ for any fixed $x, y \in X$ and $t_1 > t_2$.

(ii) Suppose that the mapping M satisfies the ⋄-triangle inequality. Subsequently, M is symmetrically non-decreasing in the sense of $M(x, y, t_1) \geq M(y, x, t_2)$ for any fixed $x, y \in X$ and $t_1 > t_2$.

(iii) Suppose that the mapping M satisfies the ▷-triangle inequality or the ◁-triangle inequality. Afterwards, M is both non-decreasing and symmetrically non-decreasing.

Let $\{x_n\}_{n=1}^\infty$ be a sequence in the fuzzy semi-metric space (X, M).

- We write $x_n \xrightarrow{M^\triangleright} x$ as $n \to \infty$ when $M(x_n, x, t) \to 1$ as $n \to \infty$ for all $t > 0$.
- We write $x_n \xrightarrow{M^\triangleleft} x$ as $n \to \infty$ when $M(x, x_n, t) \to 1$ as $n \to \infty$ for all $t > 0$.
- We write $x_n \xrightarrow{M} x$ as $n \to \infty$ when $x_n \xrightarrow{M^\triangleright} x$ and $x_n \xrightarrow{M^\triangleleft} x$ as $n \to \infty$.

Proposition 3 (Wu [17]). *Let (X, M) be a fuzzy semi-metric space, and let $\{x_n\}_{n=1}^\infty$ be a sequence in X. Suppose that the t-norm $*$ is left-continuous at 1 with respect to the first or second component. Afterwards, we have the following results.*

(i) Assume that the mapping M satisfies the ⋈-triangle inequality or ⋄-triangle inequality. Subsequently, we have the following properties.

- If $x_n \xrightarrow{M^\triangleleft} x$ and $x_n \xrightarrow{M^\triangleright} y$ as $n \to \infty$, then $x = y$.
- If $x_n \xrightarrow{M^\triangleright} x$ and $x_n \xrightarrow{M^\triangleleft} y$ as $n \to \infty$, then $x = y$.

(ii) Assume that M satisfies the ◁-triangle inequality. If $x_n \xrightarrow{M^\triangleright} x$ and $x_n \xrightarrow{M^\triangleright} y$ as $n \to \infty$, then $x = y$.

(iii) Assume that M satisfies the ▷-triangle inequality. If $x_n \xrightarrow{M^\triangleleft} x$ and $x_n \xrightarrow{M^\triangleleft} y$ as $n \to \infty$, then $x = y$.

Proposition 4 (Wu [18]). *Let (X, M) be a fuzzy semi-metric space, and let $\{(x_n, y_n, t_n)\}_{n=1}^\infty$ be a sequence in $X \times X \times (0, \infty)$. Assume that the t-norm $*$ is left-continuous with respect to the first or second component. For any sequences $\{a_n\}_{n=1}^\infty$ and $\{b_n\}_{n=1}^\infty$ in $[0, 1]$, we also assume that the following inequality is satisfied*

$$\sup_n (a_n * b_n) \geq \left(\sup_n a_n\right) * \left(\sup_n b_n\right).$$

(i) Suppose that M satisfies the ⋈-triangle inequality, and that $t_n \to t^\circ$, $x_n \xrightarrow{M} x^\circ$ and $y_n \xrightarrow{M} y^\circ$ as $n \to \infty$. Subsequently, the following statements hold true.

- If M is continuous with respect to the distance at t°, then $M(x_n, y_n, t_n) \to M(x^\circ, y^\circ, t^\circ)$ as $n \to \infty$.

- If M is symmetrically continuous with respect to the distance at t°, then $M(x_n, y_n, t_n) \to M(y^\circ, x^\circ, t^\circ)$ as $n \to \infty$.

(ii) Suppose that M satisfies the $\circ$-triangle inequality for $\circ \in \{\triangleright, \triangleleft\}$, and that $t_n \to t^\circ$, $x_n \xrightarrow{M} x^\circ$ and $y_n \xrightarrow{M} y^\circ$ as $n \to \infty$. If M is continuous or symmetrically continuous with respect to the distance at t°, then $M(x_n, y_n, t_n) \to M(x^\circ, y^\circ, t^\circ) = M(y^\circ, x^\circ, t^\circ)$ as $n \to \infty$.

(iii) Suppose that M satisfies the $\diamond$-triangle inequality, and that $t_n \to t^\circ$ as $n \to \infty$, $x_n \xrightarrow{M^\triangleright} x^\circ$, and $y_n \xrightarrow{M^\triangleright} y^\circ$ as $n \to \infty$ simultaneously, or $x_n \xrightarrow{M^\triangleleft} x^\circ$ and $y_n \xrightarrow{M^\triangleleft} y^\circ$ as $n \to \infty$ simultaneously. If M is continuous or symmetrically continuous with respect to the distance at t°, then $M(x_n, y_n, t_n) \to M(y^\circ, x^\circ, t^\circ) = M(x^\circ, y^\circ, t^\circ)$ as $n \to \infty$.

Definition 4. *Let $\{x_n\}_{n=1}^\infty$ be a sequence in the fuzzy semi-metric space (X, M).*

- *We say that $\{x_n\}_{n=1}^\infty$ is a $>$-Cauchy sequence when, given any pair (r, t) with $t > 0$ and $0 < r < 1$, there exists $n_{r,t} \in \mathbb{N}$ satisfying $M(x_m, x_n, t) > 1 - r$ for all pairs (m, n) of integers m and n with $m > n \geq n_{r,t}$.*
- *We say that $\{x_n\}_{n=1}^\infty$ is a $<$-Cauchy sequence when, given any pair (r, t) with $t > 0$ and $0 < r < 1$, there exists $n_{r,t} \in \mathbb{N}$ satisfying $M(x_n, x_m, t) > 1 - r$ for all pairs (m, n) of integers m and n with $m > n \geq n_{r,t}$.*
- *We say that $\{x_n\}_{n=1}^\infty$ is a Cauchy sequence when, given any pair (r, t) with $t > 0$ and $0 < r < 1$, there exists $n_{r,t} \in \mathbb{N}$ satisfying $M(x_m, x_n, t) > 1 - r$ and $M(x_n, x_m, t) > 1 - r$ for all pairs (m, n) of integers m and n with $m, n \geq n_{r,t}$ and $m \neq n$.*
- *We say that (X, M) is $(>, \triangleright)$-complete when each $>$-Cauchy sequence $\{x_n\}_{n=1}^\infty$ is convergent in the sense of $x_n \xrightarrow{M^\triangleright} x$.*
- *We say that (X, M) is $(>, \triangleleft)$-complete when each $>$-Cauchy sequence $\{x_n\}_{n=1}^\infty$ is convergent in the sense of $x_n \xrightarrow{M^\triangleleft} x$.*
- *We say that (X, M) is $(<, \triangleright)$-complete when each $<$-Cauchy sequence $\{x_n\}_{n=1}^\infty$ is convergent in the sense of $x_n \xrightarrow{M^\triangleright} x$.*
- *We say that (X, M) is $(<, \triangleleft)$-complete when each $<$-Cauchy sequence $\{x_n\}_{n=1}^\infty$ is convergent in the sense of $x_n \xrightarrow{M^\triangleleft} x$.*

Definition 5. *Let (X, M) be a fuzzy semi-metric space. Four types of continuities are defined below.*

- *We say that the function $f : X \to X$ is $(\triangleright, \triangleright)$-continuous with respect to M when, given any sequence $\{x_n\}_{n=1}^\infty$ in X, $x_n \xrightarrow{M^\triangleright} x$, as $n \to \infty$ implies $f(x_n) \xrightarrow{M^\triangleright} f(x)$ as $n \to \infty$.*
- *We say that the function $f : X \to X$ is $(\triangleright, \triangleleft)$-continuous with respect to M when, given any sequence $\{x_n\}_{n=1}^\infty$ in X, $x_n \xrightarrow{M^\triangleright} x$, as $n \to \infty$ implies $f(x_n) \xrightarrow{M^\triangleleft} f(x)$ as $n \to \infty$.*
- *We say that the function $f : X \to X$ is $(\triangleleft, \triangleright)$-continuous with respect to M when, given any sequence $\{x_n\}_{n=1}^\infty$ in X, $x_n \xrightarrow{M^\triangleleft} x$, as $n \to \infty$ implies $f(x_n) \xrightarrow{M^\triangleright} f(x)$ as $n \to \infty$.*
- *We say that the function $f : X \to X$ is $(\triangleleft, \triangleleft)$-continuous with respect to M when, given any sequence $\{x_n\}_{n=1}^\infty$ in X, $x_n \xrightarrow{M^\triangleleft} x$, as $n \to \infty$ implies $f(x_n) \xrightarrow{M^\triangleleft} f(x)$ as $n \to \infty$.*

3. Auxiliary Functions Based on the Supremum

The concept of auxiliary function based on X^2 was proposed by Wu [17] to study the common coincidence point. In this paper, we are going to consider the auxiliary function that is based on X^4 to study the common coupled coincidence point.

Definition 6. *Let (X, M) be a fuzzy semi-metric space. We say that the mapping M satisfies the rational condition when $M(x, y, t) \to 0$, as $t \to 0+$ for any fixed $x, y \in X$.*

Let (X, M) be a fuzzy semi-metric space along with a t-norm $*$. We define the mapping $\eta : X^4 \times [0, \infty) \to [0, 1]$ on the product space $X^4 \times [0, \infty)$, as follows

$$\eta(x, y, u, v, t) = M(x, y, t) * M(u, v, t).$$

Subsequently, we have the following interesting result that will be used to define the auxiliary functions.

Proposition 5. *Let (X, M) be a fuzzy semi-metric space, such that the mapping M satisfies the rational condition. Suppose that the t-norm $*$ is right-continuous at 0 with respect to the first or second component. Subsequently, we have*

$$\lim_{t \to 0+} \eta(x, y, u, v, t) = 0. \quad (1)$$

The following definition of auxiliary functions are based on X^4. This new concept extends the auxiliary functions based on X^2, as proposed by Wu [17].

Definition 7. *Let (X, M) be a fuzzy semi-metric space, such that M satisfies the rational condition in which the t-norm $*$ is also right-continuous at 0 with respect to the first or second component. For any fixed $x, y, u, v \in X$ and $\lambda \in [0, 1)$ with $x \neq y$ or $u \neq v$, we define a function $\Phi : X^4 \to [0, \infty)$ on the product space X^4 by*

$$\Phi(\lambda, x, y, u, v) = \sup\{t > 0 : \eta(x, y, u, v, t) \leq 1 - \lambda\}$$

and $\Phi(\lambda, x, x, u, u) = 0$ for $\lambda \in [0, 1)$.

For $x \neq y$ or $u \neq v$, we need to claim that the set $\{t > 0 : \eta(x, y, u, v, t) \leq 1 - \lambda\}$ is not empty. Suppose that $\{t > 0 : \eta(x, y, u, v, t) \leq 1 - \lambda\} = \emptyset$. By definition, we must have $\eta(x, y, u, v, t) > 1 - \lambda$ for all $t > 0$. This says that

$$\lim_{t \to 0+} \eta(x, y, u, v, t) \geq 1 - \lambda,$$

which contradicts (1). Therefore, we indeed have $\{t > 0 : \eta(x, y, u, v, t) \leq 1 - \lambda\} \neq \emptyset$, which says that the function Φ is well-defined.

Proposition 6. *Let (X, M) be a fuzzy semi-metric space such that the mapping M satisfies the rational condition in which the t-norm $*$ is right-continuous at 0 with respect to the first or second component. Given any fixed $x, y, u, v \in X$ and $\lambda \in (0, 1)$, we have the following properties.*

(i) *Suppose that $\Phi(\lambda, x, y, u, v) < \infty$. For any $\epsilon > 0$, we have*

$$\eta(x, y, u, v, \Phi(\lambda, x, y, u, v) + \epsilon) > 1 - \lambda$$

(ii) *Assume that $\epsilon > 0$ is sufficiently small satisfying $\Phi(\lambda, x, y, u, v) > \epsilon$. Subsequently, we have the following properties.*

- *If the mapping M satisfies the $\bowtie$-triangle inequality or the $\triangleright$-triangle inequality or the $\triangleleft$-triangle inequality, then*

$$\eta(x, y, u, v, \Phi(\lambda, x, y, u, v) - \epsilon) \leq 1 - \lambda.$$

- *If the mapping M satisfies the $\triangleright$-triangle inequality or the $\triangleleft$-triangle inequality, then*

$$\eta(y, x, u, v, \Phi(\lambda, x, y, u, v) - \epsilon) \leq 1 - \lambda \text{ and } \eta(x, y, v, u, \Phi(\lambda, x, y, u, v) - \epsilon) \leq 1 - \lambda.$$

- If the mapping M satisfies the $\triangleright$-triangle inequality or the $\triangleleft$-triangle inequality or the $\diamond$-triangle inequality, then

$$\eta(y, x, v, u, \Phi(\lambda, x, y, u, v) - \epsilon) \leq 1 - \lambda.$$

Proof. The proof is similar to the argument in Wu [17] by considering X^4 instead of X^2. □

Proposition 7. *Let (X, M) be a fuzzy semi-metric space, such that the mapping M satisfies the rational condition in which the t-norm $*$ is right-continuous at 0 with respect to the first or second component. Given any fixed $x, y, u, v \in X$ and $\lambda \in (0, 1)$, we have the following properties.*

(i)   Assume that $\eta(x, y, u, v, t) \leq 1 - \lambda$. Then, we have the following results.
- *If the mapping M satisfies the $\bowtie$-triangle inequality or the $\triangleright$-triangle inequality or the $\triangleleft$-triangle inequality, then $t \leq \Phi(\lambda, x, y, u, v)$.*
- *If the mapping M satisfies the $\triangleright$-triangle inequality or the $\triangleleft$-triangle inequality, then $t \leq \Phi(\lambda, y, x, u, v)$ and $t \leq \Phi(\lambda, x, y, v, u)$.*
- *If the mapping M satisfies the $\triangleright$-triangle inequality or the $\triangleleft$-triangle inequality or the $\diamond$-triangle inequality, then $t \leq \Phi(\lambda, y, x, v, u)$.*

(ii)   We have the following results.
- *Suppose that the mapping M satisfies the $\bowtie$-triangle inequality or the $\triangleright$-triangle inequality or the $\triangleleft$-triangle inequality. If $\eta(x, y, u, v, t) > 1 - \lambda$, then $\Phi(\lambda, x, y, u, v) < \infty$ and $t \geq \Phi(\lambda, x, y, u, v)$.*
- *Suppose that the mapping M satisfies the $\triangleright$-triangle inequality or the $\triangleleft$-triangle inequality.*
  - *If $\eta(x, y, u, v, t) > 1 - \lambda$, then $\Phi(\lambda, y, x, u, v) < \infty$ and $\Phi(\lambda, x, y, v, u) < \infty$.*
  - *If $\eta(x, y, u, v, t) > 1 - \lambda$ and $\Phi(\lambda, x, y, u, v) < \infty$, then $t \geq \Phi(\lambda, x, y, u, v)$.*
- *Suppose that the mapping M satisfies the $\triangleright$-triangle inequality or the $\triangleleft$-triangle inequality or the $\diamond$-triangle inequality.*
  - *If $\eta(x, y, u, v, t) > 1 - \lambda$, then $\Phi(\lambda, y, x, v, u) < \infty$.*
  - *If $\eta(x, y, u, v, t) > 1 - \lambda$ and $\Phi(\lambda, x, y, u, v) < \infty$, then $t \geq \Phi(\lambda, x, y, u, v)$.*

Proof. The proof is similar to the argument in Wu [17] by considering X^4 instead of X^2. □

Proposition 8. *Let (X, M) be a fuzzy semi-metric space, such that M satisfies the rational condition, in which the t-norm $*$ is right-continuous at 0 and left-continuous at 1 with respect to the first or second component.*

(i)   *Suppose that M satisfies the $\bowtie$-triangle inequality. Given any fixed $x_1, x_2, \cdots, x_p, y_1, y_2, \cdots, y_p \in X$ and any fixed $\mu \in (0, 1]$, there exists $\lambda \in (0, 1)$, such that*

$$\Phi(\mu, x_1, x_p, y_1, y_p) \leq \Phi(\lambda, x_1, x_2, y_1, y_2) + \Phi(\lambda, x_2, x_3, y_2, y_3) + \cdots$$
$$+ \Phi(\lambda, x_{p-2}, x_{p-1}, y_{p-2}, y_{p-1}) + \Phi(\lambda, x_{p-1}, x_p, y_{p-1}, y_p)$$

$$\Phi(\mu, x_1, x_p, y_p, y_1) \leq \Phi(\lambda, x_1, x_2, y_2, y_1) + \Phi(\lambda, x_2, x_3, y_3, y_2) + \cdots$$
$$+ \Phi(\lambda, x_{p-2}, x_{p-1}, y_{p-1}, y_{p-2}) + \Phi(\lambda, x_{p-1}, x_p, y_p, y_{p-1})$$

$$\Phi(\mu, x_p, x_1, y_1, y_p) \leq \Phi(\lambda, x_p, x_{p-1}, y_{p-1}, y_p) + \Phi(\lambda, x_{p-1}, x_{p-2}, y_{p-2}, y_{p-1})$$
$$+ \cdots + \Phi(\lambda, x_3, x_2, y_2, y_3) + \Phi(\lambda, x_2, x_1, y_1, y_2)$$

$$\Phi(\mu, x_p, x_1, y_p, y_1) \leq \Phi(\lambda, x_p, x_{p-1}, y_p, y_{p-1}) + \Phi(\lambda, x_{p-1}, x_{p-2}, y_{p-1}, y_{p-2})$$
$$+ \cdots + \Phi(\lambda, x_3, x_2, y_3, y_2) + \Phi(\lambda, x_2, x_1, y_2, y_1).$$

(ii) *Suppose that M satisfies the ▷-triangle inequality. Given any fixed* $x_1, x_2, \cdots, x_p, y_1, y_2, \cdots, y_p \in X$ *and any fixed* $\mu \in (0,1]$, *there exists* $\lambda \in (0,1)$ *such that*

$$\max\{\Phi(\mu, x_1, x_p, y_1, y_p), \Phi(\mu, x_1, x_p, y_p, y_1), \Phi(\mu, x_p, x_1, y_1, y_p), \Phi(\mu, x_p, x_1, y_p, y_1)\}$$
$$\leq \Phi(\lambda, x_1, x_2, y_1, y_2) + \Phi(\lambda, x_3, x_2, y_3, y_2) + \Phi(\lambda, x_4, x_3, y_4, y_3)$$
$$+ \cdots + \Phi(\lambda, x_p, x_{p-1}, y_p, y_{p-1})$$

(iii) *Suppose that M satisfies the ◁-triangle inequality. Given any fixed* $x_1, x_2, \cdots, x_p, y_1, y_2, \cdots, y_p \in X$ *and any fixed* $\mu \in (0,1]$, *there exists* $\lambda \in (0,1)$, *such that*

$$\max\{\Phi(\mu, x_1, x_p, y_1, y_p), \Phi(\mu, x_1, x_p, y_p, y_1), \Phi(\mu, x_p, x_1, y_1, y_p), \Phi(\mu, x_p, x_1, y_p, y_1)\}$$
$$\leq \Phi(\lambda, x_2, x_1, y_2, y_1) + \Phi(\lambda, x_2, x_3, y_2, y_3) + \Phi(\lambda, x_3, x_4, y_3, y_4)$$
$$+ \cdots + \Phi(\lambda, x_{p-1}, x_p, y_{p-1}, y_p)$$

(iv) *Suppose that M satisfies the ⋄-triangle inequality. Given any fixed* $x_1, x_2, \cdots, x_p, y_1, y_2, \cdots, y_p \in X$ *and any fixed* $\mu \in (0,1]$, *there exists* $\lambda \in (0,1)$, *such that the following inequalities are satisfied.*

- *If p is even and* $\Phi(\mu, x_1, x_p, y_1, y_p) < \infty$, *then*

$$\Phi(\mu, x_1, x_p, y_1, y_p) \leq \Phi(\lambda, x_1, x_2, y_1, y_2) + \Phi(\lambda, x_2, x_3, y_2, y_3) + \Phi(\lambda, x_4, x_3, y_4, y_3)$$
$$+ \Phi(\lambda, x_4, x_5, y_4, y_5) + \Phi(\lambda, x_6, x_5, y_6, y_5) + \Phi(\lambda, x_6, x_7, y_6, y_7)$$
$$+ \cdots + \Phi(\lambda, x_p, x_{p-1}, y_p, y_{p-1}) \quad (2)$$

- *If p is even and* $\Phi(\mu, x_1, x_p, y_p, y_1) < \infty$, *then*

$$\Phi(\mu, x_1, x_p, y_p, y_1) \leq \Phi(\lambda, x_1, x_2, y_2, y_1) + \Phi(\lambda, x_2, x_3, y_3, y_2) + \Phi(\lambda, x_4, x_3, y_3, y_4)$$
$$+ \Phi(\lambda, x_4, x_5, y_5, y_4) + \Phi(\lambda, x_6, x_5, y_5, y_6) + \Phi(\lambda, x_6, x_7, y_7, y_6)$$
$$+ \cdots + \Phi(\lambda, x_p, x_{p-1}, y_{p-1}, y_p) \quad (3)$$

- *If p is even and* $\Phi(\mu, x_p, x_1, y_1, y_p) < \infty$, *then*

$$\Phi(\mu, x_p, x_1, y_1, y_p) \leq \Phi(\lambda, x_2, x_1, y_1, y_2) + \Phi(\lambda, x_3, x_2, y_2, y_3) + \Phi(\lambda, x_3, x_4, y_4, y_3)$$
$$+ \Phi(\lambda, x_5, x_4, y_4, y_5) + \Phi(\lambda, x_5, x_6, y_6, y_5) + \Phi(\lambda, x_7, x_6, y_6, y_7)$$
$$+ \cdots + \Phi(\lambda, x_{p-1}, x_p, y_p, y_{p-1}) \quad (4)$$

- *If p is even and* $\Phi(\mu, x_p, x_1, y_p, y_1) < \infty$, *then*

$$\Phi(\mu, x_p, x_1, y_p, y_1) \leq \Phi(\lambda, x_2, x_1, y_2, y_1) + \Phi(\lambda, x_3, x_2, y_3, y_2) + \Phi(\lambda, x_3, x_4, y_3, y_4)$$
$$+ \Phi(\lambda, x_5, x_4, y_5, y_4) + \Phi(\lambda, x_5, x_6, y_5, y_6) + \Phi(\lambda, x_7, x_6, y_7, y_6)$$
$$+ \cdots + \Phi(\lambda, x_{p-1}, x_p, y_{p-1}, y_p) \quad (5)$$

- *If p is odd and* $\Phi(\mu, x_1, x_p, y_1, y_p) < \infty$, *then*

$$\Phi(\mu, x_1, x_p, y_1, y_p) \leq \Phi(\lambda, x_2, x_1, y_2, y_1) + \Phi(\lambda, x_3, x_2, y_3, y_2) + \Phi(\lambda, x_3, x_4, y_3, y_4)$$
$$+ \Phi(\lambda, x_5, x_4, y_5, y_4) + \Phi(\lambda, x_5, x_6, y_5, y_6) + \Phi(\lambda, x_7, x_6, y_7, y_6)$$
$$+ \cdots + \Phi(\lambda, x_{p-1}, x_p, y_{p-1}, y_p) \quad (6)$$

- *If p is odd and* $\Phi(\mu, x_1, x_p, y_p, y_1) < \infty$, *then*

$$\Phi(\mu, x_1, x_p, y_p, y_1) \leq \Phi(\lambda, x_2, x_1, y_1, y_2) + \Phi(\lambda, x_3, x_2, y_2, y_3) + \Phi(\lambda, x_3, x_4, y_4, y_3)$$
$$+ \Phi(\lambda, x_5, x_4, y_4, y_5) + \Phi(\lambda, x_5, x_6, y_6, y_5) + \Phi(\lambda, x_7, x_6, y_6, y_7)$$
$$+ \cdots + \Phi(\lambda, x_{p-1}, x_p, y_p, y_{p-1}) \quad (7)$$

- If p is odd and $\Phi(\mu, x_p, x_1, y_1, y_p) < \infty$, then

$$\Phi(\mu, x_p, x_1, y_1, y_p) \leq \Phi(\lambda, x_1, x_2, y_2, y_1) + \Phi(\lambda, x_2, x_3, y_3, y_2) + \Phi(\lambda, x_4, x_3, y_3, y_4)$$
$$+ \Phi(\lambda, x_4, x_5, y_5, y_4) + \Phi(\lambda, x_6, x_5, y_5, y_6) + \Phi(\lambda, x_6, x_7, y_7, y_6)$$
$$+ \cdots + \Phi(\lambda, x_p, x_{p-1}, y_{p-1}, y_p) \tag{8}$$

- If p is odd and $\Phi(\mu, x_p, x_1, y_p, y_1) < \infty$, then

$$\Phi(\mu, x_p, x_1, y_p, y_1) \leq \Phi(\lambda, x_1, x_2, y_1, y_2) + \Phi(\lambda, x_2, x_3, y_2, y_3) + \Phi(\lambda, x_4, x_3, y_4, y_3)$$
$$+ \Phi(\lambda, x_4, x_5, y_4, y_5) + \Phi(\lambda, x_6, x_5, y_6, y_5) + \Phi(\lambda, x_6, x_7, y_6, y_7)$$
$$+ \cdots + \Phi(\lambda, x_p, x_{p-1}, y_p, y_{p-1}) \tag{9}$$

Proof. The proof is similar to the argument put foward in Wu [17] by considering X^4 instead of X^2. □

Proposition 9. *Let (X, M) be a fuzzy semi-metric space, such that M satisfies the rational condition, in which the t-norm $*$ is right-continuous at 0 with respect to the first or second component. Let $\{x_n\}_{n=1}^{\infty}$ and $\{y_n\}_{n=1}^{\infty}$ be two sequences in X.*

(i) *Assume that M satisfies the $\bowtie$-triangle inequality or the $\triangleright$-triangle inequality or the $\triangleleft$-triangle inequality. Subsequently, we have the following results.*

- $\{x_n\}_{n=1}^{\infty}$ and $\{y_n\}_{n=1}^{\infty}$ are two >-Cauchy sequences if and only if, given any $\epsilon > 0$ and $\lambda \in (0, 1)$, there exists $n_{\epsilon, \lambda} \in \mathbb{N}$ satisfying $\Phi(\lambda, x_m, x_n, y_m, y_n) < \epsilon$ for $m > n \geq n_{\epsilon, \lambda}$.
- $\{x_n\}_{n=1}^{\infty}$ is a >-Cauchy sequences and $\{y_n\}_{n=1}^{\infty}$ is a <-Cauchy sequences if and only if, given any $\epsilon > 0$ and $\lambda \in (0, 1)$, there exists $n_{\epsilon, \lambda} \in \mathbb{N}$ satisfying $\Phi(\lambda, x_m, x_n, y_n, y_m) < \epsilon$ for $m > n \geq n_{\epsilon, \lambda}$.
- $\{x_n\}_{n=1}^{\infty}$ is a <-Cauchy sequences and $\{y_n\}_{n=1}^{\infty}$ is a >-Cauchy sequences if and only if, given any $\epsilon > 0$ and $\lambda \in (0, 1)$, there exists $n_{\epsilon, \lambda} \in \mathbb{N}$ satisfying $\Phi(\lambda, x_n, x_m, y_m, y_n) < \epsilon$ for $m > n \geq n_{\epsilon, \lambda}$.
- $\{x_n\}_{n=1}^{\infty}$ and $\{y_n\}_{n=1}^{\infty}$ are two <-Cauchy sequences if and only if, given any $\epsilon > 0$ and $\lambda \in (0, 1)$, there exists $n_{\epsilon, \lambda} \in \mathbb{N}$ satisfying $\Phi(\lambda, x_n, x_m, y_n, y_m) < \epsilon$ for $m > n \geq n_{\epsilon, \lambda}$.

(ii) *Assume that M satisfies the $\diamond$-triangle inequality. Then, we have the following results.*

- Let $\{x_n\}_{n=1}^{\infty}$ and $\{y_n\}_{n=1}^{\infty}$ be two >-Cauchy sequences. Suppose that $\Phi(\lambda, x_m, x_n, y_m, y_n) < \infty$ for all $\lambda \in (0, 1)$ and $m > n$. Subsequently, given any $\epsilon > 0$, there exists $n_{\epsilon, \lambda} \in \mathbb{N}$ satisfying $\Phi(\lambda, x_m, x_n, y_m, y_n) < \epsilon$ for $m > n \geq n_{\epsilon, \lambda}$.
- Let $\{x_n\}_{n=1}^{\infty}$ be a >-Cauchy sequence and let $\{y_n\}_{n=1}^{\infty}$ be a <-Cauchy sequence. Suppose that $\Phi(\lambda, x_m, x_n, y_n, y_m) < \infty$ for any all $\lambda \in (0, 1)$ and $m > n$. Afterwards, given any $\epsilon > 0$, there exists $n_{\epsilon, \lambda} \in \mathbb{N}$ satisfying $\Phi(\lambda, x_m, x_n, y_n, y_m) < \epsilon$ for $m > n \geq n_{\epsilon, \lambda}$.
- Let $\{x_n\}_{n=1}^{\infty}$ be a <-Cauchy sequence and let $\{y_n\}_{n=1}^{\infty}$ be a >-Cauchy sequence. Suppose that $\Phi(\lambda, x_n, x_m, y_m, y_n) < \infty$ for all $\lambda \in (0, 1)$ and $m > n$. Subsequently, given any $\epsilon > 0$, there exists $n_{\epsilon, \lambda} \in \mathbb{N}$ satisfying $\Phi(\lambda, x_n, x_m, y_m, y_n) < \epsilon$ for $m > n \geq n_{\epsilon, \lambda}$.
- Let $\{x_n\}_{n=1}^{\infty}$ and $\{y_n\}_{n=1}^{\infty}$ be two <-Cauchy sequences. Suppose that $\Phi(\lambda, x_n, x_m, y_n, y_m) < \infty$ for all $\lambda \in (0, 1)$ and $m > n$. Subsequently, given any $\epsilon > 0$, there exists $n_{\epsilon, \lambda} \in \mathbb{N}$ satisfying $\Phi(\lambda, x_n, x_m, y_n, y_m) < \epsilon$ for $m > n \geq n_{\epsilon, \lambda}$.
- Suppose that, given any $\epsilon > 0$ and $\lambda \in (0, 1)$, there exists $n_{\epsilon, \lambda} \in \mathbb{N}$ satisfying $\Phi(\lambda, x_m, x_n, y_n, y_n) < \epsilon$ for $m > n \geq n_{\epsilon, \lambda}$. Then $\{x_n\}_{n=1}^{\infty}$ and $\{y_n\}_{n=1}^{\infty}$ are two <-Cauchy sequences.
- Suppose that, given any $\epsilon > 0$ and $\lambda \in (0, 1)$, there exists $n_{\epsilon, \lambda} \in \mathbb{N}$ satisfying $\Phi(\lambda, x_m, x_n, y_n, y_n) < \epsilon$ for $m > n \geq n_{\epsilon, \lambda}$. Then $\{x_n\}_{n=1}^{\infty}$ is a <-Cauchy sequences and $\{y_n\}_{n=1}^{\infty}$ is a >-Cauchy sequences.

- Suppose that, given any $\epsilon > 0$ and $\lambda \in (0,1)$, there exists $n_{\epsilon,\lambda} \in \mathbb{N}$ satisfying $\Phi(\lambda, x_n, x_m, y_m, y_n) < \epsilon$ for $m > n \geq n_{\epsilon,\lambda}$. Subsequently, $\{x_n\}_{n=1}^{\infty}$ is a $>$-Cauchy sequences and $\{y_n\}_{n=1}^{\infty}$ is a $<$-Cauchy sequences.
- Suppose that, given any $\epsilon > 0$ and $\lambda \in (0,1)$, there exists $n_{\epsilon,\lambda} \in \mathbb{N}$ satisfying $\Phi(\lambda, x_n, x_m, y_n, y_m) < \epsilon$ for $m > n \geq n_{\epsilon,\lambda}$. Afterwards, $\{x_n\}_{n=1}^{\infty}$ and $\{y_n\}_{n=1}^{\infty}$ are two $>$-Cauchy sequences.

Proof. The proof is similar to the argument in Wu [17] by considering X^4 instead of X^2. □

4. Cauchy Sequences

Given any $a \in [0,1]$, for convenience, we write

$$(*a)^n = \overbrace{a * a * \cdots * a}^{n \text{ times}}$$

and

$$\left[*\eta\left(a,b,c,d,\frac{t}{k^n}\right) \right]^{2^n} = \overbrace{\eta\left(a,b,c,d,\frac{t}{k^n}\right) * \eta\left(a,b,c,d,\frac{t}{k^n}\right) * \cdots * \eta\left(a,b,c,d,\frac{t}{k^n}\right)}^{2^n \text{ times}}.$$

The following results will be used for further discussion.

Proposition 10. *Let (X, M) be a fuzzy semi-metric space such that M satisfies the rational condition in which the t-norm is right-continuous at 0 and left-continuous at 1 in the first or second component. Let $0 < k < 1$ be any fixed constant, and let $\{x_n\}_{n=1}^{\infty}$ and $\{y_n\}_{n=1}^{\infty}$ be two sequences in X.*

(i) *Suppose that M satisfies the $\bowtie$-triangle inequality. Subsequently, we have the following results.*

- *Assume that there exist fixed elements $a_1, b_1, c_1, d_1 \in X$ satisfying*

$$\sup_{\lambda \in [0,1)} \Phi(\lambda, a_1, b_1, c_1, d_1) < \infty \tag{10}$$

and

$$\eta(x_n, x_{n+1}, y_n, y_{n+1}, t) \geq \left[*\eta\left(a_1, b_1, c_1, d_1, \frac{t}{k^n}\right) \right]^{2^n} \text{ for each } n \in \mathbb{N}. \tag{11}$$

Afterwards, $\{x_n\}_{n=1}^{\infty}$ and $\{y_n\}_{n=1}^{\infty}$ are $<$-Cauchy sequences.

- *Assume that there exist fixed elements $a_2, b_2, c_2, d_2 \in X$ satisfying*

$$\sup_{\lambda \in [0,1)} \Phi(\lambda, a_2, b_2, c_2, d_2) < \infty \tag{12}$$

and

$$\eta(x_n, x_{n+1}, y_{n+1}, y_n, t) \geq \left[*\eta\left(a_2, b_2, c_2, d_2, \frac{t}{k^n}\right) \right]^{2^n} \text{ for each } n \in \mathbb{N}. \tag{13}$$

Subsequently, $\{x_n\}_{n=1}^{\infty}$ is a $<$-Cauchy sequence and $\{y_n\}_{n=1}^{\infty}$ is a $>$-Cauchy sequence.

- *Assume that there exist fixed elements $a_3, b_3, c_3, d_3 \in X$ satisfying*

$$\sup_{\lambda \in [0,1)} \Phi(\lambda, a_3, b_3, c_3, d_3) < \infty \tag{14}$$

and

$$\eta(x_{n+1}, x_n, y_n, y_{n+1}, t) \geq \left[*\eta\left(a_3, b_3, c_3, d_3, \frac{t}{k^n}\right) \right]^{2^n} \text{ for each } n \in \mathbb{N}. \quad (15)$$

Subsequently, $\{x_n\}_{n=1}^{\infty}$ is a $>$-Cauchy sequence and $\{y_n\}_{n=1}^{\infty}$ is a $<$-Cauchy sequence.
- Assume that there exist fixed elements $a_4, b_4, c_4, d_4 \in X$ satisfying

$$\sup_{\lambda \in [0,1)} \Phi(\lambda, a_4, b_4, c_4, d_4) < \infty \quad (16)$$

and

$$\eta(x_{n+1}, x_n, y_{n+1}, y_n, t) \geq \left[*\eta\left(a_4, b_4, c_4, d_4, \frac{t}{k^n}\right) \right]^{2^n} \text{ for each } n \in \mathbb{N}. \quad (17)$$

Afterwards, $\{x_n\}_{n=1}^{\infty}$ and $\{y_n\}_{n=1}^{\infty}$ are $>$-Cauchy sequences.

(ii) Suppose that the mapping M satisfies the $\triangleright$-triangle inequality or the $\triangleleft$-triangle inequality, and that the conditions (10), (11), (16) and (17) are satisfied. Subsequently, $\{x_n\}_{n=1}^{\infty}$ and $\{y_n\}_{n=1}^{\infty}$ are both $>$-Cauchy and $<$-Cauchy sequences. In other words, $\{x_n\}_{n=1}^{\infty}$ and $\{y_n\}_{n=1}^{\infty}$ are Cauchy sequences.

(iii) Suppose that the mapping M satisfies the $\diamond$-triangle inequality, and that any one of the following two conditions is satisfied:
- conditions (10), (11), (16) and (17) are satisfied;
- conditions (12), (13), (14) and (15) are satisfied.

Afterwards, $\{x_n\}_{n=1}^{\infty}$ and $\{y_n\}_{n=1}^{\infty}$ are both $>$-Cauchy and $<$-Cauchy sequences.

Proof. To prove part (i), if

$$\left[*\eta\left(a_1, b_1, c_1, d_1, \frac{t}{k^n}\right) \right]^{2^n} \leq 1 - \lambda,$$

then, using Proposition 1, there exists $\bar{\lambda}(t) \in (0,1)$ satisfying

$$\eta\left(a_1, b_1, c_1, d_1, \frac{t}{k^n}\right) \leq 1 - \bar{\lambda}(t).$$

Let

$$\lambda_0 \equiv \inf_t \bar{\lambda}(t) \in [0,1).$$

Then λ_0 depends only on λ and

$$\eta\left(a_1, b_1, c_1, d_1, \frac{t}{k^n}\right) \leq 1 - \inf_t \bar{\lambda}(t) \equiv 1 - \lambda_0.$$

It follows that

$$\left\{ t > 0 : \left[*\eta\left(a_1, b_1, c_1, d_1, \frac{t}{k^n}\right) \right]^{2^n} \leq 1 - \lambda \right\} \subseteq \left\{ t > 0 : \eta\left(a_1, b_1, c_1, d_1, \frac{t}{k^n}\right) \leq 1 - \lambda_0 \right\}. \quad (18)$$

Therefore, we obtain

$$\Phi(\lambda, x_n, x_{n+1}, y_n, y_{n+1})$$
$$= \sup\{t > 0 : \eta(x_n, x_{n+1}, y_n, y_{n+1}, t) \leq 1 - \lambda\}$$
$$\leq \sup\left\{t > 0 : \left[*\eta\left(a_1, b_1, c_1, d_1, \frac{t}{k^n}\right)\right]^{2^n} \leq 1 - \lambda\right\} \text{ (by (11))}$$
$$\leq \sup\left\{t > 0 : \eta\left(a_1, b_1, c_1, d_1, \frac{t}{k^n}\right) \leq 1 - \lambda_0\right\} \text{ (by (18))}$$
$$= \sup\{k^n \cdot t > 0 : \eta(a_1, b_1, c_1, d_1, t) \leq 1 - \lambda_0\}$$
$$= k^n \cdot \sup\{t > 0 : \eta(a_1, b_1, c_1, d_1, t) \leq 1 - \lambda_0\}$$
$$= k^n \cdot \Phi(\lambda_0, a_1, b_1, c_1, d_1), \tag{19}$$

where λ_0 only depends on λ. Now, we assume that $m, n \in \mathbb{N}$ with $m > n$. Given any $\mu \in (0, 1]$, by part (i) of Proposition 8, there exists $\lambda \in (0, 1)$, such that

$$\Phi(\mu, x_n, x_m, y_n, y_m)$$
$$\leq \Phi(\lambda, x_n, x_{n+1}, y_n, y_{n+1}) + \Phi(\lambda, x_{n+1}, x_{n+2}, y_{n+1}, y_{n+2}) + \cdots + \Phi(\lambda, x_{m-1}, x_m, y_{m-1}, y_m)$$
$$\leq k^n \cdot \Phi(\lambda_0, a_1, b_1, c_1, d_1) + k^{n+1} \cdot \Phi(\lambda_0, a_1, b_1, c_1, d_1) + \cdots + k^{m-1} \cdot \Phi(\lambda_0, a_1, b_1, c_1, d_1) \text{ (by (19))}$$
$$= \Phi(\lambda_0, a_1, b_1, c_1, d_1) \cdot \frac{k^n \cdot (1 - k^{m-n})}{1-k} \leq \Phi(\lambda_0, a_1, b_1, c_1, d_1) \cdot \frac{k^n}{1-k}$$
$$\leq \left[\sup_{\lambda \in [0,1)} \Phi(\lambda, a_1, b_1, c_1, d_1)\right] \cdot \frac{k^n}{1-k} \to 0 \text{ as } n \to \infty, \tag{20}$$

which also says that, given any $\epsilon \in (0,1)$ and $\mu \in (0,1)$, there exists $n_{\mu,\epsilon} \in \mathbb{N}$ such that $m > n \geq n_{\mu,\epsilon}$ implies $\Phi(\mu, x_n, x_m, y_n, y_m) < \epsilon$. By the fourth case of part (i) of Proposition 9, it follows that $\{x_n\}_{n=1}^\infty$ and $\{y_n\}_{n=1}^\infty$ are $<$-Cauchy sequences. The other results can be similarly obtained by using the corresponding cases of Proposition 9 and part (i) of Proposition 8.

To prove part (ii), we consider two cases below.

- Suppose that the mapping M satisfies the $\triangleright$-triangle inequality. While using part (ii) of Proposition 8, we have

$$\max\{\Phi(\mu, x_n, x_m, y_n, y_m), \Phi(\mu, x_m, x_n, y_n, y_m), \Phi(\mu, x_n, x_m, y_m, y_n), \Phi(\mu, x_m, x_n, y_m, y_n)\}$$
$$\leq \Phi(\lambda, x_m, x_{m-1}, y_m, y_{m-1}) + \Phi(\lambda, x_{m-1}, x_{m-2}, y_{m-1}, y_{m-2})$$
$$+ \cdots + \Phi(\lambda, x_{n+2}, x_{n+1}, y_{n+2}, y_{n+1}) + \Phi(\lambda, x_n, x_{n+1}, y_n, y_{n+1}). \tag{21}$$

By referring to (19), we can similarly obtain

$$\Phi(\lambda, x_{n+1}, x_n, y_{n+1}, y_n) \leq k^n \cdot \Phi(\lambda_0, a_4, b_4, c_4, d_4). \tag{22}$$

By using (19), (22), (21) and referring to (20), we have

$$\max\{\Phi(\mu, x_n, x_m, y_n, y_m), \Phi(\mu, x_m, x_n, y_n, y_m), \Phi(\mu, x_n, x_m, y_m, y_n), \Phi(\mu, x_m, x_n, y_m, y_n)\}$$
$$\leq \max\left\{\left[\sup_{\lambda \in [0,1)} \Phi(\lambda, a_1, b_1, c_1, d_1)\right], \left[\sup_{\lambda \in [0,1)} \Phi(\lambda, a_4, b_4, c_4, d_4)\right]\right\} \cdot \frac{k^n}{1-k} \to 0 \text{ as } n \to \infty.$$

Using the above argument, we can show that $\{x_n\}_{n=1}^\infty$ and $\{y_n\}_{n=1}^\infty$ are both $>$-Cauchy and $<$-Cauchy sequences in metric sense.

- Suppose that the mapping M satisfies the $\triangleleft$-triangle inequality. While using part (iii) of Proposition 8, we have

$$\max\{\Phi(\mu, x_n, x_m, y_n, y_m), \Phi(\mu, x_m, x_n, y_n, y_m), \Phi(\mu, x_n, x_m, y_m, y_n), \Phi(\mu, x_m, x_n, y_m, y_n)\}$$
$$\leq \Phi(\lambda, x_{n+1}, x_n, y_{n+1}, y_n) + \Phi(\lambda, x_{n+1}, x_{n+2}, y_{n+1}, y_{n+2})$$
$$+ \cdots + \Phi(\lambda, x_{m-2}, x_{m-1}, y_{m-2}, y_{m-1}) + \Phi(\lambda, x_{m-1}, x_m, y_{m-1}, y_m).$$

Using the above argument, we can show that $\{x_n\}_{n=1}^{\infty}$ and $\{y_n\}_{n=1}^{\infty}$ are both $>$-Cauchy and $<$-Cauchy sequences in metric sense.

To prove part (iii), we consider two cases below.

- Assume that the conditions (10), (11), (16) and (17) are satisfied. If p is even, then, using (2) and (5) in part (iv) of Proposition 8, we can similarly show that $\{x_n\}_{n=1}^{\infty}$ and $\{y_n\}_{n=1}^{\infty}$ are both $>$-Cauchy and $<$-Cauchy sequences in metric sense. If p is odd, then, using (6) and (9) in Proposition 8, we can similarly obtain the desired results.
- Assume that the conditions (12), (13), (14) and (15) are satisfied. If p is even, then, using (3) and (4) in part (iv) of Proposition 8, we can similarly show that $\{x_n\}_{n=1}^{\infty}$ and $\{y_n\}_{n=1}^{\infty}$ are both $>$-Cauchy and $<$-Cauchy sequences in the metric sense. If p is odd, then, using (7) and (8) in Proposition 8, we can similarly obtain the desired results.

This completes the proof. □

5. Common Coupled Coincidence Points

In this section, we are going to investigate the common coupled coincidence points in fuzzy semi-metric space under some suitable conditions. We consider two mappings $T : X \times X \to X$ and $f : X \to X$.

- Recall that the mappings T and f commute when $f(T(x,y)) = T(f(x), f(y))$ for all $x, y \in X$.
- Recall that an element $(x^*, y^*) \in X \times X$ is called a coupled coincidence point of mappings T and f when $T(x^*, y^*) = f(x^*)$ and $T(y^*, x^*) = f(y^*)$. In particular, if $x^* = f(x^*) = T(x^*, y^*)$ and $y^* = f(y^*) = T(y^*, x^*)$, then (x^*, y^*) is called a common coupled fixed point of T and f.

Let $\{T_n\}_{n=1}^{\infty}$ be a sequence of mappings from the product space $X \times X$ into X, and let f be a mapping from X into itself satisfying $T_n(X, X) \subseteq f(X)$ for all $n \in \mathbb{N}$. Given any two initial elements $x_0, y_0 \in X$, since $T_n(X, X) \subseteq f(X)$, there exist $x_1, y_1 \in X$ satisfying

$$f(x_1) = T_1(x_0, y_0) \text{ and } f(y_1) = T_1(y_0, x_0).$$

Similarly, there also exist $x_2, y_2 \in X$, satisfying

$$f(x_2) = T_2(x_1, y_1) \text{ and } f(y_2) = T_2(y_1, x_1).$$

Continuing this process, we can construct two sequences $\{x_n\}_{n=1}^{\infty}$ and $\{y_n\}_{n=1}^{\infty}$, satisfying

$$f(x_n) = T_n(x_{n-1}, y_{n-1}) \text{ and } f(y_n) = T_n(y_{n-1}, x_{n-1}) \tag{23}$$

for $n \in \mathbb{N}$.

In the sequel, the common coupled coincidence points will be separately studied by considering the four different types of triangle inequalities.

Theorem 1 (Satisfying the $\bowtie$-Triangle Inequality). *Let (X, M) be a fuzzy semi-metric space, such that the mapping M satisfies the rational condition and the $\bowtie$-triangle inequality. Suppose that the following conditions are satisfied.*

- *The t-norm $*$ is left-continuous with respect to the first or second component.*
- *Given any fixed $x, y \in X$, the mapping $M(x, y, \cdot) : (0, \infty) \to [0, 1]$ is left-continuous at each point $t \in (0, \infty)$.*

- The mappings $T_n : X \times X \to X$ and $f : X \to X$ satisfy the inclusions $T_n(X, X) \subseteq f(X)$ for all $n \in \mathbb{N}$.
- The mappings f and T_n commute; that is, $f(T_n(x, y)) = T_n(f(x), f(y))$ for all $x, y \in X$ and all $n \in \mathbb{N}$.
- Given any $x, y, u, v \in X$, the following contractive inequality is satisfied:

$$M(T_i(x, y), T_j(u, v), k_{ij} \cdot t) \geq M(f(x), f(u), t) * M(f(y), f(v), t), \qquad (24)$$

where k_{ij} satisfies $0 < k_{ij} \leq k < 1$ for all $i, j \in \mathbb{N}$ and for some constant k.

Subsequently, we have the following results.

(i) Suppose that there exist $x^*, y^* \in X$ satisfying

$$\sup_{\lambda \in (0,1)} \Phi(\lambda, f(x^*), T_1(x^*, y^*), f(y^*), T_1(y^*, x^*)) < \infty,$$

and that any one of the following conditions is satisfied:

(a) (X, M) is $(<, \triangleright)$-complete and f is simultaneously $(\triangleright, \triangleright)$-continuous and $(\triangleright, \triangleleft)$-continuous with respect to M;

(b) (X, M) is $(<, \triangleleft)$-complete and f is simultaneously $(\triangleleft, \triangleright)$-continuous and $(\triangleleft, \triangleleft)$-continuous with respect to M.

Afterwards, the mappings $\{T_n\}_{n=1}^{\infty}$ and f have a common coupled coincidence point (x°, y°). We further assume that the following conditions are satisfied.

- The inequality (24) is replaced by the following inequality

$$M(T_i(x, y), T_j(u, v), k_{ij} \cdot t) \geq M(f(x), f(u), t) \cdot M(f(y), f(v), t), \qquad (25)$$

where the t-norm $*$ is replaced by the product of real numbers.
- The mapping M satisfies the distance condition in Definition 2.
- For any fixed $x, y \in X$ and $t > 0$, the following mapping

$$\varrho(\alpha) = M\left(x, y, k^{\log_2 \alpha} \cdot t\right) \qquad (26)$$

is differentiable on $(0, \infty)$.

Afterwards, we have the following results.

(A) Suppose that $(\bar{x}, \bar{y})$ is another coupled coincidence point of mappings f and T_{n_0} for some $n_0 \in \mathbb{N}$. Subsequently, $f(x^{\circ}) = f(\bar{x})$ and $f(y^{\circ}) = f(\bar{y})$.

(B) There exists $(x^{\circ}, y^{\circ}) \in X \times X$ such that $(f(x^{\circ}), f(y^{\circ})) \in X \times X$ is the common coupled fixed point of the mappings $\{T_n\}_{n=1}^{\infty}$.

Moreover, the point $(x^{\circ}, y^{\circ}) \in X \times X$ can be obtained, as follows.

- Suppose that condition (a) is satisfied. Afterwards, the point $(x^{\circ}, y^{\circ}) \in X \times X$ can be obtained by taking the limit $f(x_n) \xrightarrow{M^{\triangleright}} x^{\circ}$ and $f(y_n) \xrightarrow{M^{\triangleright}} y^{\circ}$.
- Suppose that condition (b) is satisfied. Subsequently, the point $(x^{\circ}, y^{\circ}) \in X \times X$ can be obtained by taking the limit $f(x_n) \xrightarrow{M^{\triangleleft}} x^{\circ}$ and $f(y_n) \xrightarrow{M^{\triangleleft}} y^{\circ}$.

The sequences $\{x_n\}_{n=1}^{\infty}$ and $\{y_n\}_{n=1}^{\infty}$ are generated from the initial element $(x_0, y_0) = (x^*, y^*) \in X \times X$ according to (23).

(ii) Suppose that there exist $x^*, y^* \in X$ satisfying

$$\sup_{\lambda \in (0,1)} \Phi(\lambda, T_1(x^*, y^*), f(x^*), T_1(y^*, x^*), f(y^*)) < \infty,$$

and that any one of the following conditions is satisfied:

(c) (X, M) is $(>, \triangleright)$-complete and f is simultaneously $(\triangleright, \triangleright)$-continuous and $(\triangleright, \triangleleft)$-continuous with respect to M;

(d) (X, M) is $(>, \triangleleft)$-complete and f is simultaneously $(\triangleleft, \triangleright)$-continuous and $(\triangleleft, \triangleleft)$-continuous with respect to M.

Afterwards, we have the same result as part (i).

Proof. We can generate two sequences $\{x_n\}_{n=1}^\infty$ and $\{y_n\}_{n=1}^\infty$ from the initial element $x_0 = x^*$ and $y_0 = y^*$ according to (23). Then we have

$$f(x^*) = f(x_0) \text{ and } f(y^*) = f(y_0)$$

and

$$T_1(x^*, y^*) = T_1(x_0, y_0) = f(x_1) \text{ and } T_1(y^*, x^*) = T_1(y_0, x_0) = f(y_1).$$

To prove part (i), from (23) and (24), we obtain

$$M(f(x_1), f(x_2), t) = M(T_1(x_0, y_0), T_2(x_1, y_1), t)$$
$$\geq M\left(f(x_0), f(x_1), \frac{t}{k_{12}}\right) * M\left(f(y_0), f(y_1), \frac{t}{k_{12}}\right)$$

and

$$M(f(y_1), f(y_2), t) = M(T_1(y_0, x_0), T_2(y_1, x_1), t)$$
$$\geq M\left(f(y_0), f(y_1), \frac{t}{k_{12}}\right) * M\left(f(x_0), f(x_1), \frac{t}{k_{12}}\right).$$

By induction, we can obtain

$$M(f(x_n), f(x_{n+1}), t) \geq \left[*M\left(f(x_0), f(x_1), \frac{t}{\prod_{i=1}^n k_{i,i+1}}\right)\right]^{2^{n-1}}$$
$$* \left[*M\left(f(y_0), f(y_1), \frac{t}{\prod_{i=1}^n k_{i,i+1}}\right)\right]^{2^{n-1}} \quad (27)$$

and

$$M(f(y_n), f(y_{n+1}), t) \geq \left[*M\left(f(x_0), f(x_1), \frac{t}{\prod_{i=1}^n k_{i,i+1}}\right)\right]^{2^{n-1}}$$
$$* \left[*M\left(f(y_0), f(y_1), \frac{t}{\prod_{i=1}^n k_{i,i+1}}\right)\right]^{2^{n-1}}. \quad (28)$$

Part (i) of Proposition 2 says that the mapping $M(x, y, \cdot)$ is nondecreasing. Because $k_{i,i+1} \leq k$ for each $i \in \mathbb{N}$, using the increasing property of t-norm to (27) and (28), we also have

$$M(f(x_n), f(x_{n+1}), t) \geq \left[*M\left(f(x_0), f(x_1), \frac{t}{k^n}\right)\right]^{2^{n-1}} * \left[*M\left(f(y_0), f(y_1), \frac{t}{k^n}\right)\right]^{2^{n-1}}$$
$$= \left[*\eta\left(f(x_0), f(x_1), f(y_0), f(y_1), \frac{t}{k^n}\right)\right]^{2^{n-1}} \quad (29)$$

and

$$M(f(y_n), f(y_{n+1}), t) \geq \left[*M\left(f(x_0), f(x_1), \frac{t}{k^n}\right)\right]^{2^{n-1}} * \left[*M\left(f(y_0), f(y_1), \frac{t}{k^n}\right)\right]^{2^{n-1}}$$
$$= \left[*\eta\left(f(x_0), f(x_1), f(y_0), f(y_1), \frac{t}{k^n}\right)\right]^{2^{n-1}}. \quad (30)$$

Using the increasing property of t-norm to (29) and (30), we have

$$\eta(f(x_n), f(x_{n+1}), f(y_n), f(y_{n+1}), t) = M(f(x_n), f(x_{n+1}), t) * M(f(y_n), f(y_{n+1}), t)$$
$$\geq \left[*\eta\left(f(x_0), f(x_1), f(y_0), f(y_1), \frac{t}{k^n}\right)\right]^{2^n}. \tag{31}$$

From part (i) of Proposition 10, it follows that $\{f(x_n)\}_{n=1}^{\infty}$ and $\{f(y_n)\}_{n=1}^{\infty}$ are $<$-Cauchy sequences. We consider the following cases

- Suppose that condition (a) is satisfied. Because (X, M) is $(<, \triangleright)$-complete, there exist $x^\circ, y^\circ \in X$, such that

$$f(x_n) \xrightarrow{M^\triangleright} x^\circ \text{ and } f(y_n) \xrightarrow{M^\triangleright} y^\circ \text{ as } n \to \infty. \tag{32}$$

Since f is simultaneously $(\triangleright, \triangleright)$-continuous and $(\triangleright, \triangleleft)$-continuous with respect to M, we have

$$f(f(x_n)) \xrightarrow{M^\triangleright} f(x^\circ) \text{ and } f(f(y_n)) \xrightarrow{M^\triangleright} f(y^\circ) \text{ as } n \to \infty$$

and

$$f(f(x_n)) \xrightarrow{M^\triangleleft} f(x^\circ) \text{ and } f(f(y_n)) \xrightarrow{M^\triangleleft} f(y^\circ) \text{ as } n \to \infty,$$

which say that, for all $t > 0$,

$$M(f(f(x_n)), f(x^\circ), t) \to 1- \text{ as } n \to \infty \tag{33}$$
$$M(f(f(y_n)), f(y^\circ), t) \to 1- \text{ as } n \to \infty \tag{34}$$
$$M(f(x^\circ), f(f(x_n)), t) \to 1- \text{ as } n \to \infty \tag{35}$$
$$M(f(y^\circ), f(f(y_n)), t) \to 1- \text{ as } n \to \infty. \tag{36}$$

- Suppose that condition (b) is satisfied. Since (X, M) is $(<, \triangleleft)$-complete, there exist $x^\circ, y^\circ \in X$, such that

$$f(x_n) \xrightarrow{M^\triangleleft} x^\circ \text{ and } f(y_n) \xrightarrow{M^\triangleleft} y^\circ \text{ as } n \to \infty. \tag{37}$$

Because f is simultaneously $(\triangleleft, \triangleright)$-continuous and $(\triangleleft, \triangleleft)$-continuous with respect to M, we can similarly obtain (33)–(36).

Using (23) and the commutativity of T_n and f, we obtain

$$f(f(x_{n+1})) = f(T_{n+1}(x_n, y_n)) = T_{n+1}(f(x_n), f(y_n))) \tag{38}$$

and

$$f(f(y_{n+1})) = f(T_{n+1}(y_n, x_n)) = T_{n+1}(f(y_n), f(x_n))).$$

We shall show that $f(x^\circ) = T_n(x^\circ, y^\circ)$ and $f(y^\circ) = T_n(y^\circ, x^\circ)$ for all $n \in \mathbb{N}$. Now we have

$$M(f(f(x_{n+1})), T_n(x^\circ, y^\circ), kt) \geq M(f(f(x_{n+1})), T_n(x^\circ, y^\circ), k_{n+1,n} \cdot t)$$
$$= M(T_{n+1}(f(x_n), f(y_n))), T_n(x^\circ, y^\circ), k_{n+1,n} \cdot t) \text{ (by (38))}$$
$$\geq M(f(f(x_n)), f(x^\circ), t) * M(f(f(y_n)), f(y^\circ), t) \text{ (by (24))}. \tag{39}$$

Using Proposition 1 and applying (33) and (34) to (39), we obtain

$$\liminf_{n \to \infty} M(f(f(x_{n+1})), T_n(x^\circ, y^\circ), t)$$
$$\geq \lim_{n \to \infty} \left[M\left(f(f(x_n)), f(x^\circ), \frac{t}{k}\right) * M\left(f(f(y_n)), f(y^\circ), \frac{t}{k}\right)\right] = 1 * 1 = 1,$$

which says that

$$1 \geq \limsup_{n \to \infty} M(f(f(x_{n+1})), T_n(x^\circ, y^\circ), t) \geq \liminf_{n \to \infty} M(f(f(x_{n+1})), T_n(x^\circ, y^\circ), t) \geq 1.$$

Therefore, we obtain

$$\lim_{n \to \infty} M(f(f(x_{n+1})), T_n(x^\circ, y^\circ), t) = 1, \text{ i.e., } M(f(f(x_{n+1})), T_n(x^\circ, y^\circ), t) \to 1-. \quad (40)$$

Using the ⋈-triangle inequality, we see that

$$M(f(x^\circ), T_n(x^\circ, y^\circ), 2t) \geq M(f(x^\circ), f(f(x_{n+1})), t) * M(f(f(x_{n+1})), T_n(x^\circ, y^\circ), t).$$

While using the left-continuity of t-norm $*$ to (35) and (40), we obtain $M(f(x^\circ), T_n(x^\circ, y^\circ), 2t) = 1$ for all $t > 0$. Therefore we must have $f(x^\circ) = T_n(x^\circ, y^\circ)$ for all $n \in \mathbb{N}$. We can similarly show that $f(y^\circ) = T_n(y^\circ, x^\circ)$ for all $n \in \mathbb{N}$.

To prove property (A), let $(\bar{x}, \bar{y})$ be another coupled coincidence point of f and T_{n_0} for some $n_0 \in \mathbb{N}$, i.e., $f(\bar{x}) = T_{n_0}(\bar{x}, \bar{y})$ and $f(\bar{y}) = T_{n_0}(\bar{y}, \bar{x})$. Because the mapping $M(x, y, \cdot)$ is non-decreasing, by (25), we have

$$M(f(x^\circ), f(\bar{x}), t) = M(T_{n_0}(x^\circ, y^\circ), T_{n_0}(\bar{x}, \bar{y}), t)$$
$$\geq M\left(f(x^\circ), f(\bar{x}), \frac{t}{k_{n_0, n_0}}\right) \cdot M\left(f(y^\circ), f(\bar{y}), \frac{t}{k_{n_0, n_0}}\right)$$
$$\geq M\left(f(x^\circ), f(\bar{x}), \frac{t}{k}\right) \cdot M\left(f(y^\circ), f(\bar{y}), \frac{t}{k}\right) \quad (41)$$

and

$$M(f(y^\circ), f(\bar{y}), t) = M(T_{n_0}(y^\circ, x^\circ), T_{n_0}(\bar{y}, \bar{x}), t)$$
$$\geq M\left(f(y^\circ), f(\bar{y}), \frac{t}{k_{n_0, n_0}}\right) \cdot M\left(f(x^\circ), f(\bar{x}), \frac{t}{k_{n_0, n_0}}\right)$$
$$\geq M\left(f(y^\circ), f(\bar{y}), \frac{t}{k}\right) \cdot M\left(f(x^\circ), f(\bar{x}), \frac{t}{k}\right). \quad (42)$$

Therefore we obtain

$$M(f(x^\circ), f(\bar{x}), t) \geq M\left(f(x^\circ), f(\bar{x}), \frac{t}{k}\right) \cdot M\left(f(y^\circ), f(\bar{y}), \frac{t}{k}\right) \text{ (by (41))}$$
$$\geq \left[M\left(f(x^\circ), f(\bar{x}), \frac{t}{k^2}\right) \cdot M\left(f(y^\circ), f(\bar{y}), \frac{t}{k^2}\right)\right] \cdot \left[M\left(f(x^\circ), f(\bar{x}), \frac{t}{k^2}\right) \cdot M\left(f(y^\circ), f(\bar{y}), \frac{t}{k^2}\right)\right]$$
(by (41) and (42))
$$= \left[M\left(f(x^\circ), f(\bar{x}), \frac{t}{k^2}\right)\right]^2 \cdot \left[M\left(f(y^\circ), f(\bar{y}), \frac{t}{k^2}\right)\right]^2$$
$$\geq \cdots \geq \left[M\left(f(x^\circ), f(\bar{x}), \frac{t}{k^n}\right)\right]^{2^{n-1}} \cdot \left[M\left(f(y^\circ), f(\bar{y}), \frac{t}{k^n}\right)\right]^{2^{n-1}}$$
(by repeating to use (41) and (42))
$$\geq \left[M\left(f(x^\circ), f(\bar{x}), \frac{t}{k^n}\right)\right]^{2^n} \cdot \left[M\left(f(y^\circ), f(\bar{y}), \frac{t}{k^n}\right)\right]^{2^n} \quad (43)$$
(since $M(x, y, t) \leq 1$ for any $x, y \in X$ and $t > 0$),

Equivalently, we have

$$M(f(x^\circ), f(\bar{x}), t) \geq \left[M\left(f(x^\circ), f(\bar{x}), \frac{t}{k^{\log_2 n}}\right)\right]^n \cdot \left[M\left(f(y^\circ), f(\bar{y}), \frac{t}{k^{\log_2 n}}\right)\right]^n.$$

which can be rewritten as

$$M\left(f(x^\circ), f(\tilde{x}), k^{\log_2 n} \cdot t\right) \geq [M(f(x^\circ), f(\tilde{x}), t)]^n \cdot [M(f(y^\circ), f(\tilde{y}), t)]^n. \quad (44)$$

We are going to claim that there exists $\bar{t} > 0$, such that $M(f(y^\circ), f(\tilde{y}), t) \neq 0$ for all $t \geq \bar{t}$. We consider the following two cases.

- If $f(y^\circ) = f(\tilde{y})$, then $M(f(y^\circ), f(\tilde{y}), t) = 1$ for all $t > 0$.
- If $f(y^\circ) \neq f(\tilde{y})$, then the distance condition says that there exits $\bar{t} > 0$ such that $M(f(y^\circ), f(\tilde{y}), \bar{t}) \neq 0$. Part (i) of Proposition 2 says that the mapping $M(x, y, \cdot)$ is nondecreasing. It follows that $M(f(y^\circ), f(\tilde{y}), t) \neq 0$ for all $t \geq \bar{t}$.

Therefore, from (44), for any fixed $t > 0$ with $t \geq \bar{t}$, we have

$$\left[M\left(f(x^\circ), f(\tilde{x}), k^{\log_2 n} \cdot t\right)\right]^{1/n} \cdot \frac{1}{M(f(y^\circ), f(\tilde{y}), t)} \geq M(f(x^\circ), f(\tilde{x}), t). \quad (45)$$

Because $0 < k < 1$ and the mapping $M(x, y, \cdot)$ is non-decreasing, the function ϱ defined in (26) is non-increasing, which says that $\varrho'(\alpha) \leq 0$ on $(0, \infty)$. Because M satisfies the rational condition, we have

$$\lim_{t \to 0+} M(x, y, t) = 0 \quad (46)$$

for any fixed $x, y \in X$ with $x \neq y$. We consider

$$\varrho(\alpha) = M\left(f(x^\circ), f(\tilde{x}), k^{\log_2 \alpha} \cdot t\right).$$

Suppose that $f(x^\circ) \neq f(\tilde{x})$. Because $0 < k < 1$, it follows that $k^{\log_2 \alpha} \cdot t \to 0+$ as $\alpha \to \infty$. Therefore, (46) says that $\varrho(\alpha) \to 0+$ as $\alpha \to \infty$. Subsequently, we obtain

$$\lim_{n \to \infty} \left[M(f(x^\circ), f(\tilde{x}), k^{\log_2 n} \cdot t)\right]^{1/n}$$
$$= \lim_{n \to \infty} [\varrho(n)]^{1/n} = \lim_{n \to \infty} \exp\left[\frac{\ln \varrho(n)}{n}\right] = \exp\left[\lim_{n \to \infty} \frac{\varrho'(n)}{\varrho(n)}\right] \text{ (using the l'Hospital's rule)}$$
$$= 0 \text{ (since } \varrho'(n) \leq 0 \text{ and } \varrho(n) \to 0+). \quad (47)$$

By taking $n \to \infty$ in (45) and using (47), it follows that $M(f(x^\circ), f(\tilde{x}), t) = 0$ for all $t \geq \bar{t}$. Because $f(x^\circ) \neq f(\tilde{x})$, the distance condition says that there exits $t_0 > 0$, such that $M(f(x^\circ), f(\tilde{x}), t_0) \neq 0$, i.e., $M(f(x^\circ), f(\tilde{x}), t) \neq 0$ for all $t \geq t_0$ by the nondecreasing property of $M(x, y, \cdot)$, which contradicts $M(f(x^\circ), f(\tilde{x}), t) = 0$ for all $t \geq \bar{t}$. Therefore, we must have $f(x^\circ) = f(\tilde{x})$. We can similarly obtain $f(y^\circ) = f(\tilde{y})$.

To prove property (B), using the commutativity of T_n and f, we have

$$f(T_n(x^\circ, y^\circ)) = T_n(f(x^\circ), f(y^\circ)) = T_n(T_n(x^\circ, y^\circ), T_n(y^\circ, x^\circ)) \quad (48)$$

and

$$f(T_n(y^\circ, x^\circ)) = T_n(f(y^\circ), f(x^\circ)) = T_n(T_n(y^\circ, x^\circ), T_n(x^\circ, y^\circ)). \quad (49)$$

By regarding $\tilde{x}$ as $T_n(x^\circ, y^\circ)$ and $\tilde{y}$ as $T_n(y^\circ, x^\circ)$, the equalities (48) and (49) say that

$$f(\tilde{x}) = T_n(\tilde{x}, \tilde{y}) \text{ and } f(\tilde{y}) = T_n(\tilde{y}, \tilde{x}).$$

Therefore, using property (A), we must have

$$f(x^\circ) = f(\tilde{x}) = f(T_n(x^\circ, y^\circ)) = T_n(f(x^\circ), f(y^\circ))$$

and

$$f(y^\circ) = f(\tilde{y}) = f(T_n(y^\circ, x^\circ)) = T_n(f(y^\circ), f(x^\circ)),$$

which says that $(f(x^\circ), f(y^\circ)) \in X \times X$ is the common coupled fixed point of the mappings $\{T_n\}_{n=1}^\infty$.

To prove part (ii), we can similarly obtain

$$\eta(f(x_{n+1}), f(x_n), f(y_{n+1}), f(y_n), t) \geq \left[*\eta\left(f(x_1), f(x_0), f(y_1), f(y_0), \frac{t}{k^n} \right) \right]^{2^n}. \quad (50)$$

From part (i) of Proposition 10, it follows that $\{f(x_n)\}_{n=1}^\infty$ and $\{f(y_n)\}_{n=1}^\infty$ are $>$-Cauchy sequences. We consider two cases below.

- Suppose that condition (c) is satisfied. Because (X, M) is $(>, \triangleright)$-complete, there exist $x^\circ, y^\circ \in X$, such that

$$f(x_n) \xrightarrow{M^\triangleright} x^\circ \text{ and } f(y_n) \xrightarrow{M^\triangleright} y^\circ \text{ as } n \to \infty.$$

Because f is simultaneously $(\triangleright, \triangleright)$-continuous and $(\triangleright, \triangleleft)$-continuous with respect to M, we can similarly obtain (33)–(36).

- Suppose that condition (d) is satisfied. Because (X, M) is $(>, \triangleleft)$-complete, there exist $x^\circ, y^\circ \in X$, such that

$$f(x_n) \xrightarrow{M^\triangleleft} x^\circ \text{ and } f(y_n) \xrightarrow{M^\triangleleft} y^\circ \text{ as } n \to \infty.$$

Because f is simultaneously $(\triangleleft, \triangleright)$-continuous and $(\triangleleft, \triangleleft)$-continuous with respect to M, we can similarly obtain (33)–(36).

The remaining proof follows from the similar argument of part (i). This completes the proof. □

In Theorem 1, since the fuzzy semi-metric M is not necessarily symmetric, if the contractive inequalities (24) and (25) are not satisfied and, alternatively, the following converse-contractive inequalities

$$M(T_i(x,y), T_j(u,v), k_{ij} \cdot t) \geq M(f(u), f(x), t) * M(f(v), f(y), t)$$

and

$$M(T_i(x,y), T_j(u,v), k_{ij} \cdot t) \geq M(f(u), f(x), t) \cdot M(f(v), f(y), t)$$

are satisfied, then we can also obtain the desired results by assuming the different conditions.

Theorem 2 (Satisfying the ⋈-Triangle Inequality: Converse-Contractive Inequality). *Let (X, M) be a fuzzy semi-metric space, such that the mapping M satisfies the rational condition and the ⋈-triangle inequality. Suppose that the following conditions are satisfied.*

- *The first four conditions in Theorem 1 are satisfied.*
- *For any $x, y, u, v \in X$, the following converse-contractive inequality is satisfied:*

$$M(T_i(x,y), T_j(u,v), k_{ij} \cdot t) \geq M(f(u), f(x), t) * M(f(v), f(y), t), \quad (51)$$

where k_{ij} satisfies $0 < k_{ij} \leq k < 1$ for all $i, j \in \mathbb{N}$ and for some constant k.

Subsequently, we have the following results.

(i) *Suppose that there exist $x^*, y^* \in X$ satisfying*

$$\sup_{\lambda \in (0,1)} \Phi(\lambda, f(x^*), T_1(x^*, y^*), f(y^*), T_1(y^*, x^*)) < \infty,$$

and that any one of the following conditions is satisfied:

(a) *(X, M) is $(<, \triangleright)$-complete and f is $(\triangleright, \triangleright)$-continuous or $(\triangleright, \triangleleft)$-continuous with respect to M;*

(b) (X, M) is $(<, \triangleleft)$-complete and f is $(\triangleleft, \triangleright)$-continuous or $(\triangleleft, \triangleleft)$-continuous with respect to M.

Subsequently, the mappings $\{T_n\}_{n=1}^{\infty}$ and f have a common coupled coincidence point (x°, y°). We further assume that the following conditions are satisfied.

- The inequality (51) is replaced by the following inequality

$$M(T_i(x,y), T_j(u,v), k_{ij} \cdot t) \geq M(f(u), f(x), t) \cdot M(f(v), f(y), t),$$

where the t-norm $*$ is replaced by the product of real numbers;
- The mapping M satisfies the distance condition in Definition 2.
- For any fixed $x, y \in X$ and $t > 0$, the following mapping

$$\varrho(\alpha) = M\left(x, y, k^{\log_2 \alpha} \cdot t\right)$$

is differentiable on $(0, \infty)$.

Afterwards, we have the following results.

(A) Suppose that $(\bar{x}, \bar{y})$ is another coupled coincidence point of f and T_{n_0} for some $n_0 \in \mathbb{N}$. Then $f(x^\circ) = f(\bar{x})$ and $f(y^\circ) = f(\bar{y})$.
(B) There exists $(x^\circ, y^\circ) \in X \times X$, such that $(f(x^\circ), f(y^\circ)) \in X \times X$ is the common coupled fixed point of the mappings $\{T_n\}_{n=1}^{\infty}$.

Moreover, the point $(x^\circ, y^\circ) \in X \times X$ can be obtained, as follows.

- Suppose that condition (a) is satisfied. Then the point $(x^\circ, y^\circ) \in X \times X$ can be obtained by taking the limit $f(x_n) \xrightarrow{M^\triangleright} x^\circ$ and $f(y_n) \xrightarrow{M^\triangleright} y^\circ$.
- Suppose that condition (b) is satisfied. Subsequently, the point $(x^\circ, y^\circ) \in X \times X$ can be obtained by taking the limit $f(x_n) \xrightarrow{M^\triangleleft} x^\circ$ and $f(y_n) \xrightarrow{M^\triangleleft} y^\circ$.

The sequences $\{x_n\}_{n=1}^{\infty}$ and $\{y_n\}_{n=1}^{\infty}$ are generated from the initial element $(x_0, y_0) = (x^*, y^*) \in X \times X$, according to (23).

(ii) Suppose that there exist $x^*, y^* \in X$ satisfying

$$\sup_{\lambda \in (0,1)} \Phi(\lambda, T_1(x^*, y^*), f(x^*), T_1(y^*, x^*), f(y^*)) < \infty,$$

and that any one of the following conditions is satisfied:

(c) (X, M) is $(>, \triangleright)$-complete and f is $(\triangleright, \triangleright)$-continuous or $(\triangleright, \triangleleft)$-continuous with respect to M;
(d) (X, M) is $(>, \triangleleft)$-complete and f is $(\triangleleft, \triangleright)$-continuous or $(\triangleleft, \triangleleft)$-continuous with respect to M;

Subsequently, we have the same result as part (i).

Theorem 3 (Satisfying the $\triangleright$-Triangle Inequality). *Let (X, M) be a fuzzy semi-metric space, such that the mapping M satisfies the rational condition and the $\triangleright$-triangle inequality. Let $(x_0, y_0) \in X \times X$ be an initial element that generates the sequences $\{x_n\}_{n=1}^{\infty}$ and $\{y_n\}_{n=1}^{\infty}$ according to (23). Suppose that the following conditions are satisfied.*

- *The first four conditions in Theorem 1 are satisfied.*
- *The following contractive inequalities is satisfied*

$$M(T_i(x,y), T_j(u,v), k_{ij} \cdot t) \geq M(f(x), f(u), t) * M(f(y), f(v), t) \tag{52}$$

or the following converse-contractive inequalities is satisfied

$$M(T_i(x,y), T_j(u,v), k_{ij} \cdot t) \geq M(f(u), f(x), t) * M(f(v), f(y), t), \tag{53}$$

where k_{ij} satisfies $0 < k_{ij} \leq k < 1$ for all $i, j \in \mathbb{N}$ and for some constant k.

- There exist $x^*, y^* \in X$ satisfying
$$\sup_{\lambda \in (0,1)} \Phi(\lambda, f(x^*), T_1(x^*, y^*), f(y^*), T_1(y^*, x^*)) < \infty$$

and
$$\sup_{\lambda \in (0,1)} \Phi(\lambda, T_1(x^*, y^*), f(x^*), T_1(y^*, x^*), f(y^*)) < \infty.$$

- Any one of the following conditions is satisfied:
  (a) (X, M) is $(<, \rhd)$-complete or $(>, \rhd)$-complete and f is $(\rhd, \lhd)$-continuous with respect to M;
  (b) (X, M) is $(<, \lhd)$-complete or $(>, \lhd)$-complete and f is $(\lhd, \lhd)$-continuous with respect to M.

Subsequently, the mappings $\{T_n\}_{n=1}^{\infty}$ and f have a common coupled coincidence point (x°, y°). We further assume that the following conditions are satisfied.

- The inequality (52) is replaced by the following inequality
$$M(T_i(x,y), T_j(u,v), k_{ij} \cdot t) \geq M(f(x), f(u), t) \cdot M(f(y), f(v), t) \quad (54)$$

and the inequality (53) is replaced by the following inequality
$$M(T_i(x,y), T_j(u,v), k_{ij} \cdot t) \geq M(f(u), f(x), t) \cdot M(f(v), f(y), t), \quad (55)$$

where the t-norm $*$ is replaced by the product of real numbers, such that any one of the inequalities (54) and (55) is satisfied.

- The mapping M satisfies the distance condition in Definition 2.
- For any fixed $x, y \in X$ and $t > 0$, the following mapping
$$\varrho(\alpha) = M\left(x, y, k^{\log_2 \alpha} \cdot t\right)$$

is differentiable on $(0, \infty)$.

Afterwards, we have the following results.

(A) Suppose that $(\bar{x}, \bar{y})$ is another coupled coincidence point of f and T_{n_0} for some $n_0 \in \mathbb{N}$. Subsequently, $f(x^\circ) = f(\bar{x})$ and $f(y^\circ) = f(\bar{y})$.

(B) There exists $(x^\circ, y^\circ) \in X \times X$, such that $(f(x^\circ), f(y^\circ)) \in X \times X$ is the common coupled fixed point of the mappings $\{T_n\}_{n=1}^{\infty}$.

Moreover, the point $(x^\circ, y^\circ) \in X \times X$ can be obtained as follows.

- Suppose that condition (a) is satisfied. Subsequently, the point $(x^\circ, y^\circ) \in X \times X$ can be obtained by taking the limit $f(x_n) \xrightarrow{M^\rhd} x^\circ$ and $f(y_n) \xrightarrow{M^\rhd} y^\circ$.
- Suppose that condition (b) is satisfied. Afterwards, the point $(x^\circ, y^\circ) \in X \times X$ can be obtained by taking the limit $f(x_n) \xrightarrow{M^\lhd} x^\circ$ and $f(y_n) \xrightarrow{M^\lhd} y^\circ$.

The sequences $\{x_n\}_{n=1}^{\infty}$ and $\{y_n\}_{n=1}^{\infty}$ are generated from the initial element $(x_0, y_0) = (x^*, y^*) \in X \times X$ according to (23).

Theorem 4 (Satisfying the $\lhd$-Triangle Inequality). *Let (X, M) be a fuzzy semi-metric space such that the mapping M satisfies the rational condition and the $\rhd$-triangle inequality. Let $(x_0, y_0) \in X \times X$ be an initial element that generates the sequences $\{x_n\}_{n=1}^{\infty}$ and $\{y_n\}_{n=1}^{\infty}$ according to (23). Suppose that the following conditions are satisfied.*

- The first four conditions in Theorem 1 are satisfied.
- The following contractive inequalities is satisfied
$$M(T_i(x,y), T_j(u,v), k_{ij} \cdot t) \geq M(f(x), f(u), t) * M(f(y), f(v), t) \quad (56)$$

or the following converse-contractive inequalities is satisfied

$$M(T_i(x,y), T_j(u,v), k_{ij} \cdot t) \geq M(f(u), f(x), t) * M(f(v), f(y), t), \tag{57}$$

where k_{ij} satisfies $0 < k_{ij} \leq k < 1$ for all $i, j \in \mathbb{N}$ and for some constant k.
- There exist $x^*, y^* \in X$ satisfying

$$\sup_{\lambda \in (0,1)} \Phi(\lambda, f(x^*), T_1(x^*, y^*), f(y^*), T_1(y^*, x^*)) < \infty$$

and

$$\sup_{\lambda \in (0,1)} \Phi(\lambda, T_1(x^*, y^*), f(x^*), T_1(y^*, x^*), f(y^*)) < \infty;$$

- Any one of the following conditions is satisfied:
  (a) (X, M) is $(<, \triangleright)$-complete or $(>, \triangleright)$-complete and f is $(\triangleright, \triangleright)$-continuous with respect to M;
  (b) (X, M) is $(<, \triangleleft)$-complete or $(>, \triangleleft)$-complete and f is $(\triangleleft, \triangleright)$-continuous with respect to M.

Subsequently, the mappings $\{T_n\}_{n=1}^{\infty}$ and f have a common coupled coincidence point (x°, y°). We further assume that the following conditions are satisfied.
- The inequality (56) is replaced by the following inequality

$$M(T_i(x,y), T_j(u,v), k_{ij} \cdot t) \geq M(f(x), f(u), t) \cdot M(f(y), f(v), t) \tag{58}$$

and the inequality (57) is replaced by the following inequality

$$M(T_i(x,y), T_j(u,v), k_{ij} \cdot t) \geq M(f(u), f(x), t) \cdot M(f(v), f(y), t), \tag{59}$$

where the t-norm $*$ is replaced by the product of real numbers, such that any one of the inequalities (58) and (59) is satisfied.
- The mapping M satisfies the distance condition in Definition 2.
- For any fixed $x, y \in X$ and $t > 0$, the following mapping

$$\varrho(\alpha) = M\left(x, y, k^{\log_2 \alpha} \cdot t\right)$$

is differentiable on $(0, \infty)$.

Subsequently, we have the following results.
(A) Suppose that $(\bar{x}, \bar{y})$ is another coupled coincidence point of f and T_{n_0} for some $n_0 \in \mathbb{N}$. Subsequently, $f(x^\circ) = f(\bar{x})$ and $f(y^\circ) = f(\bar{y})$.
(B) There exists $(x^\circ, y^\circ) \in X \times X$ such that $(f(x^\circ), f(y^\circ)) \in X \times X$ is the common coupled fixed point of the mappings $\{T_n\}_{n=1}^{\infty}$.

Moreover, the point $(x^\circ, y^\circ) \in X \times X$ can be obtained, as follows.
- Suppose that condition (a) is satisfied. Afterwards, the point $(x^\circ, y^\circ) \in X \times X$ can be obtained by taking the limit $f(x_n) \xrightarrow{M^{\triangleright}} x^\circ$ and $f(y_n) \xrightarrow{M^{\triangleright}} y^\circ$.
- Suppose that condition (b) is satisfied. Subsequently, the point $(x^\circ, y^\circ) \in X \times X$ can be obtained by taking the limit $f(x_n) \xrightarrow{M^{\triangleleft}} x^\circ$ and $f(y_n) \xrightarrow{M^{\triangleleft}} y^\circ$.

The sequences $\{x_n\}_{n=1}^{\infty}$ and $\{y_n\}_{n=1}^{\infty}$ are generated from the initial element $(x_0, y_0) = (x^*, y^*) \in X \times X$ according to (23).

Theorem 5 (Satisfying the $\diamond$-Triangle Inequality). *Let (X, M) be a fuzzy semi-metric space, such that the mapping M satisfies the rational condition and the $\diamond$-triangle inequality. Suppose that the following conditions are satisfied.*
- *All five conditions in Theorem 1 are satisfied.*

- There exist $x^*, y^* \in X$ satisfying
$$\sup_{\lambda \in (0,1)} \Phi(\lambda, f(x^*), T_1(x^*, y^*), f(y^*), T_1(y^*, x^*)) < \infty$$

and
$$\sup_{\lambda \in (0,1)} \Phi(\lambda, T_1(x^*, y^*), f(x^*), T_1(y^*, x^*), f(y^*)) < \infty.$$

- Any one of the following conditions is satisfied:
  (a) (X, M) is $(<, \triangleright)$-complete or $(>, \triangleright)$-complete, and f is $(\triangleright, \triangleright)$-continuous or $(\triangleright, \triangleleft)$-continuous with respect to M;
  (b) (X, M) is $(<, \triangleleft)$-complete or $(>, \triangleleft)$-complete, and f is $(\triangleleft, \triangleright)$-continuous or $(\triangleleft, \triangleleft)$-continuous with respect to M.

Subsequently, the mappings $\{T_n\}_{n=1}^{\infty}$ and f have a common coupled coincidence point (x°, y°). Moreover, the point $(x^\circ, y^\circ) \in X \times X$ can be obtained, as follows.

- Suppose that condition (a) is satisfied. Afterwards, the point $(x^\circ, y^\circ) \in X \times X$ can be obtained by taking the limit $f(x_n) \xrightarrow{M^\triangleright} x^\circ$ and $f(y_n) \xrightarrow{M^\triangleright} y^\circ$.
- Suppose that condition (b) is satisfied. Subsequently, the point $(x^\circ, y^\circ) \in X \times X$ can be obtained by taking the limit $f(x_n) \xrightarrow{M^\triangleleft} x^\circ$ and $f(y_n) \xrightarrow{M^\triangleleft} y^\circ$.

The sequences $\{x_n\}_{n=1}^{\infty}$ and $\{y_n\}_{n=1}^{\infty}$ are generated from the initial element $(x_0, y_0) = (x^*, y^*) \in X \times X$ according to (23).

Theorem 6 (Satisfying the $\diamond$-Triangle Inequality: Converse-Contractive Inequality). *Let (X, M) be a fuzzy semi-metric space, such that the mapping M satisfies the rational condition and the $\diamond$-triangle inequality. Let $(x_0, y_0) \in X \times X$ be an initial element that generates the sequences $\{x_n\}_{n=1}^{\infty}$ and $\{y_n\}_{n=1}^{\infty}$ according to (23). Suppose that the following conditions are satisfied.*

- *The first four conditions in Theorem 1 are satisfied.*
- *For any $x, y, u, v \in X$, the following converse-contractive inequality is satisfied:*
$$M(T_i(x, y), T_j(u, v), k_{ij} \cdot t) \geq M(f(u), f(x), t) * M(f(v), f(y), t),$$

where k_{ij} satisfies $0 < k_{ij} \leq k < 1$ for all $i, j \in \mathbb{N}$ and for some constant k.
- *There exist $x^*, y^* \in X$ satisfying*
$$\sup_{\lambda \in (0,1)} \Phi(\lambda, f(x^*), T_1(x^*, y^*), f(y^*), T_1(y^*, x^*)) < \infty$$

and
$$\sup_{\lambda \in (0,1)} \Phi(\lambda, T_1(x^*, y^*), f(x^*), T_1(y^*, x^*), f(y^*)) < \infty.$$

- *Any one of the following conditions is satisfied:*
  (a) (X, M) is $(<, \triangleright)$-complete or $(>, \triangleright)$-complete and f is simultaneously $(\triangleright, \triangleright)$-continuous and $(\triangleright, \triangleleft)$-continuous with respect to M;
  (b) (X, M) is $(<, \triangleleft)$-complete or $(>, \triangleleft)$-complete and f is simultaneously $(\triangleleft, \triangleright)$-continuous and $(\triangleleft, \triangleleft)$-continuous with respect to M.

Subsequently, the mappings $\{T_n\}_{n=1}^{\infty}$ and f have a common coupled coincidence point (x°, y°). Moreover, the point $(x^\circ, y^\circ) \in X \times X$ can be obtained as follows.

- *Suppose that condition (a) is satisfied. Afterwards, the point $(x^\circ, y^\circ) \in X \times X$ can be obtained by taking the limit $f(x_n) \xrightarrow{M^\triangleright} x^\circ$ and $f(y_n) \xrightarrow{M^\triangleright} y^\circ$.*
- *Suppose that condition (b) is satisfied. Subsequently, the point $(x^\circ, y^\circ) \in X \times X$ can be obtained by taking the limit $f(x_n) \xrightarrow{M^\triangleleft} x^\circ$ and $f(y_n) \xrightarrow{M^\triangleleft} y^\circ$.*

The sequences $\{x_n\}_{n=1}^{\infty}$ and $\{y_n\}_{n=1}^{\infty}$ are generated from the initial element $(x_0, y_0) = (x^*, y^*) \in X \times X$ according to (23).

6. Common Coupled Fixed Points

Consider the mappings $T : X \times X \to X$ and $f : X \to X$. Recall that an element $(x^*, y^*) \in X \times X$ is called a common coupled fixed point when

$$x^* = f(x^*) = T(x^*, y^*) \text{ and } y^* = f(y^*) = T(y^*, x^*).$$

The common coupled fixed points are also the common coupled coincidence points. The uniqueness of common coupled coincidence points presented above was not guaranteed. In this section, we shall investigate the uniqueness of common coupled fixed points.

The contractive inequality and converse-contractive inequality should consider the product of real numbers instead of t-norm $*$ in order to obtain the unique common coupled fixed point.

Theorem 7 (Satisfying the $\bowtie$-Triangle Inequality). *Let (X, M) be a fuzzy semi-metric space, such that the mapping M satisfies the rational condition and the $\bowtie$-triangle inequality. Suppose that the following conditions are satisfied:*

- *For any sequences $\{a_n\}_{n=1}^{\infty}$ and $\{b_n\}_{n=1}^{\infty}$ in $[0, 1]$, the following inequality is satisfied:*

$$\sup_n (a_n * b_n) \geq \left(\sup_n a_n\right) * \left(\sup_n b_n\right).$$

- *The t-norm $*$ is left-continuous with respect to the first or second component.*
- *Given any fixed $x, y \in X$, the mapping $M(x, y, \cdot) : (0, \infty) \to [0, 1]$ is continuous on $(0, \infty)$.*
- *The mapping M satisfies the distance condition in Definition 2.*
- *Given any fixed $x, y \in X$ and $t > 0$, the following mapping*

$$\varrho(\alpha) = M\left(x, y, k^{\log_2 \alpha} \cdot t\right)$$

is differentiable on $(0, \infty)$.
- *The mappings $T_n : X \to X$ and $f : X \to X$ satisfy the inclusion $T_n(X, X) \subseteq f(X)$ for all $n \in \mathbb{N}$.*
- *The mappings f and T_n commute.*
- *Any one of the following conditions is satisfied:*
  - *the mapping f is simultaneously $(\triangleright, \triangleright)$-continuous and $(\triangleright, \triangleleft)$-continuous with respect to M;*
  - *the mapping f is simultaneously $(\triangleleft, \triangleright)$-continuous and $(\triangleleft, \triangleleft)$-continuous with respect to M.*
- *for any $x, y, u, v \in X$, the following contractive inequality is satisfied:*

$$M(T_i(x, y), T_j(u, v), k_{ij} \cdot t) \geq M(f(x), f(u), t) \cdot M(f(y), f(v), t), \tag{60}$$

where k_{ij} satisfies $0 < k_{ij} \leq k < 1$ for all $i, j \in \mathbb{N}$ and for some constant k.

Subsequently, we have the following results.

(i) *Suppose that the space (X, M) is simultaneously $(<, \triangleright)$-complete and $(<, \triangleleft)$-complete. We also assume that there exist $x^*, y^* \in X$ satisfying*

$$\sup_{\lambda \in (0,1)} \Phi(\lambda, f(x^*), T_1(x^*, y^*), f(y^*), T_1(y^*, x^*)) < \infty.$$

Afterwards, the mappings $\{T_n\}_{n=1}^{\infty}$ and f have a unique common coupled fixed point (x°, y°).

(ii) Suppose that the space (X, M) is simultaneously $(>, \triangleright)$-complete and $(>, \triangleleft)$-complete. We also assume that there exist $x^*, y^* \in X$ satisfying

$$\sup_{\lambda \in (0,1)} \Phi(\lambda, T_1(x^*, y^*), f(x^*), T_1(y^*, x^*), f(y^*)) < \infty.$$

Then the mappings $\{T_n\}_{n=1}^{\infty}$ and f have a unique common coupled fixed point (x°, y°). Moreover, the point $(x^{\circ}, y^{\circ}) \in X \times X$ can be obtained as follows.

- The point x° can be obtained by taking the limit $f(x_n) \xrightarrow{M^{\triangleright}} x^{\circ}$ or the limit $f(x_n) \xrightarrow{M^{\triangleleft}} x^{\circ}$;
- The point y° can be obtained by taking the limit $f(y_n) \xrightarrow{M^{\triangleright}} y^{\circ}$ or the limit $f(y_n) \xrightarrow{M^{\triangleleft}} y^{\circ}$.

The sequences $\{x_n\}_{n=1}^{\infty}$ and $\{y_n\}_{n=1}^{\infty}$ are generated from the initial element $(x_0, y_0) = (x^*, y^*) \in X \times X$ according to (23).

Proof. According to (23), we can generate two sequences $\{x_n\}_{n=1}^{\infty}$ and $\{y_n\}_{n=1}^{\infty}$ from the initial element $x_0 = x^*$ and $y_0 = y^*$. To prove part (i), while using part (i) of Theorem 1, we have $f(x^{\circ}) = T_n(x^{\circ}, y^{\circ})$ and $f(y^{\circ}) = T_n(y^{\circ}, x^{\circ})$ for all $n \in \mathbb{N}$. According to the proof of part (i) of Theorem 1, we see that $\{f(x_n)\}_{n=1}^{\infty}$ and $\{f(y_n)\}_{n=1}^{\infty}$ are $<$-Cauchy sequences. Since (X, M) is simultaneously $(<, \triangleright)$-complete and $(<, \triangleleft)$-complete, using part (i) of Proposition 3, there exists $x^{\circ}, y^{\circ} \in X$ satisfying $f(x_n) \xrightarrow{M^{\triangleright}} x^{\circ}$, $f(x_n) \xrightarrow{M^{\triangleleft}} x^{\circ}$, $f(y_n) \xrightarrow{M^{\triangleright}} y^{\circ}$ and $f(y_n) \xrightarrow{M^{\triangleleft}} y^{\circ}$ as $n \to \infty$, which also says that $f(x_n) \xrightarrow{M} x^{\circ}$ and $f(y_n) \xrightarrow{M} y^{\circ}$ as $n \to \infty$.

Next, we are going to claim that x° is a fixed point of f. While using (60) and the nondecreasing property of $M(x, y, \cdot)$ by part (i) of Proposition 2, we have

$$M(f(x_{n+1}), f(x^{\circ}), t) = M(T_{n+1}(x_n, y_n), T_n(x^{\circ}, y^{\circ}), t)$$
$$\geq M\left(f(x_n), f(x^{\circ}), \frac{t}{k_{n+1,n}}\right) \cdot M\left(f(y_n), f(y^{\circ}), \frac{t}{k_{n+1,n}}\right)$$
$$\geq M\left(f(x_n), f(x^{\circ}), \frac{t}{k}\right) \cdot M\left(f(y_n), f(y^{\circ}), \frac{t}{k}\right) \quad (61)$$

and

$$M(f(y_{n+1}), f(y^{\circ}), t) = M(T_{n+1}(y_n, x_n), T_n(y^{\circ}, x^{\circ}), t)$$
$$\geq M\left(f(y_n), f(y^{\circ}), \frac{t}{k_{n+1,n}}\right) \cdot M\left(f(x_n), f(x^{\circ}), \frac{t}{k_{n+1,n}}\right)$$
$$\geq M\left(f(y_n), f(y^{\circ}), \frac{t}{k}\right) \cdot M\left(f(x_n), f(x^{\circ}), \frac{t}{k}\right) \quad (62)$$

Because $f(x_n) \xrightarrow{M} x^{\circ}$ and $f(y_n) \xrightarrow{M} y^{\circ}$ as $n \to \infty$, applying part (i) of Proposition 4 to (61) and (62), we obtain

$$M(x^{\circ}, f(x^{\circ}), t) \geq M\left(x^{\circ}, f(x^{\circ}), \frac{t}{k}\right) \cdot M\left(y^{\circ}, f(y^{\circ}), \frac{t}{k}\right).$$

and

$$M(y^{\circ}, f(y^{\circ}), t) \geq M\left(x^{\circ}, f(x^{\circ}), \frac{t}{k}\right) \cdot M\left(y^{\circ}, f(y^{\circ}), \frac{t}{k}\right).$$

By referring to (43), we can obtain

$$M(x^{\circ}, f(x^{\circ}), t) \geq \left[M\left(x^{\circ}, f(x^{\circ}), \frac{t}{k^n}\right)\right]^{2^n} \cdot \left[M\left(y^{\circ}, f(y^{\circ}), \frac{t}{k^n}\right)\right]^{2^n},$$

which is equivalent to
$$M(x^\circ, f(x^\circ), k^{\log_2 n} \cdot t) \geq [M(x^\circ, f(x^\circ), t)]^n \cdot [M(y^\circ, f(y^\circ), t)]^n. \tag{63}$$

We are going to claim that there exists $\bar{t} > 0$, such that $M(y^\circ, f(y^\circ), t) \neq 0$ for all $t \geq \bar{t}$. We consider the following cases.

- If $f(y^\circ) = y^\circ$, then $M(y^\circ, f(y^\circ), t) = 1$ for all $t > 0$.
- If $f(y^\circ) \neq y^\circ$, then the distance condition says that there exits $\bar{t} > 0$, such that $M(y^\circ, f(y^\circ), \bar{t}) \neq 0$. Part (i) of Proposition 2 says that the mapping $M(x, y, \cdot)$ is nondecreasing. Therefore, we have $M(y^\circ, f(y^\circ), t) \neq 0$ for all $t \geq \bar{t}$.

From (63), for any fixed $t > 0$ with $t \geq \bar{t}$, we have

$$\left[M\left(x^\circ, f(x^\circ), k^{\log_2 n} \cdot t\right)\right]^{1/n} \cdot \frac{1}{M(y^\circ, f(y^\circ), t)} \geq M(x^\circ, f(x^\circ), t). \tag{64}$$

Because $0 < k < 1$ and the mapping $M(x, y, \cdot)$ is nondecreasing, the function ϱ that is defined in (26) is non-increasing, which says that $\varrho'(\alpha) \leq 0$ on $(0, \infty)$. Because M satisfies the rational condition, we have

$$\lim_{t \to 0+} M(x, y, t) = 0 \tag{65}$$

for any fixed $x, y \in X$ with $x \neq y$. We consider

$$\varrho(\alpha) = M\left(x^\circ, f(x^\circ), k^{\log_2 \alpha} \cdot t\right).$$

Suppose that $f(x^\circ) \neq x^\circ$. Since $0 < k < 1$, it follows that $k^{\log_2 \alpha} \cdot t \to 0+$ as $\alpha \to \infty$. Therefore, (65) says that $\varrho(\alpha) \to 0+$ as $\alpha \to \infty$. Subsequently, we obtain

$$\lim_{n \to \infty} \left[M(x^\circ, f(x^\circ), k^{\log_2 n} \cdot t)\right]^{1/n}$$
$$= \lim_{n \to \infty} [\varrho(n)]^{1/n} = \lim_{n \to \infty} \exp\left[\frac{\ln \varrho(n)}{n}\right] = \exp\left[\lim_{n \to \infty} \frac{\varrho'(n)}{\varrho(n)}\right] \text{ (using the l'Hospital's rule)}$$
$$= 0 \text{ (since } \varrho'(n) \leq 0 \text{ and } \varrho(n) \to 0+). \tag{66}$$

Applying (66) to (64), we obtain $M(x^\circ, f(x^\circ), t) = 0$ for all $t \geq \bar{t}$. Because $f(x^\circ) \neq x^\circ$, the distance condition says that there exits $t_0 > 0$ such that $M(x^\circ, f(x^\circ), t_0) \neq 0$, i.e., $M(x^\circ, f(x^\circ), t) \neq 0$ for all $t \geq t_0$ by the nondecreasing property of $M(x, y, \cdot)$, which contradicts $M(x^\circ, f(x^\circ), t) = 0$ for all $t \geq \bar{t}$. Therefore we must have $f(x^\circ) = x^\circ$. We can similarly obtain $f(y^\circ) = y^\circ$; that is,

$$x^\circ = f(x^\circ) = T_n(x^\circ, y^\circ) \text{ and } y^\circ = f(y^\circ) = T_n(y^\circ, x^\circ)$$

for all $n \in \mathbb{N}$.

In order to prove the uniqueness, let $(\bar{x}, \bar{y})$ be another common coupled fixed point of f and $\{T_n\}_{n=1}^\infty$, i.e., $\bar{x} = f(\bar{x}) = T_n(\bar{x}, \bar{y})$ and $\bar{y} = f(\bar{y}) = T_n(\bar{y}, \bar{x})$ for all $n \in \mathbb{N}$. The inequality (43) is equivalent to

$$M(x^\circ, \bar{x}, k^{\log_2 n} \cdot t) \geq [M(x^\circ, \bar{x}, t)]^n \cdot [M(y^\circ, \bar{y}, t)]^n. \tag{67}$$

We can similarly show that there exists $\hat{t} > 0$, such that $M(y^\circ, \bar{y}, t) \neq 0$ for all $t \geq \hat{t}$. Therefore, from (67), for any fixed $t > 0$ with $t \geq \hat{t}$, we have

$$\left[M\left(x^\circ, \bar{x}, k^{\log_2 n} \cdot t\right)\right]^{1/n} \cdot \frac{1}{M(y^\circ, \bar{y}, t)} \geq M(x^\circ, \bar{x}, t). \tag{68}$$

By referring to (68), it follows that $M(x^\circ, \tilde{x}, t) = 0$ for all $t \geq \hat{t}$. Because $\tilde{x} \neq x^\circ$, the distance condition says that there exits $t_0 > 0$, such that $M(x^\circ, \tilde{x}, t_0) \neq 0$, i.e., $M(x^\circ, \tilde{x}, t) \neq 0$ for all $t \geq t_0$ by the non-decreasing property of $M(x, y, \cdot)$, which contradicts $M(x^\circ, \tilde{x}, t) = 0$ for all $t \geq \hat{t}$. Therefore, we must have $\tilde{x} = x^\circ$. We can similarly obtain $\tilde{y} = y^\circ$. This proves the uniqueness. Finally, part (ii) can be obtained by applying part (ii) of Theorem 1 to the above argument. This completes the proof. □

Theorem 8 (Satisfying the ⋈-Triangle Inequality: Converse-Contractive Inequality). *Let (X, M) be a fuzzy semi-metric space such that the mapping M satisfies the rational condition and the ⋈-triangle inequality. Suppose that the following conditions are satisfied.*

- *The first eight conditions of Theorem 7 are satisfied.*
- *For any $x, y, u, v \in X$, the following converse-contractive inequality is satisfied:*

$$M(T_i(x,y), T_j(u,v), k_{ij} \cdot t) \geq M(f(u), f(x), t) \cdot M(f(v), f(y), t), \qquad (69)$$

where k_{ij} satisfies $0 < k_{ij} \leq k < 1$ for all $i, j \in \mathbb{N}$ and for some constant k.

Subsequently, we have the following results.

(i) *Suppose that the space (X, M) is simultaneously $(<, \triangleright)$-complete and $(<, \triangleleft)$-complete. We also assume that there exist $x^*, y^* \in X$ satisfying*

$$\sup_{\lambda \in (0,1)} \Phi(\lambda, f(x^*), T_1(x^*, y^*), f(y^*), T_1(y^*, x^*)) < \infty.$$

Afterwards, the mappings $\{T_n\}_{n=1}^\infty$ and f have a unique common coupled fixed point (x°, y°).

(ii) *Suppose that the space (X, M) is simultaneously $(>, \triangleright)$-complete and $(>, \triangleleft)$-complete. We also assume that there exist $x^*, y^* \in X$ satisfying*

$$\sup_{\lambda \in (0,1)} \Phi(\lambda, T_1(x^*, y^*), f(x^*), T_1(y^*, x^*), f(y^*)) < \infty.$$

Subsequently, the mappings $\{T_n\}_{n=1}^\infty$ and f have a unique common coupled fixed point (x°, y°).

Moreover, the point $(x^\circ, y^\circ) \in X \times X$ can be obtained, as follows.

- *The point x° can be obtained by taking the limit $f(x_n) \xrightarrow{M^\triangleright} x^\circ$ or the limit $f(x_n) \xrightarrow{M^\triangleleft} x^\circ$.*
- *The point y° can be obtained by taking the limit $f(y_n) \xrightarrow{M^\triangleright} y^\circ$ or the limit $f(y_n) \xrightarrow{M^\triangleleft} y^\circ$.*

The sequences $\{x_n\}_{n=1}^\infty$ and $\{y_n\}_{n=1}^\infty$ are generated from the initial element $(x_0, y_0) = (x^*, y^*) \in X \times X$, according to (23).

Theorem 9 (Satisfying the ▷-Triangle Inequality). *Let (X, M) be a fuzzy semi-metric space, such that the mapping M satisfies the rational condition and the ▷-triangle inequality. Suppose that the following conditions are satisfied.*

- *The first eight conditions of Theorem 7 are satisfied.*
- *The following contractive inequalities is satisfied*

$$M(T_i(x,y), T_j(u,v), k_{ij} \cdot t) \geq M(f(x), f(u), t) \cdot M(f(y), f(v), t) \qquad (70)$$

or the following converse-contractive inequalities is satisfied

$$M(T_i(x,y), T_j(u,v), k_{ij} \cdot t) \geq M(f(u), f(x), t) \cdot M(f(v), f(y), t), \qquad (71)$$

where k_{ij} satisfies $0 < k_{ij} \leq k < 1$ for all $i, j \in \mathbb{N}$ and for some constant k.
- *There exist $x^*, y^* \in X$, satisfying*

$$\sup_{\lambda \in (0,1)} \Phi(\lambda, f(x^*), T_1(x^*, y^*), f(y^*), T_1(y^*, x^*)) < \infty$$

and
$$\sup_{\lambda \in (0,1)} \Phi(\lambda, T_1(x^*, y^*), f(x^*), T_1(y^*, x^*), f(y^*)) < \infty.$$

- The mapping f is $(\triangleright, \triangleleft)$-continuous or $(\triangleleft, \triangleleft)$-continuous with respect to M.
- Any one of the following conditions is satisfied:
  - (X, M) is $(<, \triangleright)$-complete and $(<, \triangleleft)$-complete simultaneously;
  - (X, M) is $(>, \triangleright)$-complete and $(>, \triangleleft)$-complete simultaneously.

Subsequently, the mappings $\{T_n\}_{n=1}^{\infty}$ and f have a unique common coupled fixed point (x°, y°). Moreover, the point $(x^\circ, y^\circ) \in X \times X$ can be obtained, as follows.

- The point x° can be obtained by taking the limit $f(x_n) \xrightarrow{M^\triangleright} x^\circ$ or the limit $f(x_n) \xrightarrow{M^\triangleleft} x^\circ$.
- The point y° can be obtained by taking the limit $f(y_n) \xrightarrow{M^\triangleright} y^\circ$ or the limit $f(y_n) \xrightarrow{M^\triangleleft} y^\circ$.

The sequences $\{x_n\}_{n=1}^{\infty}$ and $\{y_n\}_{n=1}^{\infty}$ are generated from the initial element $(x_0, y_0) = (x^*, y^*) \in X \times X$ according to (23).

Theorem 10 (Satisfying the $\triangleleft$-Triangle Inequality). *Let (X, M) be a fuzzy semi-metric space, such that the mapping M satisfies the rational condition and the $\triangleleft$-triangle inequality. Suppose that the following conditions are satisfied.*

- The first eight conditions of Theorem 7 are satisfied.
- The following contractive inequalities is satisfied

$$M(T_i(x,y), T_j(u,v), k_{ij} \cdot t) \geq M(f(x), f(u), t) \cdot M(f(y), f(v), t)$$

or the following converse-contractive inequalities are satisfied

$$M(T_i(x,y), T_j(u,v), k_{ij} \cdot t) \geq M(f(u), f(x), t) \cdot M(f(v), f(y), t),$$

where k_{ij} satisfies $0 < k_{ij} \leq k < 1$ for all $i, j \in \mathbb{N}$ and for some constant k.
- There exist $x^*, y^* \in X$ satisfying

$$\sup_{\lambda \in (0,1)} \Phi(\lambda, f(x^*), T_1(x^*, y^*), f(y^*), T_1(y^*, x^*)) < \infty$$

and
$$\sup_{\lambda \in (0,1)} \Phi(\lambda, T_1(x^*, y^*), f(x^*), T_1(y^*, x^*), f(y^*)) < \infty.$$

- The mapping f is $(\triangleright, \triangleright)$-continuous or $(\triangleleft, \triangleright)$-continuous with respect to M.
- Any one of the following conditions is satisfied:
  - (X, M) is $(<, \triangleright)$-complete and $(<, \triangleleft)$-complete simultaneously;
  - (X, M) is $(>, \triangleright)$-complete and $(>, \triangleleft)$-complete simultaneously.

Subsequently, the mappings $\{T_n\}_{n=1}^{\infty}$ and f have a unique common coupled fixed point (x°, y°). Moreover, the point $(x^\circ, y^\circ) \in X \times X$ can be obtained, as follows.

- The point x° can be obtained by taking the limit $f(x_n) \xrightarrow{M^\triangleright} x^\circ$ or the limit $f(x_n) \xrightarrow{M^\triangleleft} x^\circ$.
- The point y° can be obtained by taking the limit $f(y_n) \xrightarrow{M^\triangleright} y^\circ$ or the limit $f(y_n) \xrightarrow{M^\triangleleft} y^\circ$.

The sequences $\{x_n\}_{n=1}^{\infty}$ and $\{y_n\}_{n=1}^{\infty}$ are generated from the initial element $(x_0, y_0) = (x^*, y^*) \in X \times X$ according to (23).

Theorem 11 (Satisfying the $\diamond$-Triangle Inequality). *Let (X, M) be a fuzzy semi-metric space such that the mapping M satisfies the rational condition and the $\diamond$-triangle inequality. Suppose that the following conditions are satisfied.*

- All nine conditions of Theorem 7 are satisfied.

- There exist $x^*, y^* \in X$ satisfying

$$\sup_{\lambda \in (0,1)} \Phi(\lambda, f(x^*), T_1(x^*, y^*), f(y^*), T_1(y^*, x^*)) < \infty$$

and

$$\sup_{\lambda \in (0,1)} \Phi(\lambda, T_1(x^*, y^*), f(x^*), T_1(y^*, x^*), f(y^*)) < \infty;$$

- Any one of the following conditions is satisfied:
  (a) (X, M) is $(<, \triangleright)$-complete or $(>, \triangleright)$-complete and f is $(\triangleright, \triangleright)$-continuous or $(\triangleright, \triangleleft)$-continuous with respect to M;
  (b) (X, M) is $(<, \triangleleft)$-complete or $(>, \triangleleft)$-complete and f is $(\triangleleft, \triangleright)$-continuous or $(\triangleleft, \triangleleft)$-continuous with respect to M.

Afterwards, the mappings T and f have a unique common coupled fixed point (x°, y°) Moreover, the point $(x^\circ, y^\circ) \in X \times X$ can be obtained, as follows.

- Suppose that condition (a) is satisfied. Afterwards, the point $(x^\circ, y^\circ) \in X \times X$ can be obtained by taking the limit $f(x_n) \xrightarrow{M^\triangleright} x^\circ$ and $f(y_n) \xrightarrow{M^\triangleright} y^\circ$.
- Suppose that condition (b) is satisfied. Subsequently, the point $(x^\circ, y^\circ) \in X \times X$ can be obtained by taking the limit $f(x_n) \xrightarrow{M^\triangleleft} x^\circ$ and $f(y_n) \xrightarrow{M^\triangleleft} y^\circ$.

The sequences $\{x_n\}_{n=1}^\infty$ and $\{y_n\}_{n=1}^\infty$ are generated from the initial element $(x_0, y_0) = (x^*, y^*) \in X \times X$, according to (23).

Theorem 12 (Satisfying the $\diamond$-Triangle Inequality: Converse-Contractive Inequality). *Let (X, M) be a fuzzy semi-metric space, such that the mapping M satisfies the rational condition and the $\diamond$-triangle inequality. Suppose that the following conditions are satisfied.*

- The first eight conditions of Theorem 7 are satisfied.
- The following converse-contractive inequalities are satisfied

$$M(T_i(x, y), T_j(u, v), k_{ij} \cdot t) \geq M(f(u), f(x), t) \cdot M(f(v), f(y), t),$$

where k_{ij} satisfies $0 < k_{ij} \leq k < 1$ for all $i, j \in \mathbb{N}$ and for some constant k.
- There exist $x^*, y^* \in X$ satisfying

$$\sup_{\lambda \in (0,1)} \Phi(\lambda, f(x^*), T_1(x^*, y^*), f(y^*), T_1(y^*, x^*)) < \infty$$

and

$$\sup_{\lambda \in (0,1)} \Phi(\lambda, T_1(x^*, y^*), f(x^*), T_1(y^*, x^*), f(y^*)) < \infty.$$

- Any one of the following conditions is satisfied:
  (a) (X, M) is $(<, \triangleright)$-complete or $(>, \triangleright)$-complete and f is $(\triangleright, \triangleright)$-continuous and $(\triangleright, \triangleleft)$-continuous with respect to M;
  (b) (X, M) is $(<, \triangleleft)$-complete or $(>, \triangleleft)$-complete and f is $(\triangleleft, \triangleright)$-continuous and $(\triangleleft, \triangleleft)$-continuous with respect to M.

Subsequently, the mappings T and f have a unique common coupled fixed point (x°, y°) Moreover, the point $(x^\circ, y^\circ) \in X \times X$ can be obtained, as follows.

- Suppose that condition (a) is satisfied. Subsequently, the point $(x^\circ, y^\circ) \in X \times X$ can be obtained by taking the limit $f(x_n) \xrightarrow{M^\triangleright} x^\circ$ and $f(y_n) \xrightarrow{M^\triangleright} y^\circ$.
- Suppose that condition (b) is satisfied. Afterwards, the point $(x^\circ, y^\circ) \in X \times X$ can be obtained by taking the limit $f(x_n) \xrightarrow{M^\triangleleft} x^\circ$ and $f(y_n) \xrightarrow{M^\triangleleft} y^\circ$.

The sequences $\{x_n\}_{n=1}^\infty$ and $\{y_n\}_{n=1}^\infty$ are generated from the initial element $(x_0, y_0) = (x^*, y^*) \in X \times X$ according to (23).

7. Conclusions

Four different kinds of triangle inequalities play the important role of studying the common coupled coincidence points and common coupled fixed points in fuzzy semi-metric spaces. We separately present the theorems of common coupled coincidence points that are based on the different kinds of triangle inequalities.

- Suppose that the fuzzy semi-metric space satisfies the ⋈-triangle inequality. Theorem 1 studies the common coupled coincidence points. Because the symmetric condition is not satisfied. Theorem 2 also studies the common coupled coincidence points by considering the so-called converse-contractive inequality.
- Theorems 3 and 4 study the common coupled coincidence points when the fuzzy semi-metric space satisfies the ▷-triangle inequality and ◁-triangle inequality, respectively.
- Suppose that the fuzzy semi-metric space satisfies the ◇-triangle inequality. Theorem 5 studies the common coupled coincidence points, and Theorem 6 studies the common coupled coincidence points by considering the so-called converse-contractive inequality.

Because the common coupled fixed points are the common coupled coincidence points, Theorems 1–6 can also be used to present the common coupled fixed points. However, the uniqueness cannot be realized from Theorems 1–6. Section 6 studies the uniqueness of common coupled fixed points.

- Suppose that the fuzzy semi-metric space satisfies the ⋈-triangle inequality. Theorem 7 studies the uniqueness of common coupled fixed points, and Theorem 8 also studies the uniqueness of common coupled fixed points by considering the so-called converse-contractive inequality.
- Theorems 9 and 10 study the uniqueness of common coupled fixed points when the fuzzy semi-metric space satisfies the ▷-triangle inequality and ◁-triangle inequality, respectively.
- Suppose that the fuzzy semi-metric space satisfies the ◇-triangle inequality. Theorem 11 studies the uniqueness of common coupled fixed points and Theorem 12 studies the uniqueness of common coupled fixed points by considering the so-called converse-contractive inequality.

Funding: This research received no external funding.

Conflicts of Interest: The author declares no conflict of interest.

References

1. Wu, H.-C. Fuzzy Semi-Metric Spaces. *Mathematics* **2018**, *6*, 106. [CrossRef]
2. Schweizer, B.; Sklar, A. Statistical Metric Spaces. *Pac. J. Math.* **1960**, *10*, 313–334. [CrossRef]
3. Schweizer, B.; Sklar, A.; Thorp, E. The Metrization of Statistical Metric Spaces. *Pac. J. Math.* **1960**, *10*, 673–675. [CrossRef]
4. Schweizer, B.; Sklar, A. Triangle Inequalities in a Class of Statistical Metric Spaces. *J. Lond. Math. Soc.* **1963**, *38*, 401–406. [CrossRef]
5. Hadžić, O.; Pap, E. *Fixed Point Theory in Probabilistic Metric Spaces*; Klumer Academic Publishers: New York, NY, USA, 2001.
6. Chang, S.S.; Cho, Y.J.; Kang, S.M. *Nonlinear Operator Theory in Probabilistic Metric Space*; Nova Science Publishers: New York, NY, USA, 2001.
7. Kramosil, I.; Michalek, J. Fuzzy Metric and Statistical Metric Spaces. *Kybernetika* **1975**, *11*, 336–344.
8. Rakić, D.; Mukheimer, A.; Došenović, T.; Mitrović, Z.D.; Radenović, S. On Some New Fixed Point Results in b-Fuzzy Metric Spaces. *J. Inequalities Appl.* **2020** *2020*, 99. [CrossRef]
9. Rakić, A.; Došenović, T.; Mitrović, Z.D.; de la Sen, M.; Radenović, S. Some Fixed Point Theorems of ĆIrić Type in Fuzzy Metric Spaces. *Mathematics* **2020**, *8*, 297. [CrossRef]
10. Mecheraoui, R.; Mukheimer, A.; Radenović, S. From G-completeness to M-completeness. *Symmetry* **2019**, *11*, 839. [CrossRef]
11. Gu, F.; Shatanawi, W. Some New Results on Common Coupled Fixed Points of Two Hybrid Pairs of Mappings in Partial Metric Spaces. *J. Nonlinear Funct. Anal.* **2019**, *2019*, 13.
12. Petruel, A. Local fixed point results for graphic contractions. *J. Nonlinear Var. Anal.* **2019**, *3*, 141–148.
13. Petruel, A.; Petruel, G. Fixed Point Results for Multi-Valued Locally Contractive Operators. *Appl. Set-Valued Anal. Optim.* **2020**, *2*, 175–181.
14. Hu, X.-Q.; Zheng, M.-X.; Damjanovi'c, B.; Shao, X.-F. Common Coupled Fixed Point Theorems for Weakly Compatible Mappings In Fuzzy Metric Spaces. *Fixed Point Theory Appl.* **2013**, *2013*, 220. [CrossRef]

15. Mohiuddine, S.A.; Alotaibi, A. Coupled Coincidence Point Theorems for Compatible Mappings in Partially Ordered Intuitionistic Generalized Fuzzy Metric Spaces. *Fixed Point Theory Appl.* **2013**, *2013*, 220. [CrossRef]
16. Qiu, Z.; Hong, S. Coupled Fixed Points For Multivalued Mappings In Fuzzy Metric Spaces. *Fixed Point Theory Appl.* **2013**, *2013*, 265. [CrossRef]
17. Wu, H.-C. Common Coincidence Points and Common Fixed Points in Fuzzy Semi-Metric Spaces. *Mathematics* **2018**, *6*, 29. [CrossRef]
18. Wu, H.-C. Convergence in Fuzzy Semi-Metric Spaces. *Mathematics* **2018**, *6*, 170. [CrossRef]

Article

e-Distance in Menger PGM Spaces with an Application

Ehsan Lotfali Ghasab [1], Hamid Majani [1,*], Manuel De la Sen [2,*] and Ghasem Soleimani Rad [3]

1. Department of Mathematics, Shahid Chamran University of Ahvaz, Ahvaz 6135783151, Iran; e-lotfali@stu.scu.ac.ir
2. Institute of Research and Development of Processes IIDP, University of the Basque Country, 48940 Leioa, Spain
3. Young Researchers and Elite Club, West Tehran Branch, Islamic Azad University, Tehran 1477893855, Iran; gha.soleimani.sci@iauctb.ac.ir
* Correspondence: h.majani@scu.ac.ir or majani.hamid@gmail.com (H.M.); manuel.delasen@ehu.eus (M.D.l.S.)

Abstract: The main purpose of the present paper is to define the concept of an e-distance (as a generalization of r-distance) on a Menger PGM space and to introduce some of its properties. Moreover, some coupled fixed point results, in terms of this distance on a complete PGM space, are proved. To support our definitions and main results, several examples and an application are considered.

Keywords: e-distance; Menger PGM space; coupled fixed point

MSC: JPrimary 47H10; Secondary 47S50

1. Introduction and Preliminaries

In 1942, Menger [1] introduced Menger probabilistic metric spaces as an extension of metric spaces. After that, Sehgal and Bharucha-Reid [2,3] studied some fixed point results for different classes of probabilistic contractions (also, see and references in the citation). Moreover, in 2009, Saadati et al. [4] introduced the concept of r-distance on this space.

Throughout this paper, the set of all Menger distance distribution functions are denoted by D^+.

Definition 1 ([5], page 1). *A binary mapping* $\mathcal{T} : [0,1] \times [0,1] \to [0,1]$ *is called t-norm if the following propertied are held:*

(a) $\mathcal{T}$ *is commutative and associative;*
(b) $\mathcal{T}$ *is continuous;*
(c) $\mathcal{T}(a,1) = a$ *if* $a \in [0,1]$;
(d) $\mathcal{T}(a,b) \leq \mathcal{T}(c,d)$ *if* $a \leq c$ *and* $b \leq d$ *for every* $a,b,c,d \in [0,1]$.

Definition 2 ([4]). *A t-norm* $\mathcal{T}$ *is called an H-type I if for* $\epsilon \in (0,1)$, *there exist* $\delta \in (0,1)$ *so that* $\mathcal{T}^m(1-\delta,...,1-\delta) > 1-\epsilon$ *for each* $m \in \mathbb{N}$, *where* $\mathcal{T}^m$ *recursively defined by* $\mathcal{T}^1 = \mathcal{T}$ *and* $\mathcal{T}^m(t_1,t_2,...,t_{m+1}) = \mathcal{T}(\mathcal{T}^{m-1}(t_1,t_2,...,t_m),t_{m+1})$ *for* $m = 2,3,\cdots$ *and* $t_i \in [0,1]$.

All t-norms in the sequel are from class of H-type I.

From another point of view, Mustafa and Sims [6] defined G-metric spaces as another extension of metric spaces, analyzed the structure of this space, and continued the theory of fixed point in such spaces. In 2014, Zhou et al. [7], by combining Menger PM-spaces and G-metric spaces, defined Menger probabilistic generalized metric space (shortly, Menger PGM space). Other researchers extended several fixed point theorems in [8–10] and references contained therein.

Definition 3 ([7]). *Assume that* $\mathcal{X}$ *is a nonempty set,* $\mathcal{T}$ *is a continuous t-norm and* $G : \mathcal{X}^3 \to D^+$ *is a mapping satisfying the following properties for all* $x,y,z,a \in \mathcal{X}$ *and* $s,t > 0$:

(PG1) $G_{x,y,z}(t) = 1$ if and only if $x = y = z$;
(PG2) $G_{x,x,y}(t) \geq G_{x,y,z}(t)$, where $z \neq y$;
(PG3) $G_{x,y,z}(t) = G_{x,z,y}(t) = G_{y,x,z}(t) = \cdots$;
(PG4) $G_{x,y,z}(t+s) \geq \mathcal{T}(G_{x,a,a}(s), G_{a,y,z}(t))$.
Then $(\mathcal{X}, G, \mathcal{T})$ is named a Menger PGM space.

For the definitions of convergent, completeness, closedness and some theorems by regarding these concepts in such spaces, one can see [7]. In 2004, Ran and Reurings [11] discussed on fixed point results for comparable elements of a metric space $(\mathcal{X}, d)$ provided with a partial order. Then, Bhaskar and Lakshmikantham [12] presented several fixed point results for a mapping having mixed monotone property in such spaces (see [13,14]).

Definition 4 ([12]). *Consider a ordered set $(\mathcal{X}, \preceq)$ and a mapping $F : \mathcal{X}^2 \to \mathcal{X}$. The mapping F is told to be have mixed monotone property if*

$$x_1 \preceq x_2 \text{ implies that } F(x_1, y) \preceq F(x_2, y) \quad \forall x_1, x_2 \in \mathcal{X},$$
$$y_1 \preceq y_2 \text{ implies that } F(x, y_1) \succeq F(x, y_2) \quad \forall y_1, y_2 \in \mathcal{X}.$$

for every $x, y \in \mathcal{X}$.

Here we introduce an *e*-distance on Menger PGM spaces and some of its properties. Then we obtain some coupled fixed point results in the quasi-ordered version of such spaces. The subject of the paper offers novelties compared to the related background literature since a new distance in Menger spaces is defined while some of its properties are revisited and extended.

2. Main Results

Here, we consider an *e*-distance on a Menger PGM space, which is an extension of *r*-distance introduced by Saadati et al. [4].

Definition 5. *Consider a Menger PGM space $(\mathcal{X}, G, \mathcal{T})$. Then the function $g : \mathcal{X}^3 \times [0, \infty] \to [0, 1]$ is called an e-distance, if for all $x, y, z, a \in \mathcal{X}$ and $s, t \geq 0$ the following are held:*
(r1) $g_{x,y,z}(t+s) \geq \mathcal{T}(g_{x,a,a}(s), g_{a,y,z}(t))$;
(r2) $g_{x,y,\cdot}(t)$ and $g_{x,\cdot,y}(t)$ are continuous;
(r3) *for each $\epsilon > 0$, there exists $\delta > 0$ provided that $g_{a,y,z}(t) \geq 1 - \delta$ and $g_{x,a,a}(s) \geq 1 - \delta$ conclude that $G_{x,y,z}(t+s) \geq 1 - \epsilon$.*

Lemma 1. *Each Menger PGM is an e-distance on $\mathcal{X}$.*

Proof. Clearly, (r1) and (r2) are true. Only, we prove that (r3) is true. Assume $\epsilon > 0$ and select $\delta > 0$ so that $\mathcal{T}(1-\delta, 1-\delta) \geq 1 - \epsilon$. Then, for $G_{a,y,z}(t) \geq 1 - \delta$ and $G_{x,a,a}(s) \geq 1 - \delta$, we get

$$G_{x,y,z}(t+s) \geq \mathcal{T}(G_{a,y,z}(t), G_{x,a,a}(s)) \geq \mathcal{T}(1-\delta, 1-\delta) \geq 1 - \epsilon.$$

□

Example 1. *Assume $(\mathcal{X}, G, \mathcal{T})$ is a Menger PGM space. Define a function $g : \mathcal{X}^3 \times [0, \infty] \to [0, 1]$ by $g_{x,y,z}(t) = 1 - c$ for each $x, y, z \in \mathcal{X}$ and $t > 0$ with $c \in (0, 1)$. Then g is an e-distance.*

Lemma 2. *Consider a Menger PGM space with a continuous mapping A on $\mathcal{X}$ and a function $g : \mathcal{X}^3 \times [0, \infty] \to [0, 1]$ by $g_{x,y,z}(t) = \min\{G_{x,y,z}(t), G_{Ax,Ay,Az}(t)\}$ for each $x, y, z \in \mathcal{X}$ and $t > 0$. Then g is an e-distance on $\mathcal{X}$.*

Proof. The condition (r2) is clearly established. To prove (r1), consider $x, y, z, a \in \mathcal{X}$ and $t, s > 0$. Then, we have two following cases:

Case 1: if $G_{x,y,z}(t) = \min\{G_{x,y,z}(t), G_{Ax,Ay,Az}(t)\}$, then

$$g_{x,y,z}(t+s) = G_{x,y,z}(t+s)$$
$$\geq \mathcal{T}(G_{x,a,a}(t), G_{a,y,z}(s))$$
$$\geq \mathcal{T}(\min\{G_{x,a,a}(t), G_{Ax,Aa,Aa}(t)\}, \min\{G_{a,y,z}(s), G_{Aa,Ay,Az}(s)\})$$
$$\geq \mathcal{T}(g_{x,a,a}(t), g_{a,y,z}(s)).$$

Case 2: if $G_{Ax,Ay,Az}(t) = \min\{G_{x,y,z}(t), G_{Ax,Ay,Az}(t)\}$, then

$$g_{x,y,z}(t+s) = G_{Ax,Ay,Az}(t+s)$$
$$\geq \mathcal{T}(G_{Ax,Aa,Aa}(t), G_{Aa,Ay,Az}(s))$$
$$\geq \mathcal{T}(\min\{G_{x,a,a}(t), G_{Ax,Aa,Aa}(t)\}, \min\{G_{a,y,z}(s), G_{Aa,Ay,Az}(s)\})$$
$$\geq \mathcal{T}(g_{x,a,a}(t), g_{a,y,z}(s)).$$

Therefore, (r1) is established. Now, assume $\epsilon > 0$ and select $\delta > 0$ so that $\mathcal{T}(1-\delta, 1-\delta) \geq 1 - \epsilon$. Using $g_{x,a,a}(t) \geq 1 - \delta$ and $g_{a,y,z}(s) \geq 1 - \delta$, we get

$$\min\{G_{x,a,a}(t), G_{Ax,Aa,Aa}(t)\} = g_{x,a,a}(t) \geq 1 - \delta,$$
$$\min\{G_{a,y,z}(s), G_{Aa,Ay,Az}(s)\} = g_{a,y,z}(s) \geq 1 - \delta,$$

which induces that

$$G_{x,y,z}(t+s) \geq \mathcal{T}(G_{x,a,a}(t), G_{a,y,z}(s))$$
$$\geq \mathcal{T}(\min\{G_{x,a,a}(t), G_{Ax,Aa,Aa}(t)\}, \min\{G_{a,y,z}(s), G_{Aa,Ay,Az}(s)\})$$
$$= \mathcal{T}(g_{x,a,a}(t), g_{a,y,z}(s)) \geq \mathcal{T}(1-\delta, 1-\delta) \geq 1 - \epsilon.$$

Thus, (r3) is established. This completes the proof. □

Lemma 3. *Consider an e-distance g on $(\mathcal{X}, G, \mathcal{T})$ with two sequences $\{x_n\}$ and $\{y_n\}$ in $\mathcal{X}$. Suppose that $\{\alpha_n\}$ and $\{\beta_n\}$ are two non-negative sequences converging to 0. Then for $x, y, z \in \mathcal{X}$ and $t, s > 0$ the following assertions are established:*

(i) $g_{z,y,x_n}(t) \geq 1 - \alpha_n$ and $g_{x,x_n,x_n}(t) \geq 1 - \beta_n$ for any $n \in \mathbb{N}$ imply $x = y = z$. Specially, $g_{x,a,a}(t) = 1$ and $g_{a,y,z}(s) = 1$ imply $x = y = z$;

(ii) $g_{y_n,x_n,x_n}(t) \geq 1 - \alpha_n$ and $g_{x_n,y_m,z}(t) \geq 1 - \beta_n$ for all $m > n$ with $m, n \in \mathbb{N}$ imply $G_{y_n,y_m,z}(t+s) \to 1$ as $n \to \infty$;

(iii) let $g_{x_n,x_m,x_l}(t) \geq 1 - \alpha_n$ for all $n, m, l \in \mathbb{N}$, where $l > m > n$. Then $\{x_n\}$ is a Cauchy sequence;

(iv) let $g_{y,y,x_l}(t) \geq 1 - \alpha_n$ for all $n \in \mathbb{N}$. Then $\{x_n\}$ is a Cauchy sequence.

Proof. To prove (ii), assume $\epsilon > 0$. By applying the definition of e-distance, there exists $\delta > 0$ so that $g_{a,y,z}(t) \geq 1 - \delta$ and $g_{x,a,a}(s) \geq 1 - \delta$ induce $G_{x,y,z}(t+s) \geq 1 - \epsilon$. Select $n_0 \in \mathbb{N}$ provided that $\alpha_n \leq \delta$ and $\beta_n \leq \delta$ for each $n \geq n_0$. Then $g_{y_n,x_n,x_n}(t) \geq 1 - \alpha_n \geq 1 - \delta$ and $g_{x_n,y_m,z}(t) \geq 1 - \beta_n \geq 1 - \delta$ for any $n \geq n_0$ and hence $G_{y_n,y_m,z}(t+s) \geq 1 - \epsilon$. Therefore, $\{y_n\}$ converges to z. Now, using (ii), (i) is established. To prove (iii), assume $\epsilon > 0$. Similar to the proof of (ii), select $\delta > 0$ and $n_0 \in \mathbb{N}$. Then, for all $n, m, l \geq n_0 + 1$, we get $g_{x_n,x_{n_0},x_{n_0}}(t) \geq 1 - \alpha_{n_0} \geq 1 - \delta$ and $g_{x_{n_0},x_l,x_m}(t) \geq 1 - \alpha_{n_0} \geq 1 - \delta$. Therefore, $G_{x_n,x_m,x_l}(t) \geq 1 - \epsilon$. Hence, $\{x_n\}$ is a Cauchy sequence. Now, it follows from (iii) that (iv) is true. □

Lemma 4. *Consider an e-distance g on $(\mathcal{X}, G, \mathcal{T})$. Suppose that $E_{\lambda,g} : \mathcal{X}^3 \to \mathbb{R}^+ \cup \{0\}$ is introduced by $E_{\lambda,g}(x,y,z) = \inf\{t > 0 : g_{x,y,z}(t) > 1 - \lambda\}$ for any $x, y, z \in \mathcal{X}$ and $\lambda \in (0,1)$. Then*

(1) for all $\mu \in (0,1)$, there exists $\lambda \in (0,1)$ so that

$$E_{\mu,g}(x_1,x_1,x_n) \leq E_{\lambda,g}(x_1,x_1,x_2) + E_{\lambda,g}(x_2,x_2,x_3) + \cdots + E_{\lambda,g}(x_{n-1},x_{n-1},x_n)$$

for each $x_1, \cdots, x_n \in \mathcal{X}$;

(2) for every sequence $\{x_n\}$ in $\mathcal{X}$, $g_{x_n,x,x}(t) \to 1$ iff $E_{\lambda,g}(x_n,x,x) \to 0$. Further, the sequence $\{x_n\}$ is Cauchy w.r.t. g iff it is Cauchy with $E_{\lambda,g}$.

Proof.

(1) For every $\mu \in (0,1)$, we can gain $\lambda \in (0,1)$ provided that $\mathcal{T}^{n-1}(1-\lambda,...,1-\lambda) \geq 1 - \mu$. Now, for every $\delta > 0$, we have

$$g_{x_1,x_1,x_n}(E_{\lambda,g}(x_1,x_1,x_2) + E_{\lambda,g}(x_2,x_2,x_3) + \cdots + E_{\lambda,g}(x_{n-1},x_{n-1},x_n) + n\delta)$$
$$\geq \mathcal{T}^{n-1}(g_{x_1,x_1,x_2}(E_{\lambda,g}(x_1,x_1,x_2) + \delta), g_{x_2,x_2,x_3}(E_{\lambda,g}(x_2,x_2,x_3) + \delta)$$
$$, \cdots, g_{x_{n-1},x_{n-1},x_n}(E_{\lambda,g}(x_{n-1},x_{n-1},x_n) + \delta))$$
$$\geq \mathcal{T}^{n-1}(1-\lambda,...,1-\lambda) \geq 1-\mu$$

which induces that

$$E_{\mu,g}(x_1,x_1,x_n) \leq E_{\lambda,g}(x_1,x_1,x_2) + E_{\lambda,g}(x_2,x_2,x_3) + \cdots + E_{\lambda,g}(x_{n-1},x_{n-1},x_n) + n\delta.$$

Since $\delta > 0$ is optional, we obtain

$$E_{\mu,g}(x_1,x_1,x_n) \leq E_{\lambda,g}(x_1,x_1,x_2) + E_{\lambda,g}(x_2,x_2,x_3) + \cdots + E_{\lambda,g}(x_{n-1},x_{n-1},x_n).$$

(2) Note that $g_{x_n,x,x}(\eta) \to 1 - \lambda$ as $n \to \infty$ iff $E_{\lambda,g}(x_n,x,x) < \eta$ for each $n \in \mathbb{N}$ and $\eta > 0$. □

In the sequel, we establish some coupled fixed point theorems by regarding an e-distance on a quasi-ordered complete PGM space.

Theorem 1. *Let $(\mathcal{X}, G, \mathcal{T}, \preceq)$ be a quasi-ordered complete Menger PGM space with $\mathcal{T}$ of Hadzić-type I, g be an e-distance and $f : \mathcal{X}^2 \to \mathcal{X}$ be a mapping having the mixed monotone property on $\mathcal{X}$. Assume that there exists a $k \in [0,1)$ such that*

$$g_{f(x,y),f(u,v),f(w,z)}(t) \geq \frac{1}{2}(g_{x,u,w}(\frac{t}{k}) + g_{y,v,z}(\frac{t}{k})) \quad (1)$$

for all $x,y,z,u,v,w \in \mathcal{X}$ with $x \succeq u \succeq w$ and $y \preceq v \preceq z$, where either $u \neq w$ or $v \neq z$ and

$$\sup\{\mathcal{T}(g_{x,y,z}(t), g_{x,y,f(x,y)}(t)) : x,y \in \mathcal{X}\} < 1. \quad (2)$$

for all $z \in \mathcal{X}$, where $z \neq f(z,q)$ for all $q \in \mathcal{X}$. If there exist $x_0, y_0 \in \mathcal{X}$ so that $x_0 \preceq f(x_0,y_0)$ and $y_0 \succeq f(y_0,x_0)$, then f have a coupled fixed point in $\mathcal{X}^2$.

Proof. Since there exist $x_0, y_0 \in \mathcal{X}$ with $x_0 \preceq f(x_0, y_0)$ and $y_0 \succeq f(y_0, x_0)$, and f has the mixed monotone property, we can construct Bhaskar-Lakshmikantham type iterative as follow:

$$x_0 \preceq x_1 \preceq x_2 \preceq \cdots \preceq x_{n+1} \preceq \cdots, \quad y_0 \succeq y_1 \succeq y_2 \succeq \cdots \succeq y_{n+1} \succeq \cdots$$

for all $n \geq 0$, where

$$x_{n+1} = f^{n+1}(x_0, y_0) = f(f^n(x_0, y_0), f^n(y_0, x_0)),$$
$$y_{n+1} = f^{n+1}(y_0, x_0) = f(f^n(y_0, x_0), f^n(x_0, y_0)).$$

If $(x_{n+1}, y_{n+1}) = (x_n, y_n)$, then f has a coupled fixed point. Otherwise, assume $(x_{n+1}, y_{n+1}) \neq (x_n, y_n)$ for each $n \geq 0$; that is, either $x_{n+1} = f(x_n, y_n) \neq x_n$ or $y_{n+1} = f(y_n, x_n) \neq y_n$. Now, by induction and (1), we obtain

$$g_{x_n, x_n, x_{n+1}}(t) \geq \frac{1}{2}(g_{x_0, x_0, x_1}(\frac{t}{k^n}) + g_{y_0, y_0, y_1}(\frac{t}{k^n})),$$

$$g_{y_n, y_n, y_{n+1}}(t) \geq \frac{1}{2}(g_{y_0, y_0, y_1}(\frac{t}{k^n}) + g_{x_0, x_0, x_1}(\frac{t}{k^n})),$$

for each $n \geq 0$ which induces that $g_{x_n, x_n, x_{n+1}}(t) \geq \frac{1}{2} g_{x_0, x_0, x_1}(\frac{t}{k^n})$ and $g_{y_n, y_n, y_{n+1}}(t) \geq \frac{1}{2} g_{y_0, y_0, y_1}(\frac{t}{k^n})$. Therefore,

$$E_{\lambda, g}(x_n, x_n, x_{n+1}) = \inf\{t > 0 : g_{x_n, x_n, x_{n+1}}(t) > 1 - \lambda\}$$
$$\leq \inf\{t > 0 : \frac{1}{2} g_{x_0, x_0, x_1}(\frac{t}{k^n}) > 1 - \lambda\}$$
$$= 2k^n E_{\lambda, g}(x_0, x_0, x_1).$$

Thus, for $m > n$ and $\lambda \in (0, 1)$, there exists $\gamma \in (0, 1)$ so that

$$E_{\lambda, g}(x_n, x_n, x_m) \leq E_{\gamma, g}(x_n, x_n, x_{n+1}) + \cdots + E_{\gamma, g}(x_{m-1}, x_{m-1}, x_m) \leq \frac{2k^n}{1-k} E_{\gamma, g}(x_0, x_0, x_1).$$

Now, there exists $n_0 \in \mathbb{N}$ so that for each $n > n_0$, $E_{\lambda, g}(x_n, x_n, x_m) \to 0$. By Lemmas 3 and 4, $\{x_n\}$ is a Cauchy sequence. Thus, using Lemma 4 (ii), there exit $n_1 \in \mathbb{N}$ and a sequence $\delta_n \to 0$ so that $g_{x_n, x_n, x_m}(t) \geq 1 - \delta_n$ for $n \geq \max\{n_0, n_1\}$. Since $\mathcal{X}$ is complete, $\{x_n\}$ converges to a point $p \in \mathcal{X}$. Similarly, $\{y_n\}$ is convergent to a point $q \in \mathcal{X}$. By (r2), we obtain $g_{x_n, x_n, p}(t) = \lim_{m \to \infty} g_{x_n, x_n, x_m}(t) \geq 1 - \delta_n$ for $n \geq \max\{n_0, n_1\}$. Moreover, we get $g_{x_n, x_{n+1}, x_{n+1}}(t) \geq 1 - \delta_n$. Now, we show that f has a coupled fixed point. Let $p \neq f(p, q)$. Then, by (2), we obtain

$$1 > \sup\{\mathcal{T}(g_{x,y,p}(t), g_{x,y,f(x,y)}(t)) : x, y \in \mathcal{X}\}$$
$$\geq \sup\{\mathcal{T}(g_{x_n, x_n, p}(t), g_{x_n, x_{n+1}, x_{n+1}}(t)) : n \in \mathbb{N}\}$$
$$\geq \sup\{\mathcal{T}(1 - \delta_n, 1 - \delta_n) : n \in \mathbb{N}\} = 1,$$

which is a contradiction. Consequently, we get $p = f(p, q)$. Similarly, we obtain $f(q, p) = q$. Here, the proof ends. □

Theorem 2. *Assume the assumptions of Theorem 1 are held and consider the continuity of f instead of relation (2). Then f has a coupled fixed point.*

Proof. As in the proof of Theorem 1, construct $\{x_n\}$ and $\{y_n\}$, where $x_n \to p$, $y_n \to q$, $x_{n+1} = f(x_n, y_n)$. Now, by the continuity of f and by taking the limit as $n \to \infty$, we get $f(p, q) = p$. Analogously, we can obtain $f(q, p) = q$. Therefore, (p, q) is a coupled fixed point of f. □

Example 2. *Assume that $\mathcal{X} = [0, \infty)$, "$\preceq$" is a quasi-ordered on $\mathcal{X}$ and $\mathcal{T}(a, b) = \min\{a, b\}$. Define a constant function $f : \mathcal{X}^2 \to \mathcal{X}$ by $f(a, b) = p$ and $G : \mathcal{X}^3 \to D^+$ by $G_{x, y, z}(t) = \frac{t}{t + G^*(x, y, z)}$ with $G^*(x, y, z) = |x - y| + |x - z| + |y - z|$ for each $x, y, z \in \mathcal{X}$. Clearly, G satisfies (PG1)-(PG4). Consider $g_{x, y, z}(t) = 1 - c$, where $c \in (0, 1)$. Then g is an e-distance on $\mathcal{X}$. Clearly, for all $x, y, z, u, v, w \in \mathcal{X}$ and for any $t > 0$, we have $g_{f(x,y), f(u,v), f(w,z)}(t) \geq \frac{1}{2}(g_{x, u, w}(\frac{t}{k}) + g_{y, v, z}(\frac{t}{k}))$. Moreover, there exist $x_0 = 0$ and $y_0 = 1$ so that $0 = x_0 \preceq f(x_0, y_0)$ and $1 = y_0 \succeq f(y_0, x_0) = 1$. Therefore, all of the hypothesis of Theorem 2 are held. Clearly, (p, p) is a coupled fixed point the function f.*

3. Application

Consider the following system of integral equations:

$$\begin{cases} x(t) = \int_a^b M(t,s)K(s,x(s),y(s))ds, \\ y(t) = \int_a^b M(t,s)K(s,y(s),x(s))ds, \end{cases} \quad (3)$$

for all $t \in I = [a,b]$, where $b > a$, $M \in C(I \times I, [0,\infty))$ and $K \in C(I \times \mathbb{R} \times \mathbb{R}, \mathbb{R})$.

Let $C(I, \mathbb{R})$ be the Banach space of every real continuous functions on I with $\|x\|_\infty = \max_{t \in I} |x(t)|$ for all $x \in C(I, \mathbb{R})$ and $C(I \times I \times C(I, \mathbb{R}), \mathbb{R})$ be the space of every continuous functions on $I \times I \times C(I, \mathbb{R})$. Define a mapping $G : C(I, \mathbb{R}) \times C(I, \mathbb{R}) \to D^+$ by $G_{x,y,z}(t) = \chi(\frac{t}{2} - (\|x - y\|_\infty + \|x - z\|_\infty + \|y - z\|_\infty))$ for all $x, y, z \in C(I, \mathbb{R})$ and $t > 0$, where

$$\chi(t) = \begin{cases} 0 & \text{if} \quad t \leq 0 \\ 1 & \text{if} \quad t > 0 \end{cases}$$

Then, $(C(I, \mathbb{R}), G, \mathcal{T})$ with $\mathcal{T}(a,b) = \min\{a,b\}$ is a complete Menger PGM space ([7]). Consider an e-distance on $\mathcal{X}$ by $g_{x,y,z}(t) = \min\{G_{x,y,z}(t), G_{Ax,Ay,Az}(t)\}$, where $A : C(I, \mathbb{R}) \to C(I, \mathbb{R})$ and $Ax = \frac{x}{2}$. Moreover, we define the relation " $\preceq$ " on $C(I, \mathbb{R})$ by $x \preceq y \Leftrightarrow \|x\|_\infty \leq \|y\|_\infty$ for all $x, y \in C(I, \mathbb{R})$. Clearly the relation " $\preceq$ " is a quasi-order relation on $C(I, \mathbb{R})$ and $(C(I, \mathbb{R}), G, \mathcal{T}, \preceq)$ is a quasi-ordered complete PGM space.

Theorem 3. Let $(C(I, \mathbb{R}), G, \mathcal{T}, \preceq)$ be a quasi-ordered complete Menger PGM space and $f : C(I, \mathbb{R}) \times C(I, \mathbb{R}) \to C(I, \mathbb{R})$ be a operator defined by $f(x,y)(t) = \int_a^b M(t,s)K(s,x(s),y(s))ds$, where $M \in C(I \times I, [0,\infty))$ and $K \in C(I \times \mathbb{R} \times \mathbb{R}, \mathbb{R})$ are two operators. Assume the following properties are held:

(i) $\|K\|_\infty = \sup_{s \in I, x,y \in C(I,\mathbb{R})} |K(s,x(s),y(s))| < \infty$;

(ii) for every $x, y \in C(I, \mathbb{R})$ and every $t, s \in I$, we have

$$\|K(s,x(s),y(s)) - K(s,u(s),v(s))\|_\infty \leq \frac{1}{4}(\max |x(s) - u(s)| + \max |y(s) - v(s)|);$$

(iii) $\max_{t \in I} \int_a^b M(t,s)ds < 1$.

Then, the system (3) have a solution in $C(I, \mathbb{R}) \times C(I, \mathbb{R})$.

Proof. For all $x, y \in C(I, \mathbb{R})$, let $\|x - y\|_\infty = \max_{t \in I}(|x(t) - y(t)|)$. Then, for all $x, y, z, u, v, w \in C(I, \mathbb{R})$, we have

$$\|f(x,y) - f(u,v)\|_\infty \leq \max_{t \in I} \int_a^b M(t,s)|K(s,x(s)y(s)) - K(s,u(s),v(s))|ds$$

$$\leq \max(\frac{1}{4}(|x(s) - u(s)| + |y(s) - v(s)|) \max_{t \in I} \int_a^b M(t,s)ds$$

$$\leq \max(\frac{1}{4}(|x(s) - u(s)| + |y(s) - v(s)|).$$

We consider two following cases:

Case 1. Let

$$g_{f(x,y),f(u,v),f(w,z)}(t) = \min\{G_{f(x,y),f(u,v),f(w,z)}(t), G_{Af(x,y),Af(u,v),Af(w,z)}(t)\}$$
$$= G_{f(x,y),f(u,v),f(w,z)}(t).$$

Then, we obtain

$$g_{f(x,y),f(u,v),f(w,z)}(t) = G_{f(x,y),f(u,v),f(w,z)}(t)$$
$$= \chi(\frac{t}{2} - (\|f(x,y) - f(u,v)\|_\infty + \|f(x,y) - f(w,z)\|_\infty + \|f(u,v) - f(w,z)\|_\infty))$$
$$\geq \chi(\frac{t}{2} - (\max(\frac{1}{4}(|x(s) - u(s)| + |y(s) - v(s)|))$$
$$+ \max(\frac{1}{4}(|x(s) - w(s)| + |y(s) - z(s)|))$$
$$+ \max(\frac{1}{4}(|u(s) - w(s)| + |v(s) - z(s)|))))$$
$$= \chi(t - \frac{1}{2}(\max((|x(s) - u(s)| + |y(s) - v(s)|))$$
$$+ \max((|x(s) - w(s)| + |y(s) - z(s)|)) + \max((|u(s) - w(s)| + |v(s) - z(s)|))))$$
$$\geq \frac{1}{2}(\chi(t - (\max(|x(s) - u(s)| + |x(s) - w(s)| + |u(s) - w(s)|)))$$
$$+ \chi(t - (\max(|y(s) - v(s)| + |y(s) - z(s)| + |v(s) - z(s)|))))$$
$$= \frac{1}{2}(G_{x,u,w}(2t) + G_{y,v,z}(2t)) \geq \frac{1}{2}(g_{x,u,w}(2t) + g_{y,v,z}(2t)).$$

Case 2. Let

$$g_{f(x,y),f(u,v),f(w,z)}(t) = \min\{G_{f(x,y),f(u,v),f(w,z)}(t), G_{Af(x,y),Af(u,v),Af(w,z)}(t)\}$$
$$= G_{Af(x,y),Af(u,v),Af(w,z)}(t).$$

By $Ax = \frac{x}{2}$, we have

$$g_{f(x,y),f(u,v),f(w,z)}(t) = G_{Af(x,y),Af(u,v),Af(w,z)}(t)$$
$$= \chi(\frac{t}{2} - \frac{1}{2}(\|f(x,y) - f(u,v)\|_\infty + \|f(x,y) - f(w,z)\|_\infty + \|f(u,v) - f(w,z)\|_\infty))$$
$$\geq \chi(\frac{t}{2} - (\|f(x,y) - f(u,v)\|_\infty + \|f(x,y) - f(w,z)\|_\infty + \|f(u,v) - f(w,z)\|_\infty))$$
$$\geq \chi(\frac{t}{2} - (\max(\frac{1}{4}(|x(s) - u(s)| + |y(s) - v(s)|))$$
$$+ \max(\frac{1}{4}(|x(s) - w(s)| + |y(s) - z(s)|))$$
$$+ \max(\frac{1}{4}(|u(s) - w(s)| + |v(s) - z(s)|))))$$
$$= \chi(t - \frac{1}{2}(\max((|x(s) - u(s)| + |y(s) - v(s)|))$$
$$+ \max((|x(s) - w(s)| + |y(s) - z(s)|)) + \max((|u(s) - w(s)| + |v(s) - z(s)|))))$$
$$\geq \frac{1}{2}(\chi(t - (\max(|x(s) - u(s)| + |x(s) - w(s)| + |u(s) - w(s)|)))$$
$$+ \chi(t - (\max(|y(s) - v(s)| + |y(s) - z(s)| + |v(s) - z(s)|))))$$
$$= \frac{1}{2}(G_{x,u,w}(2t) + G_{y,v,z}(2t)) \geq \frac{1}{2}(g_{x,u,w}(2t) + g_{y,v,z}(2t))$$

for all $x, y, z, u, v, w \in C(I, \mathbb{R})$. Therefore, by Theorem 2 with $k = \frac{1}{2}$ for all $x, y, z, u, v, w \in C(I, \mathbb{R})$ and $t > 0$, we deduce that the operator f has a coupled fixed point which is the solution of the system of the integral equations. □

4. Conclusions

The new concept of e-distance, which is a generalization of r-distance in PGM space has been introduced. Moreover, some of properties of e-distance have been discussed. In addition, we obtained several new coupled fixed point results. Ultimately, to illustrate the usability of the main theorem, the existence of a solution for a system of integral equations is proved.

Author Contributions: All authors contributed equally and significantly in writing this paper. All authors have read and agree to the published version of the manuscript.

Funding: The authors are very grateful to the Basque Government by its support through Grant IT1207-19.

Acknowledgments: The first and the second authors are grateful to the Research Council of Shahid Chamran University of Ahvaz for financial support (Grant Number: SCU.MM99.25894). Moreover, the authors are very grateful to the Basque Government by its support through Grant IT1207-19.

Conflicts of Interest: The authors declare no conflict of interest.

References

1. Menger, K. Statistical metrics. *Proc. Natl. Acad. Sci. USA* **1942**, *28*, 535–537. [CrossRef] [PubMed]
2. Sehgal, V.M. Some Fixed Point Theorems in Functional Analysis and Probability. Ph.D. Dissertation, Wayne State University, Detroit, MI, USA, 1966.
3. Sehgal, V.M.; Bharucha-Reid, A.T. Fixed points of contraction mappings on probabilistic metric spaces. *Math. Syst. Theory* **1972**, *6*, 97–102. [CrossRef]
4. Saadati, R.; O'Regan, D.; Vaezpour, S.M.; Kim, J.K. Generalized distance and common fixed point theorems in Menger probabilistic metric spaces. *Bull. Iran. Math. Soc.* **2009**, *35*, 97–117.
5. Hadzic, O.; Pap, E. *Fixed Point Theory in Probabilistic Metric Spaces*; Kluwer Academic: Dordrecht, The Netherlands, 2001.
6. Mustafa, Z.; Sims, B. A new approach to generalized metric spaces. *J. Nonlinear Convex Anal.* **2006**, *6*, 289–297.
7. Zhou, C.; Wang, S.; Ćirić, L.; Alsulami, S. Generalized probabilistic metric spaces and fixed point theorems. *Fixed Point Theory Appl.* **2014**, *2014*, 91. [CrossRef]
8. Tiwari, V.; Som, T. Fixed points for φ-contraction in Menger probabilistic generalized metric spaces. *Annals. Fuzzy Math. Inform.* **2017**, *14*, 393–405.
9. Wang, G.; Zhu, C.; Wu, Z. Some new coupled fixed point theorems in partially ordered complete Menger probabilistic G-metric spaces. *J. Comput. Anal. Appl.* **2019**, *27*, 326–344.
10. Karapinar, E.; Czerwik, S.; Aydi, H. (α, ψ)-Meir-Keeler contraction mappings in generalized b-metric spaces. *J. Func. Space* **2018**, *2018*, 3264620. [CrossRef]
11. Ran, A.C.M.; Reurings, M.C.B. A fixed point theorem in partially ordered sets and some applications to matrix equations. *Proc. Am. Math. Soc.* **2004**, *132*, 1435–1443. [CrossRef]
12. Bhaskar, T.G.; Lakshmikantham, V. Fixed point theorems in partially ordered metric spaces and applications. *Nonlinear Anal.* **2006**, *65*, 1379–1393. [CrossRef]
13. Petrusel, A.; Rus, I.A. Fixed point theory in terms of a metric and of an order relation. *Fixed Point Theory* **2019**, *20*, 601–622. [CrossRef]
14. Soleimani Rad, G.; Shukla, S.; Rahimi, H. Some relations between n-tuple fixed point and fixed point results. *Rev. Real Acad. Cienc. Exactas Físicas Nat. Ser. A Matemáticas* **2015**, *109*, 471–481. [CrossRef]

Article

Fixed Points of g-Interpolative Ćirić–Reich–Rus-Type Contractions in b-Metric Spaces

Youssef Errai *, El Miloudi Marhrani * and Mohamed Aamri

Laboratory of Algebra, Analysis and Applications (L3A), Faculty of Sciences Ben M'Sik,
Hassan II University of Casablanca, B.P 7955, Sidi Othmane, Casablanca 20700, Morocco;
aamrimohamed82@gmail.com
* Correspondence: yousseferrai1@gmail.com (Y.E.); marhrani@gmail.com (E.M.)

Received: 15 October 2020; Accepted: 12 November 2020; Published: 16 November 2020

Abstract: We use interpolation to obtain a common fixed point result for a new type of Ćirić–Reich–Rus-type contraction mappings in metric space. We also introduce a new concept of g-interpolative Ćirić–Reich–Rus-type contractions in b-metric spaces, and we prove some fixed point results for such mappings. Our results extend and improve some results on the fixed point theory in the literature. We also give some examples to illustrate the given results.

Keywords: fixed point; Ćirić–Reich–Rus-type contractions; interpolation; b-metric space

MSC: 46T99; 47H10; 54H25

1. Introduction and Preliminaries

Banach's contraction principle [1] has been applied in several branches of mathematics. As a result, researching and generalizing this outcome has proven to be a research area in nonlinear analysis (see [2–6]). It is a well-known fact that a map that satisfies the Banach contraction principle is necessarily continuous. Therefore, it was natural to wonder if in a complete metric space, a discontinuous map satisfying somewhat similar contractual conditions may have a fixed point. Kannan [7] answered yes to this question by introducing a new type of contraction. The concept of the interpolation Kannan-type contraction appeared with Karapinar [8] in 2018; this concept appealed to many researchers [8–14], making them invest in various types of contractions: interpolative Ćirić–Reich–Rus-type contraction [9–11,13], interpolative Hardy–Rogers [15]; and they used it on various spaces: metric space, b-metric space, and the Branciari distance.

In this paper, we will generalize some of the related findings to the interpolation Ćirić–Reich–Rus-type contraction in Theorems 1 and 2. In addition, we use a new concept of interpolative weakly contractive mapping to generalize some findings about the interpolation Kannan-type contraction in Theorem 3.

Now, we recall the concept of b-metric spaces as follows:

Definition 1 ([16,17])**.** *Let X be a nonempty set and $s \geq 1$ be a given real number. A function $d : X \times X \to \mathbb{R}^+$ is a b-metric if for all $x, y, z \in X$, the following conditions are satisfied:*

(b_1) $d(x,y) = 0$ if and only if $x = y$;
(b_2) $d(x,y) = d(y,x)$;
(b_3) $d(x,z) \leq s[d(x,y) + d(y,z)]$.

The pair (X,d) is called a b-metric space.

Note that the class of b-metric spaces is larger than that of metric spaces.

The notions of b-convergent and b-Cauchy sequences, as well as of b-complete b-metric spaces are defined exactly the same way as in the case of usual metric spaces (see, e.g., [18]).

Definition 2 ([19,20]). *Let $\{x_n\}$ be a sequence in a b-metric space (X,d). $g, h: X \to X$, are self-mappings, and $x \in X$. x is said to be the coincidence point of pair $\{g,h\}$ if $gx = hx$.*

Definition 3 ([10,11]). *Let Ψ be denoted as the set of all non-decreasing functions $\psi: [0, \infty) \to [0, \infty)$, such that $\sum_{k=0}^{\infty} \psi^k(t) < \infty$ for each $t > 0$. Then:*

(i) $\psi(0) = 0$,
(ii) $\psi(t) < t$ for each $t > 0$.

Remark 1 ([18]). *In a b-metric space (X,d), the following assertions hold:*

1. A b-convergent sequence has a unique limit.
2. Each b-convergent sequence is a b-Cauchy sequence.
3. In general, a b-metric is not continuous.

The fact in the last remark requires the following lemma concerning the b-convergent sequences to prove our results:

Lemma 1 ([19]). *Let (X,d) be a b-metric space with $s \geq 1$, and suppose that $\{x_n\}$ and $\{y_n\}$ are b-convergent to x, y, respectively, then we have:*

$$\frac{1}{s^2}d(x,y) \leq \liminf_{n \to \infty} d(x_n, y_n) \leq \limsup_{n \to \infty} d(x_n, y_n) \leq s^2 d(x,y).$$

In particular, if $x = y$, then we have $\lim_{n \to \infty} d(x_n, y_n) = 0$. Moreover, for each $z \in X$, we have:

$$\frac{1}{s}d(x,z) \leq \liminf_{n \to \infty} d(x_n, z) \leq \limsup_{n \to \infty} d(x_n, z) \leq s d(x,z).$$

2. Results

We denote by Φ the set of functions $\phi: [0, \infty) \to [0, \infty)$ such that $\phi(t) < t$ for every $t > 0$. Our main result is the following theorem:

Theorem 1. *Let (X,d) be a complete metric space, and T is a self-mapping on X such that:*

$$d(Tx, Ty) \leq \phi([d(x,y)]^\alpha [d(x,Tx)]^\beta [d(y,Ty)]^\gamma) \tag{1}$$

is satisfied for all $x, y \in X \setminus \text{Fix}(T)$; where $\text{Fix}(T) = \{a \in X | Ta = a\}$, $\alpha, \beta, \gamma \in (0,1)$ such that $\alpha + \beta + \gamma > 1$, and $\phi \in \Phi$.

If there exists $x \in X$ such that $d(x, Tx) < 1$, then T has a fixed point in X.

Proof. We define a sequence $\{x_n\}$ by $x_0 = x$ and $x_{n+1} = Tx_n$ for all integers n, and we assume that $x_n \neq Tx_n$, for all n.

We have:

$$d(x_n, x_{n+1}) \leq \phi([d(x_{n-1}, x_n)]^\alpha [d(x_{n-1}, x_n)]^\beta [d(x_n, x_{n+1})]^\gamma). \tag{2}$$

Using the fact $\phi(t) < t$ for each $t > 0$, from (2), we obtain:

$$d(x_n, x_{n+1}) < [d(x_{n-1}, x_n)]^\alpha [d(x_{n-1}, x_n)]^\beta [d(x_n, x_{n+1})]^\gamma.$$

which implies:
$$[d(x_n, x_{n+1})]^{1-\gamma} < [d(x_{n-1}, x_n)]^{\alpha+\beta}. \tag{3}$$

We have $d(x_0, x_1) < 1$, so that there exists a real $\lambda \in (0,1)$ such that $d(x_0, x_1) \leq \lambda$ and $\lambda = \frac{d(x_0,x_1)+1}{2}$.

By (3), we obtain:
$$d(x_1, x_2) < [d(x_0, x_1)]^{\frac{\alpha+\beta}{1-\gamma}} \leq \lambda^{\frac{\alpha+\beta}{1-\gamma}}.$$

By (3), we find:
$$d(x_{n+1}, x_n) \leq d(x_n, x_{n-1})^{1+\epsilon}$$

for all n, with $\epsilon = \frac{\alpha+\beta}{1-\gamma} - 1 > 0$.

Now, we prove by induction that for all n,
$$d(x_{n+1}, x_n) \leq \lambda^{(1+\epsilon)^n}$$

where $0 < \lambda < 1$. For $n = 1$, this is the inequality at the bottom of page 3. The induction step is:
$$d(x_{n+2}, x_{n+1}) \leq d(x_{n+1}, x_n)^{1+\epsilon} \leq \left(\lambda^{(1+\epsilon)^n}\right)^{1+\epsilon} = \lambda^{(1+\epsilon)^{n+1}}$$

Since $(1+\epsilon)^n \geq 1 + n\epsilon$ by Bernoulli's inequality and since $\lambda < 1$, this implies:
$$d(x_{n+1}, x_n) \leq \lambda^{1+n\epsilon} = \lambda \rho^n$$

for all n, where $\rho = \lambda^\epsilon < 1$. This implies:
$$d(x_{n+k}, x_n) \leq \lambda(\rho^{n+k-1} + \rho^{n+k-2} + \cdots + \rho^n) = \lambda \rho^n \left(\frac{1-\rho^k}{1-\rho}\right) = C\rho^n,$$

where $C = \lambda \left(\frac{1-\rho^k}{1-\rho}\right)$ for some integer k, from which it follows that $\{x_n\}$ forms a Cauchy sequence in (X, d), and then, it converges to some $z \in X$. Assume that $z \neq Tz$.

By letting $x = x_n$ and $y = z$ in (1), we obtain:
$$\begin{aligned} d(x_{n+1}, Tz) &\leq \phi([d(x_n, z)]^\alpha [d(x_n, x_{n+1})]^\beta [d(z, Tz)]^\gamma) \\ &< [d(x_n, z)]^\alpha [d(x_n, x_{n+1})]^\beta [d(z, Tz)]^\gamma \end{aligned}$$

for all n, which leads to $d(z, Tz) = 0$, which is a contradiction. Then, $Tz = z$. □

Example 1. Let $X = [0, 2]$ be endowed with metric $d : X \times X \to [0, \infty)$, defined by:
$$d(x, y) = \begin{cases} 0, & \text{if } x = y; \\ \frac{2}{3}, & \text{if } x, y \in [0, 1] \text{ and } x \neq y; \\ 2, & \text{otherwise.} \end{cases}$$

Consider that the self-mapping $T : X \to X$ is defined by:
$$Tx = \begin{cases} \frac{1}{2}, & \text{if } x \in [0, 1]; \\ \frac{x}{2}, & \text{if } x \in (1, 2]; \end{cases}$$

and the function $\phi(t) = 0.4t^2$ for all $t \in [0, \infty)$.

For $\alpha = 0.8$, $\beta = 0.2$, and $\gamma = 0.25$.

We discus the following cases:

Case 1. If $x, y \in [0, 1]$ or $x = y$ for all $x, y \in [0, 2]$; it is obvious.

Case 2. If $x, y \in (1, 2]$ and $x \neq y$.

We have:
$$d(Tx, Ty) = \frac{2}{3}$$

and:
$$\phi([d(x,y)]^\alpha [d(x, Tx)]^\beta [d(y, Ty)]^\gamma) = \phi(2^{\alpha+\beta+\gamma}) = \frac{2^{3,5}}{5} \geq \frac{2}{3}.$$

Then:
$$d(Tx, Ty) \leq \phi([d(x,y)]^\alpha [d(x, Tx)]^\beta [d(y, Ty)]^\gamma)$$

for all $x, y \in (1, 2]$.

Case 3. If $x \in [0, 1]$ and $y \in (1, 2]$ with $x \neq \frac{1}{2}$.

We have:
$$d(Tx, Ty) = \frac{2}{3}$$

and:
$$\phi([d(x,y)]^\alpha [d(x, Tx)]^\beta [d(y, Ty)]^\gamma) = \phi\left(2^{\alpha+\gamma} \left(\frac{2}{3}\right)^\beta\right) = \frac{2^{3,5}}{5 \cdot 3^{0,2}} \geq \frac{2}{3}.$$

Then:
$$d(Tx, Ty) \leq \phi([d(x,y)]^\alpha [d(x, Tx)]^\beta [d(y, Ty)]^\gamma)$$

for all $x \in [0, 1] \setminus \{\frac{1}{2}\}$ and $y \in (1, 2]$.

Case 4. If $x \in (1, 2]$ and $y \in [0, 1]$ with $y \neq \frac{1}{2}$.

We have:
$$d(Tx, Ty) = \frac{2}{3}$$

and:
$$\phi([d(x,y)]^\alpha [d(x, Tx)]^\beta [d(y, Ty)]^\gamma) = \phi\left(2^{\alpha+\beta} \left(\frac{2}{3}\right)^\gamma\right) = \frac{2^{3,5}}{5 \cdot 3^{0,25}} \geq \frac{2}{3}.$$

Then:
$$d(Tx, Ty) \leq \phi([d(x,y)]^\alpha [d(x, Tx)]^\beta [d(y, Ty)]^\gamma)$$

for all $x \in (1, 2]$ and $y \in [0, 1] \setminus \{\frac{1}{2}\}$.

Therefore, all the conditions of Theorem 1 are satisfied, and T has a fixed point, $x = \frac{1}{2}$.

Example 2. Let $X = \{a, q, r, s\}$ be endowed with the metric defined by the following table of values:

$d(x,y)$	a	q	r	s
a	0	$\frac{1}{3}$	$\frac{10}{3}$	$\frac{5}{3}$
q	$\frac{1}{3}$	0	3	2
r	$\frac{10}{3}$	3	0	5
s	$\frac{5}{3}$	2	5	0

Consider the self-mapping T on X as:

$$T: \begin{pmatrix} a & q & r & s \\ a & a & q & s \end{pmatrix}.$$

For $\psi(t) = \frac{2^t-1}{2^t+1}$ for all $t \in [0, \infty)$; $\alpha = 0,6$; $\beta = 0,9$; and $\gamma = 0,7$.

We have:
$$d(Tu, Tv) \leq \psi([d(u,v)]^\alpha [d(u,Tu)]^\beta [d(v,Tv)]^\gamma)$$

for all $u,v \in X \setminus \{a,s\}$.

Then, T has two fixed points, which are a and s.

If we take $\psi(t) = kt$ in Theorem (1) with $k \in (0,1)$, then we have the following corollary:

Corollary 1. *Let (X,d) be a complete metric space, and T is a self-mapping on X such that:*

$$d(Tx, Ty) \leq k[d(x,y)]^\alpha [d(x,Tx)]^\beta [d(y,Ty)]^\gamma$$

is satisfied for all $x, y \in X \setminus Fix(T)$; where $Fix(T) = \{a \in X | Ta = a\}$, and $\alpha, \beta, \gamma, k \in (0,1)$ such that $\alpha + \beta + \gamma > 1$.

If there exists $x \in X$ such that $d(x, Tx) < 1$, then T has a fixed point in X.

Example 3. *It is enough to take in Example 1: $\phi(t) = \frac{57}{58}t$ for all $t \in [0, +\infty)$.*

Example 4. *Let $X = \{a, q, r, s\}$ be endowed with the metric defined by the following table of values:*

$d(x,y)$	a	q	r	s
a	0	0,1	3,1	4
q	0,1	0	3	3,9
r	3,1	3	0	0,9
s	4	3,9	0,9	0

Consider the self-mapping T on X as:

$$T: \begin{pmatrix} a & q & r & s \\ a & a & q & s \end{pmatrix}.$$

For $k = \frac{3}{10}$; $\alpha = 0,7$; $\beta = 0,1$; and $\gamma = 0,8$.

We have:
$$d(Tu, Tv) \leq k[d(u,v)]^\alpha [d(u,Tu)]^\beta [d(v,Tv)]^\gamma$$

for all $u, v \in X \setminus \{a,s\}$.

Then, T has two fixed points, which are a and s.

Definition 4. Let (X, d, s) be a b-metric space and $T, g : X \to X$ be self-mappings on X. We say that T is a g-interpolative Ćirić–Reich–Rus-type contraction, if there exists a continuous $\psi \in \Psi$ and $\alpha, \beta \in (0,1)$ such that:

$$d(Tx, Ty) \leq \psi([d(gx, gy)]^\alpha [d(gx, Tx)]^\beta [d(gy, Ty)]^{1-\alpha-\beta}) \quad (4)$$

is satisfied for all $x, y \in X$ such that $Tx \neq gx$, $Ty \neq gy$, and $gx \neq gy$.

Theorem 2. Let (X, d, s) be a b-complete b-metric space, and T is a g-interpolative Ćirić–Reich–Rus-type contraction. Suppose that $TX \subseteq gX$ such that gX is closed. Then, T and g have a coincidence point in X.

Proof. Let $x \in X$; since $TX \subseteq gX$, we can define inductively a sequence $\{x_n\}$ such that:

$$x_0 = x, \text{ and } gx_{n+1} = Tx_n, \text{ for all integer } n.$$

If there exists $n \in \{0, 1, 2, \ldots\}$ such that $gx_n = Tx_n$, then x_n is a coincidence point of g and T. Assume that $gx_n \neq Tx_n$, for all n. By (4), we obtain:

$$\begin{aligned} d(Tx_{n+1}, Tx_n) &\leq \psi([d(gx_{n+1}, gx_n)]^\alpha [d(gx_{n+1}, Tx_{n+1})]^\beta [d(gx_n, Tx_n)]^{1-\alpha-\beta}) \\ &= \psi([d(Tx_n, Tx_{n-1})]^\alpha [d(Tx_n, Tx_{n+1})]^\beta [d(Tx_{n-1}, Tx_n)]^{1-\alpha-\beta}) \\ &= \psi([d(Tx_n, Tx_{n-1})]^{1-\beta} [d(Tx_n, Tx_{n+1})]^\beta). \end{aligned}$$

Using the fact $\psi(t) < t$ for each $t > 0$,

$$\begin{aligned} d(Tx_{n+1}, Tx_n) &\leq \psi([d(Tx_n, Tx_{n-1})]^{1-\beta} [d(Tx_n, Tx_{n+1})]^\beta) \\ &< [d(Tx_n, Tx_{n-1})]^{1-\beta} [d(Tx_n, Tx_{n+1})]^\beta. \end{aligned} \quad (5)$$

which implies:

$$[d(Tx_{n+1}, Tx_n)]^{1-\beta} < [d(Tx_n, Tx_{n-1})]^{1-\beta}.$$

Thus,

$$d(Tx_{n+1}, Tx_n) < d(Tx_n, Tx_{n-1}) \text{ for all } n \geq 1. \quad (6)$$

That is, the positive sequence $\{d(Tx_{n+1}, Tx_n)\}$ is monotone decreasing, and consequently, there exists $c \geq 0$ such that $\lim_{n \to \infty} d(Tx_{n+1}, Tx_n) = c$. From (6), we obtain:

$$\begin{aligned} [d(Tx_n, Tx_{n-1})]^{1-\beta} [d(Tx_n, Tx_{n+1})]^\beta &\leq [d(Tx_n, Tx_{n-1})]^{1-\beta} [d(Tx_n, Tx_{n-1})]^\beta \\ &= d(Tx_n, Tx_{n-1}). \end{aligned}$$

Therefore, with (5) together with the nondecreasing character of ψ, we get:

$$\begin{aligned} d(Tx_{n+1}, Tx_n) &\leq \psi([d(Tx_n, Tx_{n-1})]^{1-\beta} [d(Tx_n, Tx_{n+1})]^\beta) \\ &\leq \psi(d(Tx_n, Tx_{n-1})). \end{aligned}$$

By repeating this argument, we get:

$$d(Tx_{n+1}, Tx_n) \leq \psi(d(Tx_n, Tx_{n-1})) \leq \psi^2(d(Tx_{n-1}, Tx_{n-2})) \leq \cdots \leq \psi^n(d(Tx_1, Tx_0)). \quad (7)$$

Taking $n \to \infty$ in (7) and using the fact $\lim_{n \to \infty} \psi^n(t) = 0$ for each $t > 0$, we deduce that $c = 0$, that is,

$$\lim_{n \to \infty} d(Tx_{n+1}, Tx_n) = 0. \quad (8)$$

Then, $\{Tx_n\}$ is a b-Cauchy sequence. Suppose on the contrary that there exists an $\epsilon > 0$ and subsequences $\{Tx_{m_k}\}$ and $\{Tx_{n_k}\}$ of $\{Tx_n\}$ such that n_k is the smallest integer for which:

$$n_k > m_k > k, \ d(Tx_{n_k}, Tx_{m_k}) \geq \epsilon, \ \text{and} \ d(Tx_{n_k-1}, Tx_{m_k}) < \epsilon.$$

Then, we have:

$$\begin{aligned} d(gx_{n_k}, gx_{m_k}) = d(Tx_{n_k-1}, Tx_{m_k-1}) &\leq sd(Tx_{n_k-1}, Tx_{m_k}) + sd(Tx_{m_k}, Tx_{m_k-1}) \\ &\leq s\epsilon + sd(Tx_{m_k}, Tx_{m_k-1}). \end{aligned}$$

Using (8) in the inequality above, we obtain:

$$\limsup_{k \to \infty} d(Tx_{n_k-1}, Tx_{m_k-1}) = \limsup_{k \to \infty} d(gx_{n_k}, gx_{m_k}) \leq s\epsilon. \tag{9}$$

Putting $x = x_{n_k}$ and $y = x_{m_k}$ in (4), we have:

$$\begin{aligned} \epsilon \leq d(Tx_{n_k}, Tx_{m_k}) &\leq \psi([d(gx_{n_k}, gx_{m_k})]^\alpha [d(gx_{n_k}, Tx_{n_k})]^\beta [d(gx_{m_k}, Tx_{m_k})]^{1-\alpha-\beta}) \\ &= \psi([d(Tx_{n_k-1}, Tx_{m_k-1})]^\alpha [d(Tx_{n_k-1}, Tx_{n_k})]^\beta [d(Tx_{m_k-1}, Tx_{m_k})]^{1-\alpha-\beta}). \tag{10} \end{aligned}$$

Taking the upper limit as $k \to \infty$ in (10) and using (8) and (9) and the property of ψ, we get:

$$\epsilon \leq \limsup_{k \to \infty} d(Tx_{n_k}, Tx_{m_k}) \leq \psi(0) = 0,$$

which implies that $\epsilon = 0$, a contradiction with $\epsilon > 0$. We deduce that $\{Tx_n\}$ is a b-Cauchy sequence, and consequently, $\{gx_n\}$ is also a b-Cauchy sequence. Let $z \in X$ such that,

$$\lim_{n \to \infty} d(Tx_n, z) = \lim_{n \to \infty} d(gx_{n+1}, z) = 0.$$

Since $z \in gX$, there exists $u \in X$ such that $z = gu$. We claim that u is a coincidence point of g and T. For this, if we assume that $gu \neq Tu$, we obtain:

$$\begin{aligned} d(Tx_n, Tu) &\leq \psi([d(gx_n, gu)]^\alpha [d(gx_n, Tx_n)]^\beta [d(gu, Tu)]^{1-\alpha-\beta}) \\ &< [d(gx_n, gu)]^\alpha [d(gx_n, Tx_n)]^\beta [d(gu, Tu)]^{1-\alpha-\beta}. \end{aligned}$$

At the limit as $n \to \infty$ and using Lemma 1, we get:

$$\begin{aligned} \frac{1}{s} d(z, Tu) \leq \liminf_{n \to \infty} d(Tx_n, Tu) &\leq \limsup_{n \to \infty} [d(gx_n, gu)]^\alpha [d(gx_n, Tx_n)]^\beta [d(gu, Tu)]^{1-\alpha-\beta} \\ &\leq [sd(z, gu)]^\alpha [s^2 d(z, z)]^\beta [d(gu, Tu)]^{1-\alpha-\beta} = 0, \end{aligned}$$

which is a contradiction, which implies that:

$$Tu = z = gu.$$

Then, u is a coincidence point in X of T and g. □

Example 5. Let $X = [0, +\infty)$ and $d : X \times X \to [0, \infty)$ be defined by:

$$d(x, y) = \begin{cases} (x+y)^2, & \text{if } x \neq y; \\ 0, & \text{if } x = y. \end{cases}$$

Then, (X, d) is a complete b-metric space.

Define two self-mappings T and g on X by $g(x) = x^2$; for all $x \in X$ and:

$$Tx = \begin{cases} 1, & \text{if } x \in [0,2]; \\ \frac{1}{x}, & \text{if } x \in (2, +\infty). \end{cases}$$

T is a g-interpolative Ćirić–Reich–Rus-type contraction for $\alpha = 0,7$, $\beta = 0,4$, and:

$$\psi(t) = \begin{cases} \frac{3}{20}t^2, & \text{if } t \in [0, \frac{89}{20}]; \\ \frac{3^{t+1}-1}{3^t+1}, & \text{if } t \in (\frac{89}{20}, +\infty). \end{cases}$$

For this, we discuss the following cases:

Case 1. If $x, y \in [0,2]$ or $x = y$ for all $x \in [0, +\infty)$. It is obvious.

Case 2. If $x, y \in (2, +\infty)$ and $x \neq y$.

We have:
$$d(Tx, Ty) = (\frac{1}{x} + \frac{1}{y})^2 \leq 1.$$

Using the property of ψ, we get:

$$\psi([d(gx,gy)]^\alpha [d(gx,Tx)]^\beta [d(gy,Ty)]^{1-\alpha-\beta}) = \psi((x^2+y^2)^{2\alpha}(x^2+\frac{1}{x})^{2\beta}(y^2+\frac{1}{y})^{2(1-\alpha-\beta)})$$
$$\geq \psi(8^{2\alpha}.(\frac{9}{2})^{2(1-\alpha)}) \geq 1.$$

Therefore,
$$d(Tx, Ty) \leq \psi([d(gx,gy)]^\alpha [d(gx,Tx)]^\beta [d(gy,Ty)]^{1-\alpha-\beta}).$$

Case 3. If $x \in [0,2] \setminus \{1\}$ and $y \in (2, +\infty)$.

We have:
$$d(Tx, Ty) = (1 + \frac{1}{y})^2 \leq (\frac{3}{2})^2 = \frac{9}{4},$$

and:
$$\psi([d(gx,gy)]^\alpha [d(gx,Tx)]^\beta [d(gy,Ty)]^{1-\alpha-\beta}) = \psi((x^2+y^2)^{2\alpha}(x^2+1)^{2\beta}(y^2+\frac{1}{y})^{2(1-\alpha-\beta)})$$
$$\geq \psi(4^{2\alpha}.1^{2\beta}.(\frac{9}{2})^{2(1-\alpha-\beta)}) \geq \frac{9}{4}.$$

Therefore,
$$d(Tx, Ty) \leq \psi([d(gx,gy)]^\alpha [d(gx,Tx)]^\beta [d(gy,Ty)]^{1-\alpha-\beta}).$$

Case 4. If $x \in (2, +\infty)$ and $y \in [0,2] \setminus \{1\}$.

We have:
$$d(Tx, Ty) = (1 + \frac{1}{x})^2 \leq \frac{9}{4},$$

and:
$$\psi([d(gx,gy)]^\alpha [d(gx,Tx)]^\beta [d(gy,Ty)]^{1-\alpha-\beta}) = \psi((x^2+y^2)^{2\alpha}(x^2+\frac{1}{x})^{2\beta}(y^2+1)^{2(1-\alpha-\beta)})$$
$$\geq \psi(4^{2\alpha}.(\frac{9}{2})^{2\beta}.1^{2(1-\alpha-\beta)}) \geq \frac{9}{4}.$$

Therefore,
$$d(Tx,Ty) \leq \psi([d(gx,gy)]^\alpha [d(gx,Tx)]^\beta [d(gy,Ty)]^{1-\alpha-\beta}).$$

Then, it is clear that g, T satisfies (4) for all $u, v \in X \setminus \{1\}$. Moreover, one is a coincidence point of g and T.

Example 6. Let the set $X = \{a, b, q, r\}$ and a function $d : X \times X \to [0, \infty)$ be defined as follows:

$d(x,y)$	a	b	q	r
a	0	1	16	$\frac{49}{4}$
b	1	0	9	$\frac{25}{4}$
q	16	9	0	$\frac{1}{4}$
r	$\frac{49}{4}$	$\frac{25}{4}$	$\frac{1}{4}$	0

By a simple calculation, one can verify that the function d is a b-metric, for $s = 2$. We define the self-mappings g, T on X, as:

$$g : \begin{pmatrix} a & b & q & r \\ a & r & q & q \end{pmatrix}, \quad T : \begin{pmatrix} a & b & q & r \\ q & r & r & q \end{pmatrix}.$$

For $\alpha = 0, 3$; $\beta = 0, 8$; and $\psi(t) = \frac{t}{1+t}$ for all $t \in [0, \infty)$.

It is clear that g, T satisfies (4) for all $u, v \in X \setminus \{b, r\}$. Moreover, b and r are two coincidence points of g and T.

Definition 5. Let (X, d) is a metric space. A self-mapping $T : X \to X$ is said to be an interpolative weakly contractive mapping if there exists a constant $\alpha \in (0, 1)$ such that:

$$\zeta(d(Tx,Ty)) \leq \zeta([d(x,Tx)]^\alpha [d(y,Ty)]^{1-\alpha}) - \varphi([d(x,Tx)]^\alpha [d(y,Ty)]^{1-\alpha}), \tag{11}$$

for all $x, y \in X \setminus Fix(T)$, where
$Fix(T) = \{a \in X | Ta = a\}$,
$\zeta : [0, \infty) \to [0, \infty)$ is a continuous monotone nondecreasing function with $\zeta(t) = 0$ if and only if $t = 0$,
$\varphi : [0, \infty) \to [0, \infty)$ is a lower semi-continuous function with $\varphi(t) = 0$ if and only if $t = 0$.

Theorem 3. Let (X, d) be a complete metric space. If $T : X \to X$ is a interpolative weakly contractive mapping, then T has a fixed point.

Proof. For any $x_0 \in X$, we define a sequence $\{x_n\}$ by $x = x_0$ and $x_{n+1} = Tx_n$, $n = 0, 1, 2, \ldots$
If there exists $n_0 \in \mathbb{N}$ such that $x_{n_0+1} = x_{n_0}$, then x_{n_0} is clearly a fixed point in X. Otherwise, $x_{n+1} \neq x_n$ for each $n \geq 0$.

Substituting $x = x_n$ and $y = x_{n-1}$ in (11), we obtain that:

$$\begin{aligned}
\zeta(d(x_{n+1}, x_n)) &\leq \zeta([d(x_n, x_{n+1})]^\alpha [d(x_{n-1}, x_n)]^{1-\alpha}) - \varphi([d(x_n, x_{n+1})]^\alpha [d(x_{n-1}, x_n)]^{1-\alpha}) \\
&\leq \zeta([d(x_n, x_{n+1})]^\alpha [d(x_{n-1}, x_n)]^{1-\alpha}).
\end{aligned} \tag{12}$$

Using property of function ζ, we get:

$$d(x_{n+1}, x_n) \leq [d(x_n, x_{n+1})]^\alpha [d(x_{n-1}, x_n)]^{1-\alpha}.$$

We derive:

$$[d(x_{n+1}, x_n)]^{1-\alpha} \leq [d(x_{n-1}, x_n)]^{1-\alpha}.$$

Therefore:

$$d(x_{n+1}, x_n) \leq d(x_{n-1}, x_n), \quad \text{for all } n \geq 1.$$

It follows that the positive sequence $\{d(x_{n+1}, x_n)\}$ is decreasing. Eventually, there exists $c \geq 0$ such that $\lim_n d(x_{n+1}, x_n) = c$.

Taking $n \to \infty$ in the inequality (12), we obtain:

$$\zeta(c) \leq \zeta(c) - \varphi(c).$$

We deduce that $c = 0$. Hence:

$$\lim_n d(x_{n+1}, x_n) = 0. \tag{13}$$

Therefore, $\{x_n\}$ is a Cauchy sequence. Suppose it is not. Then, there exists a real number $\epsilon > 0$, for any $k \in \mathbb{N}, \exists m_k \geq n_k \geq k$ such that:

$$d(x_{m_k}, x_{n_k}) \geq \epsilon. \tag{14}$$

Putting $x = x_{n_k-1}$ and $y = x_{m_k-1}$ in (11) and using (14), we get:

$$\zeta(\epsilon) \leq \zeta(d(x_{m_k}, x_{n_k})) \leq \zeta([d(x_{m_k-1}, x_{m_k})]^\alpha [d(x_{n_k-1}, x_{n_k})]^{1-\alpha}) - \varphi([d(x_{m_k-1}, x_{m_k})]^\alpha [d(x_{n_k-1}, x_{n_k})]^{1-\alpha}).$$

Letting $k \to \infty$ and using (13), we conclude:

$$\zeta(\epsilon) \leq \zeta(0) - \varphi(0) = 0,$$

which is contradiction with $\epsilon > 0$; thus, $\{x_n\}$ is a Cauchy sequence; since (X, d) is complete, we obtain $z \in X$ such that $\lim_n d(x_n, z) = 0$, and assuming that $Tz \neq z$, we have:

$$\zeta(d(x_{n+1}, Tz)) \leq \zeta([d(x_n, x_{n+1})]^\alpha [d(z, Tz)]^{1-\alpha}) - \varphi([d(x_n, x_{n+1})]^\alpha [d(z, Tz)]^{1-\alpha}) \quad \text{for all } n.$$

Letting $n \to \infty$, we get:

$$\zeta(d(z, Tz)) \leq \zeta([d(z,z)]^\alpha [d(z, Tz)]^{1-\alpha}) - \varphi([d(z,z)]^\alpha [d(z, Tz)]^{1-\alpha}) = \zeta(0) - \varphi(0) = 0,$$

which is a contradiction; thus, $Tz = z$. □

Example 7. *Let the set $X = [0, 3]$ and a function $\delta : X \times X \to [0, \infty)$ be defined as follows:*

$$\delta(x, y) = \begin{cases} 0, & \text{if } x = y; \\ 3, & \text{if } x, y \in [0, 1) \text{ and } x \neq y; \\ 2, & \text{otherwise}. \end{cases}$$

Then, (X, δ) is a complete metric space.

Let $T: X \to X$ be defined as:

$$Tx = \begin{cases} 0, & \text{if } x \in [0, 1); \\ 1, & \text{if } x \in [1, 3]. \end{cases}$$

For $\zeta(t) = t^2$, $\varphi(t) = \frac{1}{2}t$ for all $t \in [0, +\infty)$ and $\alpha = 0, 6$.

We discuss the following cases.

Case 1. *If $x = y$ or $x, y \in (0, 1)$, or $x, y \in (1, 3]$ with $x \neq y$. It is obvious.*

Case 2. *If $x \in (0, 1)$ and $y \in (1, 3]$.*

We have:

$$\zeta(\delta(Tx, Ty)) = \zeta(\delta(0, 1)) = \zeta(2) = 4,$$

and:
$$[\delta(x,Tx)]^\alpha [\delta(y,Ty)]^{1-\alpha} = [\delta(x,0)]^\alpha [\delta(y,1)]^{1-\alpha} = 2.(\frac{3}{2})^\alpha.$$

Therefore:
$$\zeta([\delta(x,Tx)]^\alpha [\delta(y,Ty)]^{1-\alpha}) - \varphi([\delta(x,Tx)]^\alpha [\delta(y,Ty)]^{1-\alpha}) = (\frac{3}{2})^\alpha [4.(\frac{3}{2})^\alpha - 1] \geq 4 = \zeta(2) = \zeta(\delta(Tx,Ty)).$$

Case 3. If $x \in (1,3]$ and $y \in (0,1)$.

We have:
$$\zeta(\delta(Tx,Ty)) = \zeta(\delta(1,0)) = \zeta(2) = 4,$$

and:
$$[\delta(x,Tx)]^\alpha [\delta(y,Ty)]^{1-\alpha} = [\delta(x,1)]^\alpha [\delta(y,0)]^{1-\alpha} = 3.(\frac{2}{3})^\alpha.$$

Therefore,
$$\zeta([\delta(x,Tx)]^\alpha [\delta(y,Ty)]^{1-\alpha}) - \varphi([\delta(x,Tx)]^\alpha [\delta(y,Ty)]^{1-\alpha}) = (\frac{2}{3})^\alpha [9.(\frac{2}{3})^\alpha - \frac{3}{2}] \geq 4 = \zeta(2) = \zeta(\delta(Tx,Ty)).$$

Thus,
$$\zeta(d(Tu,Tv)) \leq \zeta([d(u,Tu)]^\alpha [d(v,Tv)]^{1-\alpha}) - \varphi([d(u,Tu)]^\alpha [d(v,Tv)]^{1-\alpha}),$$

for all $u,v \in X \setminus \{0,1\}$.

Then, T has two fixed points, which are zero and one.

Example 8. Let $X = \{a,b,r,s\}$ be endowed with the metric defined by the following table of values:

$d(x,y)$	a	b	r	s
a	0	1	4	1
b	1	0	5	2
r	4	5	0	3
s	1	2	3	0

Consider the self-mapping T on X as:
$$T: \begin{pmatrix} a & b & r & s \\ a & s & a & s \end{pmatrix}.$$

For $\zeta(t) = e^t - 1$ and $\varphi(t) = 2^t - 1$ for all $t \in [0,\infty)$; $\alpha = 0,3$.

We have:
$$\zeta(d(Tu,Tv)) \leq \zeta([d(u,Tu)]^\alpha [d(v,Tv)]^{1-\alpha}) - \varphi([d(u,Tu)]^\alpha [d(v,Tv)]^{1-\alpha}),$$

for all $u,v \in X \setminus \{a,s\}$.

Then, T has two fixed points, which are a and s.

If $\zeta(t) = t$ in Theorem (3), then we have the following corollary:

Corollary 2. *Let (X, d) be a complete metric space and $T : X \to X$ a self-mapping on X. If there exists a constant $\alpha \in (0, 1)$ such that:*

$$d(Tx, Ty) \leq [d(x, Tx)]^{\alpha}[d(y, Ty)]^{1-\alpha} - \varphi([d(x, Tx)]^{\alpha}[d(y, Ty)]^{1-\alpha}),$$

for all $x, y \in X$ and $x \neq Tx, y \neq Ty$.
$\varphi : [0, \infty) \to [0, \infty)$ is a lower semi-continuous function with $\varphi(t) = 0$ if and only if $t = 0$.

Then, T has a fixed point.

Remark 2. *In Corollary 2, if we take $\varphi(t) = (1 - \lambda)t$ for a constant $\lambda \in (0, 1)$, then the result of Theorem [8] is obtained.*

Author Contributions: All authors contributed equally and significantly to the writing of this article. All authors read and approved the final manuscript.

Funding: This research received no external funding.

Conflicts of Interest: The authors declare no conflict of interest.

References

1. Banach, S. Sur les opérations dans les ensembles abstraits et leur application aux équations intégrales. *Fund. Math.* **1922**, *3*, 133–181. [CrossRef]
2. Petruşel, A.; Petruşel, G. On Reich's strict fixed point theorem for multi-valued operators in complete metric spaces. *J. Nonlinear Var. Anal.* **2018**, *2*, 103–112.
3. Suzuki, T. Edelstein's fixed point theorem in semimetric spaces. *J. Nonlinear Var. Anal.* **2018**, *2*, 165–175.
4. Park, S. Some general fixed point theorems on topological vector spaces. *Appl. Set-Valued Anal. Optim.* **2019**, *1*, 19–28.
5. Đorić, D. Common fixed point for generalized (ψ, ϕ)-weak contractions. *Appl. Math. Lett.* **2009**, *22*, 1896–1900. [CrossRef]
6. Dutta, P.; Choudhury, B.S. A generalisation of contraction principle in metric spaces. *Fixed Point Theory Appl.* **2008**, *2008*, 406368. [CrossRef]
7. Kannan, R. Some results on fixed points. *Bull. Cal. Math. Soc.* **1968**, *60*, 71–76.
8. Karapinar, E. Revisiting the Kannan type contractions via interpolation. *Adv. Theory Nonlinear Anal. Appl.* **2018**, *2*, 85–87. [CrossRef]
9. Aydi, H.; Chen, C.M.; Karapınar, E. Interpolative Ćirić–Reich–Rus type contractions via the Branciari distance. *Mathematics* **2019**, *7*, 84. [CrossRef]
10. Aydi, H.; Karapinar, E.; Roldán López de Hierro, A.F. ω-interpolative Ćirić–Reich–Rus-type contractions. *Mathematics* **2019**, *7*, 57. [CrossRef]
11. Debnath, P.; de La Sen, M. Fixed-points of interpolative Ćirić-Reich-Rus-type contractions in b-metric Spaces. *Symmetry* **2020**, *12*, 12. [CrossRef]
12. Gaba, Y.U.; Karapınar, E. A New Approach to the Interpolative Contractions. *Axioms* **2019**, *8*, 110. [CrossRef]
13. Karapinar, E.; Agarwal, R.; Aydi, H. Interpolative Reich–Rus–Ćirić type contractions on partial metric spaces. *Mathematics* **2018**, *6*, 256. [CrossRef]
14. Noorwali, M. Common fixed point for Kannan type contractions via interpolation. *J. Math. Anal.* **2018**, *9*, 92–94.
15. Karapınar, E.; Alqahtani, O.; Aydi, H. On interpolative Hardy–Rogers type contractions. *Symmetry* **2019**, *11*, 8. [CrossRef]
16. Bakhtin, I. The contraction mapping principle in quasimetric spaces. *Func. Anal. Gos. Ped. Inst. Unianowsk* **1989**, *30*, 26–37.
17. Czerwik, S. Contraction mappings in b-metric spaces. *Acta Math. Inform. Univ. Ostrav.* **1993**, *1*, 5–11.
18. Boriceanu, M.; Bota, M.; Petruşel, A. Multivalued fractals in b-metric spaces. *Cent. Eur. J. Math.* **2010**, *8*, 367–377. [CrossRef]

19. Aghajani, A.; Abbas, M.; Roshan, J. Common fixed point of generalized weak contractive mappings in partially ordered *b*-metric spaces. *Math. Slovaca* **2014**, *64*, 941–960. [CrossRef]
20. Alqahtani, B.; Fulga, A.; Karapınar, E.; Özturk, A. Fisher-type fixed point results in *b*-metric spaces. *Mathematics* **2019**, *7*, 102. [CrossRef]

Publisher's Note: MDPI stays neutral with regard to jurisdictional claims in published maps and institutional affiliations.

© 2020 by the authors. Licensee MDPI, Basel, Switzerland. This article is an open access article distributed under the terms and conditions of the Creative Commons Attribution (CC BY) license (http://creativecommons.org/licenses/by/4.0/).

Article

Strong Convergence of Extragradient-Type Method to Solve Pseudomonotone Variational Inequalities Problems

Nopparat Wairojjana [1], Nuttapol Pakkaranang [2], Habib ur Rehman [2], Nattawut Pholasa [3,*] and Tiwabhorn Khanpanuk [4,*]

[1] Applied Mathematics Program, Faculty of Science and Technology, Valaya Alongkorn Rajabhat University under the Royal Patronage (VRU), 1 Moo 20 Phaholyothin Road, Klong Neung, Klong Luang, Pathumthani 13180, Thailand; nopparat@vru.ac.th

[2] Department of Mathematics, Faculty of Science, King Mongkut's University of Technology Thonburi (KMUTT), Bangkok 10140, Thailand; nuttapol.pak@mail.kmutt.ac.th (N.P.); habib.rehman@mail.kmutt.ac.th (H.u.R.)

[3] School of Science, University of Phayao, Phayao 56000, Thailand

[4] Department of Mathematics, Faculty of Science and Technology, Phetchabun Rajabhat University, Phetchabun 67000, Thailand

* Correspondence: nattawut_math@hotmail.com (N.P.); mathiproof@gmail.com (T.K.)

Received: 23 August 2020; Accepted: 30 September 2020; Published: 13 October 2020

Abstract: A number of applications from mathematical programmings, such as minimax problems, penalization methods and fixed-point problems can be formulated as a variational inequality model. Most of the techniques used to solve such problems involve iterative algorithms, and that is why, in this paper, we introduce a new extragradient-like method to solve the problems of variational inequalities in real Hilbert space involving pseudomonotone operators. The method has a clear advantage because of a variable stepsize formula that is revised on each iteration based on the previous iterations. The key advantage of the method is that it works without the prior knowledge of the Lipschitz constant. Strong convergence of the method is proved under mild conditions. Several numerical experiments are reported to show the numerical behaviour of the method.

Keywords: pseudomonotone mapping; subgradient extragradient method; strong convergence; Hilbert spaces; variational inequality problems

---

1. Introduction

In this article, we consider the classic variational inequalities problems (VIPs) [1,2] for an operator $\mathcal{F} : \mathcal{E} \to \mathcal{E}$ is formulated in the following way:

$$\text{Find } u^* \in \mathcal{K} \text{ such that } \langle \mathcal{F}(u^*), y - u^* \rangle \geq 0, \ \forall y \in \mathcal{K}, \tag{1}$$

where $\mathcal{K}$ is a nonempty, convex and closed subset of a real Hilbert space $\mathcal{E}$. The inner product and induced norm on $\mathcal{E}$ are denoted by $\langle .,. \rangle$ and $\|.\|$, respectively. Moreover, the set of real and natural numbers are denoted by $\mathcal{R}$ and $\mathcal{N}$, respectively. It is important to note that solving the problem (1) is equivalent to solving the following problem:

$$\text{Find an element } u^* \in \mathcal{K} \text{ such that } u^* = P_\mathcal{K}[u^* - \zeta \mathcal{F}(u^*)].$$

We assume that the following requirements have been fulfilled:

(B1) The solution set of the problem (1), represented by SVIP is nonempty.

(B2) A mapping $\mathcal{F}: \mathcal{E} \to \mathcal{E}$ is called to be pseudomonotone, i.e.,

$$\langle \mathcal{F}(y_1), y_2 - y_1 \rangle \geq 0 \implies \langle \mathcal{F}(y_2), y_1 - y_2 \rangle \leq 0, \ \forall y_1, y_2 \in \mathcal{K}.$$

(B3) A mapping $\mathcal{F}: \mathcal{E} \to \mathcal{E}$ is said to be Lipschitz continuous, i.e., there exists $L > 0$ such that

$$\|\mathcal{F}(y_1) - \mathcal{F}(y_2)\| \leq L\|y_1 - y_2\|, \ \forall y_1, y_2 \in \mathcal{K}.$$

(B4) A mapping $\mathcal{F}: \mathcal{E} \to \mathcal{E}$ is called to be sequentially weakly continuous, i.e., $\{\mathcal{F}(u_n)\}$ converges weakly to $\mathcal{F}(u)$, where $\{u_n\}$ weakly converges to u.

The concept of variational inequalities has been used as a powerful tool to study different subjects, i.e., physics, engineering, economics and optimization theory. The problem (1) was firstly introduced by Stampacchia [1] in 1964 and also provided that this problem (1) is a crucial problem in nonlinear analysis. This is an efficient mathematical technique that integrates several key elements of applied mathematics, i.e., the problems of network equilibrium, the necessary optimality conditions, the complementarity problems and the systems of non-linear equations (for more details [3–9]). On the other hand, the projection methods are important to find the numerical solution of variational inequalities. Many authors have proposed and studied different projection methods to solve the problem of variational inequalities (see for more details [10–20]) and others in [21–32]. In particular, Karpelevich [10] and Antipin [33] introduced the following extragradient method:

$$\begin{cases} u_n \in \mathcal{K}, \\ v_n = P_\mathcal{K}[u_n - \zeta \mathcal{F}(u_n)], \\ u_{n+1} = P_\mathcal{K}[u_n - \zeta \mathcal{F}(v_n)]. \end{cases} \quad (2)$$

Recently, the subgradient extragradient algorithm was established by Censor et al. [12] for solving problem (1) in real Hilbert space. Their method has the form

$$\begin{cases} u_n \in \mathcal{K}, \\ v_n = P_\mathcal{K}[u_n - \zeta \mathcal{F}(u_n)], \\ u_{n+1} = P_{\mathcal{E}_n}[u_n - \zeta \mathcal{F}(v_n)]. \end{cases} \quad (3)$$

where $\mathcal{E}_n = \{z \in \mathcal{E} : \langle u_n - \zeta \mathcal{F}(u_n) - v_n, z - v_n \rangle \leq 0\}$. Migorski et al. [34] proposed a viscosity-type subgradient extragradient method to solve monotone variational inequalities problems. The main contribution is the presence of a viscosity scheme in the algorithm that was used to improve the convergence rate of the iterative sequence and provide strong convergence theorem. The iterative sequence $\{u_n\}$ was generated in the following way: (i) Let $u_0 \in \mathcal{K}, \mu \in (0,1), \zeta_0 > 0$ and a sequence $\gamma_n \subset (0,1)$ with $\gamma_n \to 0$ and $\sum_n^\infty \gamma_n = +\infty$. (ii) Compute

$$\begin{cases} v_n = P_\mathcal{K}[u_n - \zeta_n \mathcal{F}(u_n)], \\ w_n = P_{\mathcal{E}_n}[u_n - \zeta_n \mathcal{F}(v_n)], \\ u_{n+1} = \gamma_n f(u_n) + (1 - \gamma_n) w_n, \end{cases} \quad (4)$$

where

$$\mathcal{E}_n = \{z \in \mathcal{E} : \langle u_n - \zeta_n \mathcal{F}(u_n) - v_n, z - v_n \rangle \leq 0\}.$$

(iii) Revised the stepsize in the following way:

$$\zeta_{n+1} = \begin{cases} \min\left\{\zeta_n, \frac{\mu \|u_n - v_n\|}{\|\mathcal{F}(u_n) - \mathcal{F}(v_n)\|}\right\} & \text{if } \mathcal{F}(u_n) \neq \mathcal{F}(v_n), \\ \zeta_n & \text{otherwise}. \end{cases}$$

In this paper, inspired by the iterative methods in [12,16,35,36], a modified subgradient extragradient algorithm is proposed for solving variational inequalities problems involving pseudomonotone mapping in real Hilbert space. It is important to note that our proposed scheme is effective. In particular, by comparing the results of Migorski et al. [34], our algorithm can solve pseudomonotone variational inequalities. Similar to the results of Migorski et al. [34] the proof of strong convergence of the proposed algorithm is proved without knowing the Lipschitz constant of the operator $\mathcal{F}$. The proposed algorithm could be seen as a modification of the methods that are appeared in [10,12,34–36]. Under mild conditions, a strong convergence theorem is proved. Numerical experiments have been shown that the new approach tends to be more successful than the existing one [34].

The rest of this article has been arranged as follows: Section 2 contains some definitions and basic results that have been used throughout the paper. Section 3 contains our main algorithm and a strong convergence theorem. Section 4 presents the numerical results showing the algorithmic efficacy of the proposed method.

2. Preliminaries

This section contains useful lemmas and basic identities that have been used throughout the article. The metric projection $P_\mathcal{K}(u_1)$ for $u_1 \in \mathcal{E}$ onto a closed and convex subset $\mathcal{K}$ of $\mathcal{E}$ is defined by

$$P_\mathcal{K}(u_1) = \arg\min\{\|u_2 - u_1\| : u_2 \in \mathcal{K}\}.$$

Lemma 1. [37,38] *Assume $\mathcal{K}$ is a nonempty, convex and closed subset of a real Hilbert space $\mathcal{E}$ and $P_\mathcal{K} : \mathcal{E} \to \mathcal{K}$ is a metric projection from $\mathcal{E}$ onto $\mathcal{K}$.*

(i) *Let $u_1 \in \mathcal{K}$ and $u_2 \in \mathcal{E}$, we have*

$$\|u_1 - P_\mathcal{K}(u_2)\|^2 + \|P_\mathcal{K}(u_2) - u_2\|^2 \leq \|u_1 - u_2\|^2.$$

(ii) *$u_3 = P_\mathcal{K}(u_1)$ if and only if*

$$\langle u_1 - u_3, u_2 - u_3 \rangle \leq 0, \ \forall u_2 \in \mathcal{K}.$$

(iii) *For $u_2 \in \mathcal{K}$ and $u_1 \in \mathcal{E}$*

$$\|u_1 - P_\mathcal{K}(u_1)\| \leq \|u_1 - u_2\|.$$

Lemma 2. [37] *Let $u, v \in \mathcal{E}$ and $\omega \in \mathcal{R}$.*

(i) $\|\omega u + (1-\omega)v\|^2 = \omega\|u\|^2 + (1-\omega)\|v\|^2 - \omega(1-\omega)\|u-v\|^2.$

(ii) $\|u + v\|^2 \leq \|u\|^2 + 2\langle v, u+v \rangle.$

Lemma 3. [39] *Assume that $\{\chi_n\}$ be a sequence of non-negative real numbers satisfying*

$$\chi_{n+1} \leq (1-\tau_n)\chi_n + \tau_n\delta_n, \ \forall n \in \mathcal{N},$$

where $\{\tau_n\} \subset (0,1)$ and $\{\delta_n\} \subset \mathcal{R}$ satisfy the following conditions:

$$\lim_{n\to\infty}\tau_n = 0, \ \sum_{n=1}^{\infty}\tau_n = \infty, \ \text{and} \ \limsup_{n\to\infty}\delta_n \leq 0.$$

Then, $\lim_{n\to\infty}\chi_n = 0.$

Lemma 4. [40] *Assume that $\{\chi_n\}$ is a sequence of real numbers such that there exists a subsequence $\{n_i\}$ of $\{n\}$ such that $\chi_{n_i} < \chi_{n_{i+1}}$ for all $i \in \mathcal{N}$. Then, there exists a non decreasing sequence $m_k \subset \mathcal{N}$ such that $m_k \to \infty$ as $k \to \infty$, and the following conditions are fulfilled by all (sufficiently large) numbers $k \in \mathcal{N}$:*

$$\chi_{m_k} \leq \chi_{m_{k+1}} \text{ and } \chi_k \leq \chi_{m_{k+1}}.$$

In fact, $m_k = \max\{j \leq k : \chi_j \leq \chi_{j+1}\}$.

Lemma 5. [41] *Assume that $\mathcal{F} : \mathcal{K} \to \mathcal{E}$ is a pseudomonotone and continuous mapping. Then, u^* is a solution of the problem* (1) *if and only if u^* is a solution of the following problem.*

$$\text{Find } x \in \mathcal{K} \text{ such that } \langle \mathcal{F}(y), y - x \rangle \geq 0, \; \forall y \in \mathcal{K}.$$

3. Main Results

We provide a method consisting of two convex minimization problems through a viscosity scheme and an explicit stepsize formula which is being used to improve the convergence rate of the iterative sequence and to make the method independent of the Lipschitz constants. The detailed method is provided in Algorithm 1.

Algorithm 1 (Explicit method for pseudomonotone variational inequalities problems).

Step 0: Let $u_0 \in \mathcal{K}$, $\mu \in (0,1)$, $\zeta_0 > 0$ and a sequence $\gamma_n \subset (0,1)$ satisfying

$$\lim_{n \to \infty} \gamma_n = 0 \text{ and } \sum_n^\infty \gamma_n = +\infty.$$

Step 1: Evaluate

$$v_n = P_\mathcal{K}[u_n - \zeta_n \mathcal{F}(u_n)].$$

If $u_n = v_n$; STOP. Otherwise, go to **Step 2**.
Step 2: Evaluate

$$w_n = P_{\mathcal{E}_n}[u_n - \zeta_n \mathcal{F}(v_n)],$$

where $\mathcal{E}_n = \{z \in \mathcal{E} : \langle u_n - \zeta_n \mathcal{F}(u_n) - v_n, z - v_n \rangle \leq 0\}$.
Step 3: Compute

$$u_{n+1} = \gamma_n f(u_n) + (1 - \gamma_n) w_n.$$

Step 4: Evaluate

$$\zeta_{n+1} = \begin{cases} \min\left\{\zeta_n, \frac{\mu \|u_n - v_n\|^2 + \mu \|w_n - v_n\|^2}{2\langle \mathcal{F}(u_n) - \mathcal{F}(v_n), w_n - v_n \rangle}\right\} & \text{if } \langle \mathcal{F}(u_n) - \mathcal{F}(v_n), w_n - v_n \rangle > 0, \\ \zeta_n & \text{else.} \end{cases}$$

Lemma 6. *The stepsize sequence $\{\zeta_n\}$ is monotonically decreasing with a lower bound $\min\left\{\frac{\mu}{L}, \zeta_0\right\}$ and converges to a fixed $\zeta > 0$.*

Proof. Let $\langle \mathcal{F}(u_n) - \mathcal{F}(v_n), w_n - v_n \rangle > 0$, such that

$$\frac{\mu(\|u_n - v_n\|^2 + \|w_n - v_n\|^2)}{2\langle \mathcal{F}(u_n) - \mathcal{F}(v_n), w_n - v_n \rangle} \geq \frac{2\mu \|u_n - v_n\| \|w_n - v_n\|}{2\|\mathcal{F}(u_n) - \mathcal{F}(v_n)\| \|w_n - v_n\|}$$

$$\geq \frac{2\mu \|u_n - v_n\| \|w_n - v_n\|}{2\|u_n - v_n\| \|w_n - v_n\|} \quad (5)$$

$$\geq \frac{\mu}{L}.$$

Clearly, from above we can conclude that $\{\zeta_n\}$ has a lower bound $\min\left\{\frac{\mu}{L}, \zeta_0\right\}$. Moreover, there exists a real number $\zeta > 0$, such that $\lim_{n \to \infty} \zeta_n = \zeta$. □

Lemma 7. *Assume that $\mathcal{F} : \mathcal{E} \to \mathcal{E}$ satisfies the conditions (B1)–(B4). For a given $u^* \in SVIP \neq \emptyset$, we have*

$$\|w_n - u^*\|^2 \leq \|u_n - u^*\|^2 - \left(1 - \frac{\mu \zeta_n}{\zeta_{n+1}}\right)\|u_n - v_n\|^2 - \left(1 - \frac{\mu \zeta_n}{\zeta_{n+1}}\right)\|w_n - v_n\|^2.$$

Proof. Consider that

$$\begin{aligned}
\| w_n - u^* \|^2 &= \| P_{\mathcal{E}_n}[u_n - \zeta_n \mathcal{F}(v_n)] - u^* \|^2 \\
&= \| P_{\mathcal{E}_n}[u_n - \zeta_n \mathcal{F}(v_n)] + [u_n - \zeta_n \mathcal{F}(v_n)] - [u_n - \zeta_n \mathcal{F}(v_n)] - u^* \|^2 \\
&= \| [u_n - \zeta_n \mathcal{F}(v_n)] - u^* \|^2 + \| P_{\mathcal{E}_n}[u_n - \zeta_n \mathcal{F}(v_n)] - [u_n - \zeta_n \mathcal{F}(v_n)] \|^2 \\
&\quad + 2 \langle P_{\mathcal{E}_n}[u_n - \zeta_n \mathcal{F}(v_n)] - [u_n - \zeta_n \mathcal{F}(v_n)], [u_n - \zeta_n \mathcal{F}(v_n)] - u^* \rangle.
\end{aligned} \quad (6)$$

Given that $u^* \in SVIP \subset \mathcal{K} \subset \mathcal{E}_n$, we get

$$\begin{aligned}
&\| P_{\mathcal{E}_n}[u_n - \zeta_n \mathcal{F}(v_n)] - [u_n - \zeta_n \mathcal{F}(v_n)] \|^2 \\
&+ \langle P_{\mathcal{E}_n}[u_n - \zeta_n \mathcal{F}(v_n)] - [u_n - \zeta_n \mathcal{F}(v_n)], [u_n - \zeta_n \mathcal{F}(v_n)] - u^* \rangle \\
&= \langle [u_n - \zeta_n \mathcal{F}(v_n)] - P_{\mathcal{E}_n}[u_n - \zeta_n \mathcal{F}(v_n)], u^* - P_{\mathcal{E}_n}[u_n - \zeta_n \mathcal{F}(v_n)] \rangle \leq 0,
\end{aligned} \quad (7)$$

which implies that

$$\begin{aligned}
&\langle P_{\mathcal{E}_n}[u_n - \zeta_n \mathcal{F}(v_n)] - [u_n - \zeta_n \mathcal{F}(v_n)], [u_n - \zeta_n \mathcal{F}(v_n)] - u^* \rangle \\
&\leq - \| P_{\mathcal{E}_n}[u_n - \zeta_n \mathcal{F}(v_n)] - [u_n - \zeta_n \mathcal{F}(v_n)] \|^2.
\end{aligned} \quad (8)$$

Using expressions (6) and (8), we obtain

$$\begin{aligned}
\| w_n - u^* \|^2 &\leq \| u_n - \zeta_n \mathcal{F}(v_n) - u^* \|^2 - \| P_{\mathcal{E}_n}[u_n - \zeta_n \mathcal{F}(v_n)] - [u_n - \zeta_n \mathcal{F}(v_n)] \|^2 \\
&\leq \|u_n - u^*\|^2 - \| u_n - w_n \|^2 + 2 \zeta_n \langle \mathcal{F}(v_n), u^* - w_n \rangle.
\end{aligned} \quad (9)$$

Since u^* is the solution of problem (1), we have

$$\langle \mathcal{F}(u^*), y - u^* \rangle \geq 0, \text{ for all } y \in \mathcal{K}.$$

Due to the pseudomonotonicity of $\mathcal{F}$ on $\mathcal{K}$, we get

$$\langle \mathcal{F}(y), y - u^* \rangle \geq 0, \text{ for all } y \in \mathcal{K}.$$

By substituting $y = v_n \in \mathcal{K}$, we get

$$\langle \mathcal{F}(v_n), v_n - u^* \rangle \geq 0.$$

Thus, we have

$$\langle \mathcal{F}(v_n), u^* - w_n \rangle = \langle \mathcal{F}(v_n), u^* - v_n \rangle + \langle \mathcal{F}(v_n), v_n - w_n \rangle \leq \langle \mathcal{F}(v_n), v_n - w_n \rangle. \quad (10)$$

Combining expressions (9) and (10), we obtain

$$\begin{aligned}
\| w_n - u^* \|^2 &\leq \|u_n - u^*\|^2 - \| u_n - w_n \|^2 + 2 \zeta_n \langle \mathcal{F}(v_n), v_n - w_n \rangle \\
&\leq \|u_n - u^*\|^2 - \| u_n - v_n + v_n - w_n \|^2 + 2 \zeta_n \langle \mathcal{F}(v_n), v_n - w_n \rangle \\
&\leq \|u_n - u^*\|^2 - \| u_n - v_n \|^2 - \| v_n - w_n \|^2 + 2 \langle u_n - \zeta_n \mathcal{F}(v_n) - v_n, w_n - v_n \rangle.
\end{aligned} \quad (11)$$

Note that $w_n = P_{\mathcal{E}_n}[u_n - \zeta_n \mathcal{F}(v_n)]$ and by the definition of ζ_{n+1}, we have

$$\begin{aligned}
& 2\langle u_n - \zeta_n \mathcal{F}(v_n) - v_n, w_n - v_n\rangle \\
& = 2\langle u_n - \zeta_n \mathcal{F}(u_n) - v_n, w_n - v_n\rangle + 2\zeta_n \langle \mathcal{F}(u_n) - \mathcal{F}(v_n), w_n - v_n\rangle \\
& \leq \frac{2\zeta_n}{\zeta_{n+1}} \zeta_{n+1} \langle \mathcal{F}(u_n) - \mathcal{F}(v_n), w_n - v_n\rangle \leq \frac{\zeta_n}{\zeta_{n+1}}\left[\mu \|u_n - v_n\|^2 + \mu \|w_n - v_n\|^2\right].
\end{aligned} \tag{12}$$

Combining expressions (11) and (12), we obtain

$$\begin{aligned}
& \|w_n - u^*\|^2 \\
& \leq \|u_n - u^*\|^2 - \|u_n - v_n\|^2 - \|v_n - w_n\|^2 + \frac{\zeta_n}{\zeta_{n+1}}\left[\mu \|u_n - v_n\|^2 + \mu \|w_n - v_n\|^2\right] \\
& \leq \|u_n - u^*\|^2 - \left(1 - \frac{\mu\zeta_n}{\zeta_{n+1}}\right)\|u_n - v_n\|^2 - \left(1 - \frac{\mu\zeta_n}{\zeta_{n+1}}\right)\|w_n - v_n\|^2.
\end{aligned} \tag{13}$$

□

Lemma 8. *Suppose that conditions (B1)–(B4) hold. Let $\{u_n\}$ be a sequence generated by Algorithm 1. If there is a subsequence $\{u_{n_k}\}$ which is weakly convergent to $\hat{u} \in \mathcal{E}$ and $\lim_{n\to\infty} \|u_n - v_n\| = 0$, then $\hat{u} \in SVIP$.*

Proof. We have

$$v_{n_k} = P_{\mathcal{K}}[u_{n_k} - \zeta_{n_k} \mathcal{F}(u_{n_k})], \tag{14}$$

which is equivalent to

$$\langle u_{n_k} - \zeta_{n_k} \mathcal{F}(u_{n_k}) - v_{n_k}, y - v_{n_k}\rangle \leq 0, \ \forall y \in \mathcal{K}. \tag{15}$$

From expression (15), we can write

$$\langle u_{n_k} - v_{n_k}, y - v_{n_k}\rangle \leq \zeta_{n_k} \langle \mathcal{F}(u_{n_k}), y - v_{n_k}\rangle, \ \forall y \in \mathcal{K}. \tag{16}$$

Therefore, we get

$$\frac{1}{\zeta_{n_k}}\langle u_{n_k} - v_{n_k}, y - v_{n_k}\rangle + \langle \mathcal{F}(u_{n_k}), v_{n_k} - u_{n_k}\rangle \leq \langle \mathcal{F}(u_{n_k}), y - u_{n_k}\rangle, \ \forall y \in \mathcal{K}. \tag{17}$$

Due to the boundedness of the sequence $\{u_{n_k}\}$ so does $\{\mathcal{F}(u_{n_k})\}$. By using the facts $\lim_{n\to\infty} \|u_{n_k} - v_{n_k}\| = 0$, and $\lim_{k\to\infty} \zeta_{n_k} = \zeta > 0$, limit as $k \to \infty$ in (17), we get

$$\liminf_{k\to\infty} \langle \mathcal{F}(u_{n_k}), y - u_{n_k}\rangle \geq 0, \ \forall y \in \mathcal{K}. \tag{18}$$

Moreover, we have

$$\langle \mathcal{F}(v_{n_k}), y - v_{n_k}\rangle = \langle \mathcal{F}(v_{n_k}) - \mathcal{F}(u_{n_k}), y - u_{n_k}\rangle + \langle \mathcal{F}(u_{n_k}), y - u_{n_k}\rangle + \langle \mathcal{F}(v_{n_k}), u_{n_k} - v_{n_k}\rangle. \tag{19}$$

Since $\lim_{n\to\infty} \|u_{n_k} - v_{n_k}\| = 0$, and $\mathcal{F}$ is L-Lipschitz continuous on $\mathcal{E}$, we get

$$\lim_{n\to\infty} \|\mathcal{F}(u_{n_k}) - \mathcal{F}(v_{n_k})\| = 0. \tag{20}$$

From (19) and (20), we obtain

$$\liminf_{k\to\infty} \langle \mathcal{F}(v_{n_k}), y - v_{n_k}\rangle \geq 0, \ \forall y \in \mathcal{K}. \tag{21}$$

Next, we show that $u^* \in SVIP$. We choose a sequence $\{\epsilon_k\}$ of positive numbers decreasing and tending to 0. For each k, we denote by m_k the smallest positive integer such that

$$\liminf_{k \to \infty} \langle \mathcal{F}(u_{n_i}), y - u_{n_i} \rangle + \epsilon_k \geq 0, \ \forall i \geq m_k. \tag{22}$$

Due to $\{\epsilon_k\}$ being decreasing, the sequence $\{m_k\}$ is increasing.

Case 1: If there is a subsequence $u_{n_{m_{k_j}}}$ of $u_{n_{m_k}}$ such that $\mathcal{F}(u_{n_{m_{k_j}}}) = 0$ $(\forall j)$. Letting $j \to \infty$, we obtain

$$\langle \mathcal{F}(u^*), y - u^* \rangle = \lim_{j \to \infty} \langle \mathcal{F}(u_{n_{m_{k_j}}}), y - u^* \rangle = 0. \tag{23}$$

Hence $u^* \in \mathcal{K}$, therefore we have $u^* \in SVIP$.

Case 2: If there exists N_0 such that for all $n_{m_k} \geq N_0$, $\mathcal{F}(u_{n_{m_k}}) \neq 0$. Suppose that

$$\Theta_{n_{m_k}} = \frac{\mathcal{F}(u_{n_{m_k}})}{\|\mathcal{F}(u_{n_{m_k}})\|^2}, \ \forall n_{m_k} \geq N_0. \tag{24}$$

Due to the above definition, we obtain

$$\langle \mathcal{F}(u_{n_{m_k}}), \mathcal{F}(\Theta_{n_{m_k}}) \rangle = 1, \ \forall n_{m_k} \geq N_0. \tag{25}$$

From (18) and (25), for all $n_{m_k} \geq N_0$, we have

$$\langle \mathcal{F}(u_{n_{m_k}}), y + \epsilon_k \Theta_{n_{m_k}} - u_{n_{m_k}} \rangle \geq 0. \tag{26}$$

Due to pseudomonotonicity of $\mathcal{F}$ for $n_{m_k} \geq N_0$, we obtain

$$\langle \mathcal{F}(y + \epsilon_k \Theta_{n_{m_k}}), y + \epsilon_k \Theta_{n_{m_k}} - u_{n_{m_k}} \rangle \geq 0. \tag{27}$$

For all $n_{m_k} \geq N_0$, we have

$$\langle \mathcal{F}(y), y - u_{n_{m_k}} \rangle \geq \langle \mathcal{F}(y) - \mathcal{F}(y + \epsilon_k \Theta_{n_{m_k}}), y + \epsilon_k \Theta_{n_{m_k}} - u_{n_{m_k}} \rangle - \epsilon_k \langle \mathcal{F}(y), \Theta_{n_{m_k}} \rangle. \tag{28}$$

Since $\{u_{n_k}\}$ converges weakly to $u^* \in \mathcal{K}$ and $\mathcal{F}$ is sequentially weakly continuous on $\mathcal{K}$, we have $\{\mathcal{F}(u_{n_k})\}$ converges weakly to $\mathcal{F}(u^*)$. We can suppose that $\mathcal{F}(u^*) \neq 0$. Since the norm mapping is sequentially weakly lower semicontinuous, we have

$$\|\mathcal{F}(u^*)\| \leq \liminf_{k \to \infty} \|\mathcal{F}(u_{n_k})\|. \tag{29}$$

Since $\{u_{n_{m_k}}\} \subset \{u_{n_k}\}$ and $\lim_{k \to \infty} \epsilon_k = 0$, we have

$$0 \leq \lim_{k \to \infty} \|\epsilon_k \Theta_{n_{m_k}}\| = \lim_{k \to \infty} \frac{\epsilon_k}{\|\mathcal{F}(u_{n_{m_k}})\|} \leq \frac{0}{\|\mathcal{F}(u^*)\|} = 0. \tag{30}$$

Now, letting $k \to \infty$ in (28), we obtain

$$\langle \mathcal{F}(y), y - u^* \rangle \geq 0, \ \forall y \in \mathcal{K}. \tag{31}$$

Applying the well-known Lemma 5, we can deduce that $u^* \in SVIP$. $\square$

Theorem 1. *Assume that $\mathcal{F} : \mathcal{K} \to \mathcal{E}$ satisfies the conditions (B1)–(B4). Moreover, assume that u^* belongs to the solution set $SVIP$. Then, the sequences $\{u_n\}$, $\{v_n\}$ and $\{w_n\}$ generated by Algorithm 1 converge strongly to u^*.*

Proof. By using Lemma 7, we have

$$\|w_n - u^*\|^2 \le \|u_n - u^*\|^2 - \left(1 - \frac{\mu\zeta_n}{\zeta_{n+1}}\right)\|u_n - v_n\|^2 - \left(1 - \frac{\mu\zeta_n}{\zeta_{n+1}}\right)\|w_n - v_n\|^2. \tag{32}$$

Due to $\zeta_n \to \zeta$, there exists a fixed number $\epsilon \in (0, 1 - \mu)$ such that

$$\lim_{n \to \infty} \left(1 - \frac{\mu\zeta_n}{\zeta_{n+1}}\right) = 1 - \mu > \epsilon > 0.$$

Then, there exists a finite number $N_1 \in \mathcal{N}$ such that

$$\left(1 - \frac{\mu\zeta_n}{\zeta_{n+1}}\right) > \epsilon > 0, \ \forall n \ge N_1. \tag{33}$$

Hence, we obtain

$$\|w_n - u^*\|^2 \le \|u_n - u^*\|^2, \ \forall n \ge N_1. \tag{34}$$

From the definition of the sequence $\{u_{n+1}\}$ and the fact that f is a contraction with constant $\rho \in [0, 1)$ and $n \ge N_1$, we obtain

$$\begin{aligned}
\|u_{n+1} - u^*\| &= \|\gamma_n f(u_n) + (1 - \gamma_n)w_n - u^*\| \\
&= \|\gamma_n[f(u_n) - u^*] + (1 - \gamma_n)[w_n - u^*]\| \\
&= \|\gamma_n[f(u_n) + f(u^*) - f(u^*) - u^*] + (1 - \gamma_n)[w_n - u^*]\| \\
&\le \gamma_n\|f(u_n) - f(u^*)\| + \gamma_n\|f(u^*) - u^*\| + (1 - \gamma_n)\|w_n - u^*\| \\
&\le \gamma_n\rho\|u_n - u^*\| + \gamma_n\|f(u^*) - u^*\| + (1 - \gamma_n)\|w_n - u^*\|.
\end{aligned} \tag{35}$$

From expressions (34) and (36) and $\gamma_n \subset (0, 1)$, we obtain

$$\begin{aligned}
\|u_{n+1} - u^*\| &\le \gamma_n\rho\|u_n - u^*\| + \gamma_n\|f(u^*) - u^*\| + (1 - \gamma_n)\|u_n - u^*\| \\
&= [1 - \gamma_n + \rho\gamma_n]\|u_n - u^*\| + \gamma_n(1 - \rho)\frac{\|f(u^*) - u^*\|}{(1 - \rho)} \\
&\le \max\left\{\|u_n - u^*\|, \frac{\|f(u^*) - u^*\|}{(1 - \rho)}\right\} \\
&\le \max\left\{\|u_{N_1} - u^*\|, \frac{\|f(u^*) - u^*\|}{(1 - \rho)}\right\}.
\end{aligned} \tag{36}$$

Hence, we conclude that the sequence $\{u_n\}$ is bounded. Next, the reflexivity of $\mathcal{E}$ and the boundedness of the sequence $\{u_n\}$ guarantee that there exists a subsequence $\{u_{n_k}\}$ such that $\{u_{n_k}\} \rightharpoonup u^* \in \mathcal{E}$ as $k \to \infty$. Now, we prove the strong convergence of the sequence iterative sequence $\{u_n\}$ generated by Algorithm 1. Due to the continuity and pseudomonotonicity of the operator $\mathcal{F}$ imply that the solution set $SVIP$ is a closed and convex set (for more details see [42,43]). Since the mapping f is a contraction, $P_{SVIP} \circ f$ is a contraction. The Banach contraction theorem guarantee the existence of a fixed point of $u^* \in SVIP$ such that

$$u^* = P_{SVIP}(f(u^*)).$$

By using Lemma 1 (ii), we have

$$\langle f(u^*) - u^*, y - u^* \rangle \le 0, \ \forall y \in SVIP. \tag{37}$$

From given $u_{n+1} = \gamma_n f(u_n) + (1 - \gamma_n) w_n$, and using Lemma 2 (i) and Lemma 7, we have

$$\|u_{n+1} - u^*\|^2$$
$$= \|\gamma_n f(u_n) + (1 - \gamma_n) w_n - u^*\|^2$$
$$= \|\gamma_n [f(u_n) - u^*] + (1 - \gamma_n)[w_n - u^*]\|^2$$
$$= \gamma_n \|f(u_n) - u^*\|^2 + (1 - \gamma_n)\|w_n - u^*\|^2 - \gamma_n(1 - \gamma_n)\|f(u_n) - w_n\|^2 \quad (38)$$
$$\leq \gamma_n \|f(u_n) - u^*\|^2 + (1 - \gamma_n)\left[\|u_n - u^*\|^2 - \left(1 - \frac{\mu \zeta_n}{\zeta_{n+1}}\right)\|u_n - v_n\|^2\right.$$
$$\left. - \left(1 - \frac{\mu \zeta_n}{\zeta_{n+1}}\right)\|w_n - v_n\|^2\right] - \gamma_n(1 - \gamma_n)\|f(u_n) - w_n\|^2$$
$$\leq \gamma_n \|f(u_n) - u^*\|^2 + \|u_n - u^*\|^2 - (1 - \gamma_n)\left(1 - \frac{\mu \zeta_n}{\zeta_{n+1}}\right)\left[\|w_n - v_n\|^2 + \|u_n - v_n\|^2\right].$$

The rest of the proof shall be divided into the following two parts:

Case 1: Assume that there exists a fixed number $N_2 \in \mathcal{N}$ ($N_2 \geq N_1$) such that

$$\|u_{n+1} - u^*\| \leq \|u_n - u^*\|, \; \forall n \geq N_2. \quad (39)$$

Thus, $\lim_{n \to \infty} \|u_n - u^*\|$ exists and let $\lim_{n \to \infty} \|u_n - u^*\| = l$. From expression (38), we have

$$(1 - \gamma_n)\left(1 - \frac{\mu \zeta_n}{\zeta_{n+1}}\right)\left[\|w_n - v_n\|^2 + \|u_n - v_n\|^2\right]$$
$$\leq \gamma_n \|f(u_n) - u^*\|^2 + \|u_n - u^*\|^2 - \|u_{n+1} - u^*\|^2. \quad (40)$$

Due to the existence of $\lim_{n \to \infty} \|u_n - u^*\| = l$, and $\gamma_n \to 0$, we deduce that

$$\lim_{n \to \infty} \|u_n - v_n\| = \lim_{n \to \infty} \|w_n - v_n\| = 0. \quad (41)$$

From expression (41), we have

$$\lim_{n \to \infty} \|u_n - w_n\| \leq \lim_{n \to \infty} \|u_n - v_n\| + \lim_{n \to \infty} \|v_n - w_n\| = 0. \quad (42)$$

It follows that

$$\|u_{n+1} - u_n\| = \|\gamma_n f(u_n) + (1 - \gamma_n) w_n - u_n\|$$
$$= \|\gamma_n [f(u_n) - u_n] + (1 - \gamma_n)[w_n - u_n]\|$$
$$\leq \gamma_n \|f(u_n) - u_n\| + (1 - \gamma_n)\|w_n - u_n\| \longrightarrow 0. \quad (43)$$

Thus, the sequences $\{u_n\}$, $\{v_n\}$ and $\{w_n\}$ are bounded. Thus, we can take a subsequence $\{u_{n_k}\}$ of $\{u_n\}$ such that $\{u_{n_k}\}$ weakly converges to some $\hat{u} \in \mathcal{E}$. Moreover, due to $\|u_n - v_n\| \to 0$ and using Lemma 8, we have $\hat{u} \in SVIP$. By following expression (37), we consider that

$$\limsup_{n \to \infty} \langle f(u^*) - u^*, u_n - u^* \rangle$$
$$= \limsup_{k \to \infty} \langle f(u^*) - u^*, u_{n_k} - u^* \rangle = \langle f(u^*) - u^*, \hat{u} - u^* \rangle \leq 0. \quad (44)$$

We have $\lim_{n \to \infty} \|u_{n+1} - u_n\| = 0$. It follows (44) that

$$\limsup_{n \to \infty} \langle f(u^*) - u^*, u_{n+1} - u^* \rangle$$
$$\leq \limsup_{k \to \infty} \langle f(u^*) - u^*, u_{n+1} - u_n \rangle + \limsup_{k \to \infty} \langle f(u^*) - u^*, u_n - u^* \rangle \leq 0. \quad (45)$$

From Lemma 2 (ii) and Lemma 7 for all $n \geq N_2$, we get

$$\begin{aligned}
&\|u_{n+1} - u^*\|^2 \\
&= \|\gamma_n f(u_n) + (1 - \gamma_n) w_n - u^*\|^2 \\
&= \|\gamma_n [f(u_n) - u^*] + (1 - \gamma_n)[w_n - u^*]\|^2 \\
&\leq (1 - \gamma_n)^2 \|w_n - u^*\|^2 + 2\gamma_n \langle f(u_n) - u^*, (1 - \gamma_n)[w_n - u^*] + \gamma_n [f(u_n) - u^*] \rangle \\
&= (1 - \gamma_n)^2 \|w_n - u^*\|^2 + 2\gamma_n \langle f(u_n) - f(u^*) + f(u^*) - u^*, u_{n+1} - u^* \rangle \\
&= (1 - \gamma_n)^2 \|w_n - u^*\|^2 + 2\gamma_n \langle f(u_n) - f(u^*), u_{n+1} - u^* \rangle + 2\gamma_n \langle f(u^*) - u^*, u_{n+1} - u^* \rangle \\
&\leq (1 - \gamma_n)^2 \|w_n - u^*\|^2 + 2\gamma_n \rho \|u_n - u^*\| \|u_{n+1} - u^*\| + 2\gamma_n \langle f(u^*) - u^*, u_{n+1} - u^* \rangle \\
&\leq (1 + \gamma_n^2 - 2\gamma_n) \|u_n - u^*\|^2 + 2\gamma_n \rho \|u_n - u^*\|^2 + 2\gamma_n \langle f(u^*) - u^*, u_{n+1} - u^* \rangle \\
&= (1 - 2\gamma_n) \|u_n - u^*\|^2 + \gamma_n^2 \|u_n - u^*\|^2 + 2\gamma_n \rho \|u_n - u^*\|^2 + 2\gamma_n \langle f(u^*) - u^*, u_{n+1} - u^* \rangle \\
&= [1 - 2\gamma_n(1 - \rho)] \|u_n - u^*\|^2 + 2\gamma_n(1 - \rho) \left[\frac{\gamma_n \|u_n - u^*\|^2}{2(1 - \rho)} + \frac{\langle f(u^*) - u^*, u_{n+1} - u^* \rangle}{1 - \rho} \right]. \quad (46)
\end{aligned}$$

It follows from expressions (45) and (46), we obtain

$$\limsup_{n \to \infty} \left[\frac{\gamma_n \|u_n - u^*\|^2}{2(1 - \rho)} + \frac{\langle f(u^*) - u^*, u_{n+1} - u^* \rangle}{1 - \rho} \right] \leq 0. \quad (47)$$

Choose $n \geq N_3 \in \mathcal{N}$ ($N_3 \geq N_2$) large enough such that $2\gamma_n(1 - \rho) < 1$. Now, using expressions (46) and (47) and applying Lemma 3, we conclude that $\|u_n - u^*\| \to 0$, as $n \to \infty$.

Case 2: Suppose that there exists a subsequence $\{n_i\}$ of $\{n\}$ such that

$$\|u_{n_i} - u^*\| \leq \|u_{n_{i+1}} - u^*\|, \ \forall i \in \mathcal{N}.$$

Thus, by Lemma 4, there exits a sequence $\{m_k\} \subset \mathcal{N}$ and $\{m_k\} \to \infty$, such that

$$\|u_{m_k} - u^*\| \leq \|u_{m_{k+1}} - u^*\| \quad \text{and} \quad \|u_k - u^*\| \leq \|u_{m_{k+1}} - u^*\|, \ \forall k \in \mathcal{N}. \quad (48)$$

Similar to Case 1, using (38), we have

$$(1 - \gamma_{m_k}) \left(1 - \frac{\mu \zeta_{m_k}}{\zeta_{m_k+1}}\right) [\|w_{m_k} - v_{m_k}\|^2 + \|u_{m_k} - v_{m_k}\|^2]$$
$$\leq \gamma_{m_k} \|f(u_{m_k}) - u^*\|^2 + \|u_{m_k} - u^*\|^2 - \|u_{m_k+1} - u^*\|^2. \quad (49)$$

Due to $\gamma_{m_k} \to 0$ and $\left(1 - \frac{\mu \zeta_{m_k}}{\zeta_{m_k+1}}\right) \to 1 - \mu$, we can deduce the following:

$$\lim_{n \to \infty} \|u_{m_k} - v_{m_k}\| = \lim_{k \to \infty} \|w_{m_k} - v_{m_k}\| = 0. \quad (50)$$

From expression (50), we have

$$\lim_{k \to \infty} \|u_{m_k} - w_{m_k}\| \leq \lim_{k \to \infty} \|u_{m_k} - v_{m_k}\| + \lim_{k \to \infty} \|v_{m_k} - w_{m_k}\| = 0. \quad (51)$$

Hence, we obtain

$$\begin{aligned}
\|u_{m_k+1} - u_{m_k}\| &= \|\gamma_{m_k} f(u_{m_k}) + (1 - \gamma_{m_k}) w_{m_k} - u_{m_k}\| \\
&= \|\gamma_{m_k}[f(u_{m_k}) - u_{m_k}] + (1 - \gamma_{m_k})[w_{m_k} - u_{m_k}]\| \\
&\leq \gamma_{m_k} \|f(u_{m_k}) - u_{m_k}\| + (1 - \gamma_{m_k}) \|w_{m_k} - u_{m_k}\| \longrightarrow 0.
\end{aligned} \qquad (52)$$

We have to use the same justification as in the Case 1, such that

$$\limsup_{k \to \infty} \langle f(u^*) - u^*, u_{m_k+1} - u^* \rangle \leq 0. \qquad (53)$$

Using (46) and (48), we have

$$\begin{aligned}
\|u_{m_k+1} - u^*\|^2 &\leq [1 - 2\gamma_{m_k}(1-\rho)] \|u_{m_k} - u^*\|^2 \\
&\quad + 2\gamma_{m_k}(1-\rho) \left[\frac{\gamma_{m_k} \|u_{m_k} - u^*\|^2}{2(1-\rho)} + \frac{\langle f(u^*) - u^*, u_{m_k+1} - u^* \rangle}{1-\rho} \right] \\
&\leq [1 - 2\gamma_{m_k}(1-\rho)] \|u_{m_k+1} - u^*\|^2 \\
&\quad + 2\gamma_{m_k}(1-\rho) \left[\frac{\gamma_{m_k} \|u_{m_k} - u^*\|^2}{2(1-\rho)} + \frac{\langle f(u^*) - u^*, u_{m_k+1} - u^* \rangle}{1-\rho} \right].
\end{aligned} \qquad (54)$$

It follows that

$$\|u_{m_k+1} - u^*\|^2 \leq \frac{\gamma_{m_k} \|u_{m_k} - u^*\|^2}{2(1-\rho)} + \frac{\langle f(u^*) - u^*, u_{m_k+1} - u^* \rangle}{1-\rho}. \qquad (55)$$

Since $\gamma_{m_k} \to 0$ and $\|u_{m_k} - u^*\|$ is a bounded sequence. Thus, expressions (53) and (55) implies that

$$\|u_{m_k+1} - u^*\|^2 \to 0, \text{ as } k \to \infty. \qquad (56)$$

From the inequality (48), we have

$$\lim_{n \to \infty} \|u_k - u^*\|^2 \leq \lim_{n \to \infty} \|u_{m_k+1} - u^*\|^2 \leq 0. \qquad (57)$$

Consequently, $u_n \to u^*$. This completes the proof of the theorem. □

4. Numerical Experiments

Numerical investigations present in this section to demonstrate the efficiency of the introduced Algorithm 1 in four test problems, all of which are pseudomonotone. The MATLAB program has been performed on a PC (with Intel(R) Core(TM)i3-4010U CPU @ 1.70 GHz, RAM 4.00 GB) in MATLAB version 9.5 (R2018b). We use the built-in MATLAB Quadratic programming to solve the minimization problems.

Example 1. *Consider the non-linear complementarity problem of Kojima—Shindo where the feasible set $\mathcal{K}$ which is defined by*

$$\mathcal{K} = \{u \in \mathcal{R}^4 : 1 \leq u_i \leq 5,\ i = 1, 2, 3, 4\}.$$

The mapping $\mathcal{F}: \mathcal{R}^4 \to \mathcal{R}^4$ is defined by

$$\mathcal{F}(u) = \begin{pmatrix} u_1 + u_2 + u_3 + u_4 - 4u_2u_3u_4 \\ u_1 + u_2 + u_3 + u_4 - 4u_1u_3u_4 \\ u_1 + u_2 + u_3 + u_4 - 4u_1u_2u_4 \\ u_1 + u_2 + u_3 + u_4 - 4u_1u_2u_3 \end{pmatrix}.$$

It is easy to see that $\mathcal{F}$ is not monotone on the set $\mathcal{K}$. By using the Monte Carlo approach [44], it can be shown that $\mathcal{F}$ is pseudomonotone on $\mathcal{K}$, This problem has a unique solution $u^* = (5,5,5,5)^T$. Generate many pairs of points u and v uniformly in $\mathcal{K}$ satisfying $\mathcal{F}(u)^T(v-u) \geq 0$ and then check if $\mathcal{F}(v)^T(v-u) \geq 0$. In this experiment, we take different initial points and $D_n = \|u_n - v_n\|$. Moreover, control parameters $\zeta_0 = 0.33$, $\mu = 0.25$, $\gamma_n = \frac{1}{100(n+2)}$ and $f(u) = \frac{u}{2}$ for Algorithm 1. Numerical investigation regarding the first example was shown in Table 1.

Table 1. Numerical behaviour of Algorithm 1 using different starting points for Example 1.

TOL u_0	10^{-2} Iter.	10^{-3} Iter.	10^{-4} Iter.	10^{-5} Iter.	10^{-2} Time	10^{-3} Time	10^{-4} Time	10^{-5} Time
$[-2,2,8,10]^T$	13	51	501	5001	0.079821	0.247776	3.251465	43.637834
$[-1,1,5,6]^T$	12	51	501	5001	0.083870	0.236924	2.684370	39.651178
$[-5,2,-1,2]^T$	9	51	501	5001	0.065422	0.235173	3.034747	43.630625
$[1,2,3,4]^T$	6	1004	1004	5001	0.040866	8.051234	6.686632	42.431705

Example 2. Consider the quadratic fractional programming problem in the following form [44]:

$$\begin{cases} \min f(u) = \dfrac{u^T Q u + a^T u + a_0}{b^T u + b_0}, \\ \text{subject to } u \in \mathcal{K} = \{u \in \mathcal{R}^4 : b^T u + b_0 > 0\}, \end{cases}$$

where

$$Q = \begin{pmatrix} 5 & -1 & 2 & 0 \\ -1 & 5 & -1 & 3 \\ 2 & -1 & 3 & 0 \\ 0 & 3 & 0 & 5 \end{pmatrix}, \quad a = \begin{pmatrix} 1 \\ -2 \\ -2 \\ 1 \end{pmatrix}, \quad b = \begin{pmatrix} 2 \\ 1 \\ 1 \\ 0 \end{pmatrix}, \quad a_0 = -2, \quad \text{and} \quad b_0 = 4.$$

It is easy to verify that Q is symmetric and positive definite on $\mathcal{R}^4$ and consequently f is pseudo-convex on $\mathcal{K}$. Therefore, ∇f is pseudomonotone. Using the quotient rule, we obtain

$$\nabla f(u) = \frac{(b^T u + b_0)(2Qu + a) - b(u^T Q + a^T u + a_0)}{(b^T u + b_0)^2}. \tag{58}$$

In this point of view, we can set $\mathcal{F} = \nabla f$ in Theorem 1. We minimize f over $\mathcal{K} = \{u \in \mathcal{R}^4 : 1 \leq u_i \leq 10, \ i = 1,2,3,4\}$. This problem has a unique solution $u^* = (1,1,1,1)^T \in \mathcal{K}$. In this experiment, we take different initial points and $D_n = \|u_n - v_n\|$. Moreover, control parameters $\zeta_0 = 0.33$, $\mu = 0.25$, $\gamma_n = \frac{1}{100(n+2)}$ and $f(u) = \frac{u}{2}$ for Algorithm 1. Numerical investigation regarding the second example is shown in Table 2.

Table 2. Numerical behaviour of Algorithm 1 using different starting points for Example 2.

TOL u_0	10^{-2} Iter.	10^{-3} Iter.	10^{-4} Iter.	10^{-5} Iter.	10^{-2} Time	10^{-3} Time	10^{-4} Time	10^{-5} Time
$[10,10,10,10]^T$	43	46	99	989	0.289149	0.249285	0.475520	8.480530
$[10,20,30,40]^T$	41	46	99	989	0.211707	0.187559	0.445240	6.898924
$[20,-20,20,-20]^T$	29	32	99	989	0.138575	0.169190	0.394654	7.168460

Example 3. *The third example was taken from [45] where* $\mathcal{F}: \mathcal{R}^2 \to \mathcal{R}^2$ *is defined by*

$$\mathcal{F}(u) = \begin{pmatrix} 0.5u_1u_2 - 2u_2 - 10^7 \\ -4u_1 - 0.1u_2^2 - 10^7 \end{pmatrix},$$

on $\mathcal{K} = \{u \in \mathcal{R}^2 : (u_1 - 2)^2 + (u_2 - 2)^2 \leq 1\}$. *It can easily see that $\mathcal{F}$ is Lipschitz continuous with $L = 5$ and $\mathcal{F}$ is not monotone on $\mathcal{K}$ but pseudomonotone. The above problem has a unique solution $u^* = (2.707, 2.707)^T$. In this experiment, we take different initial points and $D_n = \|u_n - v_n\|$. Moreover, control parameters $\zeta_0 = 0.33$, $\mu = 0.25$, $\gamma_n = \frac{1}{100(n+2)}$ and $f(u) = \frac{u}{3}$ for Algorithm 1. Numerical investigations regarding the third example is shown in Table 3.*

Table 3. Numerical behaviour of Algorithm 1 using different starting points for Example 3.

TOL u_0	10^{-2} Iter.	10^{-3} Iter.	10^{-4} Iter.	10^{-5} Iter.	10^{-2} Time	10^{-3} Time	10^{-4} time	10^{-5} Time
$[0,0]^T$	8	27	265	2566	0.606917	1.907212	14.120655	107.506926
$[10,10]^T$	7	27	265	2591	0.286659	1.057623	10.764532	116.258335
$[-5,-5]^T$	8	26	258	2596	0.388227	1.190191	11.424257	107.584978

Example 4. *The fourth example was taken from [45] where* $\mathcal{F}: \mathcal{R}^2 \to \mathcal{R}^2$ *is defined by*

$$\mathcal{F}(u) = \begin{pmatrix} (u_1^2 + (u_2 - 1)^2)(1 + u_2) \\ -u_1^3 - u_1(u_2 - 1)^2 \end{pmatrix},$$

where $\mathcal{K} = \{u \in \mathcal{R}^2 : -10 \leq u_i \leq 10, i = 1,2\}$. *It can easily see that $\mathcal{F}$ is Lipschitz continuous with $L = 5$ and $\mathcal{F}$ is not monotone on $\mathcal{K}$ but pseudomonotone. In this experiment, we take different initial points and $D_n = \|u_n - v_n\|$. Moreover, control parameters $\zeta_0 = 0.33$, $\mu = 0.25$, $\gamma_n = \frac{1}{100(n+2)}$ and $f(u) = \frac{u}{4}$ for Algorithm 1. Numerical investigations regarding the fourth example is shown in Table 4.*

Table 4. Numerical behaviour of Algorithm 1 using different starting points for Example 4.

TOL u_0	10^{-2} Iter.	10^{-3} Iter.	10^{-4} Iter.	10^{-5} Iter.	10^{-2} Time	10^{-3} Time	10^{-4} Time	10^{-5} Time
$[0,0]^T$	16	220	2231	29253	0.21543	2.35401	29.86562	224.95083
$[10,10]^T$	27	190	2072	25762	0.25322	2.64742	26.84528	198.26446
$[-5,-5]^T$	43	411	3801	47891	0.78262	4.77116	42.41738	427.904781

5. Conclusions

We have developed an extragradient-like method to solve pseudomonotone variational inequalities in real Hilbert space. The method had an explicit formula for an appropriate and effective stepsize evaluation on each step. For each iteration, the stepsize formula is modified based on the previous iterations. The numerical investigation was presented to explain the numerical effectiveness of our algorithm relative to other methods. These numerical studies suggest that viscosity schemes in this sense generally improve the effectiveness of the iterative sequence.

Author Contributions: Data curation, N.W.; formal analysis, T.K.; funding acquisition, N.P. (Nuttapol Pakkaranang), N.P. (Nattawut Pholasa) and T.K.; investigation, N.W., N.P. (Nuttapol Pakkaranang) and T.K.; methodology, T.K.; project administration, H.u.R., N.P. (Nattawut Pholasa) and T.K.; resources, N.P. (Nattawut Pholasa) and T.K.; software, H.u.R.; supervision, H.u.R. and N.P. (Nattawut Pholasa); Writing—original draft, N.W. and H.u.R.; Writing—review and editing, N.P. (Nuttapol Pakkaranang). All authors have read and agreed to the published version of the manuscript.

Funding: This research was funded by University of Phayao and Phetchabun Rajabhat University.

Acknowledgments: We are very grateful to the editor and the anonymous referees for their valuable and useful comments, which helps in improving the quality of this work. N. Wairojjana would like to thank by Valaya

Alongkorn Rajabhat University under the Royal Patronage (VRU). N. Pholasa would like to thank by University of Phayao. T. Khanpanuk would like to thanks Phetchabun Rajabhat University.

Conflicts of Interest: The authors declare no conflict of interest.

References

1. Stampacchia, G. Formes bilinéaires coercitives sur les ensembles convexes. *Comptes Rendus Hebd. Seances Acad. Sci.* **1964**, *258*, 4413.
2. Konnov, I.V. On systems of variational inequalities. *Russ. Math. C/C Izv. Vyss. Uchebnye Zaved. Mat.* **1997**, *41*, 77–86.
3. Kassay, G.; Kolumbán, J.; Páles, Z. On Nash stationary points. *Publ. Math.* **1999**, *54*, 267–279.
4. Kassay, G.; Kolumbán, J.; Páles, Z. Factorization of Minty and Stampacchia variational inequality systems. *Eur. J. Oper. Res.* **2002**, *143*, 377–389. [CrossRef]
5. Kinderlehrer, D.; Stampacchia, G. *An Introduction to Variational Inequalities and Their Applications*; Society for Industrial and Applied Mathematics: Philadelphia, PA, USA, 2000. [CrossRef]
6. Konnov, I. *Equilibrium Models and Variational Inequalities*; Elsevier: Amsterdam, Netherlands, 2007; Volume 210.
7. Elliott, C.M. Variational and Quasivariational Inequalities Applications to Free—Boundary ProbLems. (Claudio Baiocchi And António Capelo). *SIAM Rev.* **1987**, *29*, 314–315. [CrossRef]
8. Nagurney, A.; Economics, E.N. *A Variational Inequality Approach*; Springer: Dordrecht, The Netherlands, 1999.
9. Takahashi, W. *Introduction to Nonlinear and Convex Analysis*; Yokohama Publishers: Yokohama, Japan, 2009.
10. Korpelevich, G. The extragradient method for finding saddle points and other problems. *Matecon* **1976**, *12*, 747–756.
11. Noor, M.A. Some iterative methods for nonconvex variational inequalities. *Comput. Math. Model.* **2010**, *21*, 97–108. [CrossRef]
12. Censor, Y.; Gibali, A.; Reich, S. The subgradient extragradient method for solving variational inequalities in Hilbert space. *J. Optim. Theory Appl.* **2010**, *148*, 318–335. [CrossRef]
13. Censor, Y.; Gibali, A.; Reich, S. Extensions of Korpelevich's extragradient method for the variational inequality problem in Euclidean space. *Optimization* **2012**, *61*, 1119–1132. [CrossRef]
14. Malitsky, Y.V.; Semenov, V.V. An Extragradient Algorithm for Monotone Variational Inequalities. *Cybern. Syst. Anal.* **2014**, *50*, 271–277. [CrossRef]
15. Tseng, P. A Modified Forward-Backward Splitting Method for Maximal Monotone Mappings. *SIAM J. Control Optim.* **2000**, *38*, 431–446. [CrossRef]
16. Moudafi, A. Viscosity Approximation Methods for Fixed-Points Problems. *J. Math. Anal. Appl.* **2000**, *241*, 46–55. [CrossRef]
17. Zhang, L.; Fang, C.; Chen, S. An inertial subgradient-type method for solving single-valued variational inequalities and fixed point problems. *Numer. Algorithms* **2018**, *79*, 941–956. [CrossRef]
18. Iusem, A.N.; Svaiter, B.F. A variant of korpelevich's method for variational inequalities with a new search strategy. *Optimization* **1997**, *42*, 309–321. doi:10.1080/02331939708844365. [CrossRef]
19. Thong, D.V.; Hieu, D.V. Modified subgradient extragradient method for variational inequality problems. *Numer. Algorithms* **2017**, *79*, 597–610. [CrossRef]
20. Thong, D.V.; Hieu, D.V. Weak and strong convergence theorems for variational inequality problems. *Numer. Algorithms* **2017**, *78*, 1045–1060. [CrossRef]
21. Censor, Y.; Gibali, A.; Reich, S. Strong convergence of subgradient extragradient methods for the variational inequality problem in Hilbert space. *Optim. Methods Softw.* **2011**, *26*, 827–845. [CrossRef]
22. Gibali, A.; Reich, S.; Zalas, R. Outer approximation methods for solving variational inequalities in Hilbert space. *Optimization* **2017**, *66*, 417–437. [CrossRef]
23. Ogbuisi, F.U.; Shehu, Y. A projected subgradient-proximal method for split equality equilibrium problems of pseudomonotone bifunctions in Banach spaces. *J. Nonlinear Var. Anal.* **2019**, *3*, 205–224.
24. Ceng, L.C. Asymptotic inertial subgradient extragradient approach for pseudomonotone variational inequalities with fixed point constraints of asymptotically nonexpansive mappings. *Commun. Optim. Theory* **2020**, *2020*, 2.

25. Wang, L.; Yu, L.; Li, T. Parallel extragradient algorithms for a family of pseudomonotone equilibrium problems and fixed point problems of nonself-nonexpansive mappings in Hilbert space. *J. Nonlinear Funct. Anal.* **2020**, *2020*, 13.
26. Ceng, L.C. Two inertial linesearch extragradient algorithms for the bilevel split pseudomonotone variational inequality with constraints. *J. Appl. Numer. Optim.* **2020**, *2*, 213–233. [CrossRef]
27. Ur Rehman, H.; Kumam, P.; Je Cho, Y.; Suleiman, Y.I.; Kumam, W. Modified Popov's explicit iterative algorithms for solving pseudomonotone equilibrium problems. *Optim. Methods Softw.* **2020**, 1–32. [CrossRef]
28. Ur Rehman, H.; Kumam, P.; Kumam, W.; Shutaywi, M.; Jirakitpuwapat, W. The inertial sub-gradient extra-gradient method for a class of pseudo-monotone equilibrium problems. *Symmetry* **2020**, *12*, 463. [CrossRef]
29. Ur Rehman, H.; Kumam, P.; Argyros, I.K.; Deebani, W.; Kumam, W. Inertial extra-gradient method for solving a family of strongly pseudomonotone equilibrium problems in real Hilbert spaces with application in variational inequality problem. *Symmetry* **2020**, *12*, 503. [CrossRef]
30. Ur Rehman, H.; Kumam, P.; Argyros, I.K.; Alreshidi, N.A.; Kumam, W.; Jirakitpuwapat, W. A self-adaptive extra-gradient methods for a family of pseudomonotone equilibrium programming with application in different classes of variational inequality problems. *Symmetry* **2020**, *12*, 523. [CrossRef]
31. Ur Rehman, H.; Kumam, P.; Shutaywi, M.; Alreshidi, N.A.; Kumam, W. Inertial optimization based two-step methods for solving equilibrium problems with applications in variational inequality problems and growth control equilibrium models. *Energies* **2020**, *13*, 3292. [CrossRef]
32. Ur Rehman, H.; Kumam, P.; Abubakar, A.B.; Cho, Y.J. The extragradient algorithm with inertial effects extended to equilibrium problems. *Comput. Appl. Math.* **2020**, *39*, 100. [CrossRef]
33. Antipin, A.S. On a method for convex programs using a symmetrical modification of the Lagrange function. *Ekonomika Matematicheskie Metody* **1976**, *12*, 1164–1173.
34. Migórski, S.; Fang, C.; Zeng, S. A new modified subgradient extragradient method for solving variational inequalities. *Appl. Anal.* **2019**, 1–10. [CrossRef]
35. Yang, J.; Liu, H.; Liu, Z. Modified subgradient extragradient algorithms for solving monotone variational inequalities. *Optimization* **2018**, *67*, 2247–2258. [CrossRef]
36. Kraikaew, R.; Saejung, S. Strong Convergence of the Halpern Subgradient Extragradient Method for Solving Variational Inequalities in Hilbert Spaces. *J. Optim. Theory Appl.* **2013**, *163*, 399–412. [CrossRef]
37. Heinz H. Bauschke, P.L.C. *Convex Analysis and Monotone Operator Theory in Hilbert Spaces*, 2nd ed.; CMS Books in Mathematics; Springer International Publishing: Cham, Switzerland, 2017.
38. Goebel, K.; Reich, S. *Uniform Convexity, Hyperbolic Geometry, and Nonexpansive Mappings*; Marcel Dekker, Inc.: New York City, NY, USA, 1984.
39. Cottle, R.W.; Yao, J.C. Pseudo-monotone complementarity problems in Hilbert space. *J. Optim. Theory Appl.* **1992**, *75*, 281–295. [CrossRef]
40. Maingé, P.E. Strong Convergence of Projected Subgradient Methods for Nonsmooth and Nonstrictly Convex Minimization. *Set-Valued Anal.* **2008**, *16*, 899–912. [CrossRef]
41. Takahashi, W. *Nonlinear Functional Analysis*; Yokohama Publishers: Yokohama, Japan, 2000.
42. Liu, Z.; Zeng, S.; Motreanu, D. Evolutionary problems driven by variational inequalities. *J. Differ. Equ.* **2016**, *260*, 6787–6799. [CrossRef]
43. Liu, Z.; Migórski, S.; Zeng, S. Partial differential variational inequalities involving nonlocal boundary conditions in Banach spaces. *J. Differ. Equ.* **2017**, *263*, 3989–4006. [CrossRef]
44. Hu, X.; Wang, J. Solving pseudomonotone variational inequalities and pseudoconvex optimization problems using the projection neural network. *IEEE Trans. Neural Networks* **2006**, *17*, 1487–1499. [CrossRef]
45. Shehu, Y.; Dong, Q.L.; Jiang, D. Single projection method for pseudo-monotone variational inequality in Hilbert spaces. *Optimization* **2018**, *68*, 385–409. [CrossRef]

© 2020 by the authors. Licensee MDPI, Basel, Switzerland. This article is an open access article distributed under the terms and conditions of the Creative Commons Attribution (CC BY) license (http://creativecommons.org/licenses/by/4.0/).

Article

The Split Various Variational Inequalities Problems for Three Hilbert Spaces

Chinda Chaichuay [†] and Atid Kangtunyakarn [*,†]

Department of Mathematics, Faculty of Science, King Mongkut's Institute of Technology Ladkrabang, Bangkok 10520, Thailand; kcchinda3@gmail.com
* Correspondence: beawrock@hotmail.com
† These authors contributed equally to this work.

Received: 10 August 2020; Accepted: 29 August 2020; Published: 7 September 2020

Abstract: There are many methods for finding a common solution of a system of variational inequalities, a split equilibrium problem, and a hierarchical fixed-point problem in the setting of real Hilbert spaces. They proved the strong convergence theorem. Many split feasibility problems are generated in real Hillbert spaces. The open problem is proving a strong convergence theorem of three Hilbert spaces with different methods from the lasted method. In this research, a new split variational inequality in three Hilbert spaces is proposed. Important tools which are used to solve classical problems will be developed. The convergence theorem for finding a common element of the set of solution of such problems and the sets of fixed-points of discontinuous mappings has been proved.

Keywords: nonspreading; pseudo-nonspreading; new split variational inequalities

MSC: 47H10; 47J20; 49J40

1. Introduction

We use the following symbols throughout this paper: let H be a real Hilbert space and C be a nonempty closed convex subset of H. We also use the symbols "$\longrightarrow$" and "$\rightharpoonup$", which represent strong and weak convergence, respectively. The variational inequality problem (VIP) is a well known problem. That is to find a point ω_* such that

$$\langle y - \omega_*, G\omega_* \rangle \geq 0, \text{ for all } y \in C, \tag{1}$$

where $G : C \longrightarrow H$ is a mapping. The set of all solutions of (1) is denoted by $Var(C, G)$.

The variational inequality problem has been applied in various fields such as industry, finance, economics, social, ecology, regional, pure and applied sciences; see [1–3]. For every $i = 1, 2$, let H_i be a real Hilbert space and C, Q be nonempty closed convex subset of H_1 and H_2, respectively. Recently, Censor [4] has introduced a new variational problem called the split inequality problem (SIP). It entails finding a solution of one variational inequality problem (VIP), the image of which, under a given bounded linear transformation, is a solution of another VIP.

The split variational inequality problem is assigned to the following formula; find a point $\omega_* \in C$ such that

$$\langle f(\omega_*), x - \omega_* \rangle \geq 0, \text{ for all } x \in C \tag{2}$$

and a point $y^* = A\omega_*$ solves

$$\langle y - y^*, g(y^*)\rangle \geq 0, \text{ for all } y \in Q, \tag{3}$$

where $A : H_1 \to H_2$ is a bounded linear operator and $f : H_1 \to H_1, g : H_2 \to H_2$ are mappings. The set of all solutions of (2) and (3) is denoted by

$$\Omega = \{x \in Var\,(C, f) : \text{ for all } x \in Var\,(Q, g)\}.$$

The split variational inequality problem can be applied to model in intensity-modulated radiation therapy (IMRT) treatment planning.

There are also a lot of authors who have introduced convergence theorem related to the split variational inequality and fixed point problems; see [5–7] for an example. In [8], they have studied the Mann implicit iterations for strongly accreative and strongly pseudo-contractive mappings and found that this implicit scheme gives a better convergence rate estimate.

The following definitions are important tools used in this research. A mapping ϑ of H into itself is called nonexpansive if $\|\vartheta x - \vartheta y\| \leq \|x - y\|$, for all $x, y \in H$. We denote by $F(\vartheta)$ the set of fixed points of ϑ (i.e., $F(\vartheta) = \{x \in C : \vartheta x = x\}$). A nonexpansive mapping ϑ is equivalent to the following inequality;

$$\langle (I - \vartheta)\,x - (I - \vartheta)\,y, \vartheta y - \vartheta x\rangle \leq \frac{1}{2}\|(I - \vartheta)x - (I - \vartheta)\,y\|^2,$$

for all $x, y \in H$. From the equation above if $y \in F(\vartheta)$ and $x \in H$, we can conclude that;

$$\langle (I - \vartheta)\,x, y - \vartheta x\rangle \leq \frac{1}{2}\|(I - \vartheta)\,x\|^2.$$

A mapping A of C into H is called inverse strongly monotonic, if there exists $\alpha > 0$ such that

$$\langle x - y, Ax - Ay\rangle \geq \alpha\|Ax - Ay\|^2,$$

for all $x, y \in C$. In [9], Kohsaka and Takahashi introduced the nonspreading mapping in Hilbert spaces H which is defined by the following inequality $2\|\vartheta x - \vartheta y\|^2 \leq \|\vartheta x - y\|^2 + \|x - \vartheta y\|^2$, for all $x, y \in C$.

Following the terminology of Browder and Petryshyn [10], in [11], Osilike and Isiogugu introduced the mapping $\vartheta : C \to C$, which is called κ−strictly pseudo-nonspreading mapping if there exists $\kappa \in [0, 1)$ such that

$$\|\vartheta x - \vartheta y\|^2 \leq \|x - y\|^2 + \kappa\|(I - \vartheta)x - (I - \vartheta)y\|^2 + 2\langle x - \vartheta x, y - \vartheta y\rangle,$$

for all $x, y \in C$. Clearly every nonspreading mapping is κ-strictly pseudo-nonspreading; see, for example, [11].

In [12], Bnouhachem modified a projection process for finding a common solution of a system of variational inequalities, a split equilibrium problem and a hierarchical fixed-point problem in the setting of real Hilbert spaces and also proved the strong convergence theorem of the sequence $\{x_n\}$ generated by

$$\begin{cases} u_n = \vartheta_{r_n}^{F_1}\left(x_n + \gamma A^*\left(T_{r_n}^{F_2} - I\right)Ax_n\right); \\ z_n = P_C\left[P_C\left[u_n - \alpha_1 B_2 u_n\right] - \alpha_1 B_1 P_C\left[u_n - \alpha_2 B_2 u_n\right]\right]; \\ y_n = \beta_n \varrho x_n + (1 - \beta_n)z_n; \\ x_{n+1} = P_C\left[\alpha_n \rho U\left(x_n\right) + (I - \alpha_n \mu F)\,\vartheta\left(y_n\right)\right], \text{ for all } n \geq 0, \end{cases}$$

where $A : H_1 \to H_2$ is a bounded linear operator. Assume that $F_1 : C \times C \to \mathbb{R}$ and $F_2 : Q \times Q \to \mathbb{R}$ are the bifunctions; $B_i : C \to H$ is a θ_i-inverse strongly monotonic mapping for each $i = 1, 2$ and $S, \vartheta : C \to C$ nonexpansive mapping; $F : C \to C$ is a k-Lipschitzian mapping and is η-strongly monotonic; $U : C \to C$ is a τ-Lipschitzian mapping; and the positive parameters are $r_n, \alpha_n, \alpha_1, \alpha_2, \rho, \mu$, for all $n \in \mathbb{N}$.

Let C and Q be nonempty closed convex subsets of the real Hilbert spaces H_1 and H_2, respectively. The split feasibility problem (SFP) is formulated as :

$$\text{to find } x^* \in C \text{ such that } Ax^* \in Q, \tag{4}$$

where $A : H_1 \to H_2$ is a bounded linear operator. The SFP are also applied in [13,14].

Recently, Moudafi [15] introduced the following new split feasibility problem, which is also called general split equality problem:

Let H_1, H_2, H_3 be real Hilbert spaces, $C \subset H_1, Q \subset H_2$ be two nonempty closed convex sets and $A : H_1 \to H_3, B : H_2 \to H_3$ be two bounded linear operators. Moudafi studied the convergence of a relaxed alternating CQ-algorithm for solving the new split feasibility problem, aiming to find

$$\omega_* \in C, y^* \in Q \text{ such that } A\omega_* = By^*. \tag{5}$$

In order to prove the weak convergence theory to solve general split equality problem (5), Moudafi defined the following iteration process $\{x_k\}$:

$$\begin{cases} x_{k+1} = P_C (x_k - \gamma_k A^* (Ax_k - By_k)), \\ y_{k+1} = P_Q (y_k + \gamma_k B^* (Ax_{k+1} - By_k)), \end{cases}$$

where A^*, B^* are adjoint operators of A, B respectively, proper conditions of the positive paramiter γ_k, for all $k \geq 1$. In order to avoid using the projection, Moudafi [16] introduced and studied the following problem: Let $T : H_1 \to H_1$ and $S : H_2 \to H_2$ be nonlinear operators such that $\text{Fix}(T) \neq \emptyset$ and $\text{Fix}(S) \neq \emptyset$, where $\text{Fix}(T)$ and $\text{Fix}(S)$ denote the sets of fixed points of T and S, respectively. If $C = \text{Fix}(T)$ and $Q = \text{Fix}(S)$; then the split equality problem reduces to

$$\text{to find } x \in \text{Fix}(T) \text{ and } y \in \text{Fix}(S) \text{ such that } Ax = By, \tag{6}$$

which is called a split equality fixed point problem (SEFPP).

Denote by Γ the solution set of split equality fixed point problem (6). There were recently SEFPP research articles in [17,18].

Question A. Can we prove a strong convergence theorem of three Hilbert spaces by different methods from Moudafi [15]?

For every $i = 1, 2, 3$, let H_i be a real Hilbert space and C_i be a nonempty closed convex subset of H_i. Let $B_i : C_i \to H_i$ be a mapping, for all $i = 1, 2, 3$, and let $A_2 : H_1 \to H_2$ and $A_3 : H_2 \to H_3$. The split various variational inequality is to find the points

$$\begin{cases} \omega_1^* \in C_1, \text{ such that } \langle B_1 \omega_1^*, x_1 - \omega_1^* \rangle \geq 0, \text{ for all } x_1 \in C_1, \text{ and} \\ \omega_2^* = A_2 \omega_1^* \in C_2, \text{ such that } \langle B_2 \omega_2^*, x_2 - \omega_2^* \rangle \geq 0, \text{ for all } x_2 \in C_2, \text{ and} \\ \omega_3^* = A_3 \omega_2^* \in C_3, \text{ such that } \langle B_3 \omega_3^*, x_3 - \omega_3^* \rangle \geq 0, \text{ for all } x_3 \in C_3. \end{cases} \tag{7}$$

The set of the solutions of (7) is denoted by $\Omega = \{\omega^* = (\omega_1^*, \omega_2^*, \omega_3^*) \in C_1 \times C_2 \times C_3 : \omega_i^* \in \text{Var}(C_i, B_i), \text{ for all } i = 1, 2, 3\}$.

To answer question A, we have created a new tool to prove a strong convergence theorem for three Hilbert spaces to be used for finding the solution of the problem (7) and the fixed points problem of nonspreading and pseudo-nonspreading mappings. Preliminaries In this section, we collect some definitions and lemmas in Hilbert space, which will be needed for proving our main results. More properties of Hilbert space can be found in [19].

Definition 1. *The (nearest point) projection P_C from H onto C assigns to each $x \in H$ the unique point $P_C x \in C$ satisfying the property*

$$\|x - P_C x\| = \min_{y \in C} \|x - y\|.$$

Lemmas 1 and 2 are properties of P_C.

Lemma 1. *([20]) For a given $x \in H$ and $y \in C$, $P_C x = y$ if and only if there holds the inequality $\langle y - x, z - y \rangle \geq 0$, for all $z \in C$.*

Lemma 2. *([21]) Let H be a Hilbert space, let C be a nonempty closed convex subset of H and let A be a mapping of C into H. Let $u \in C$. Then for $\lambda > 0$, $u = P_C(I - \lambda A)u$ if and only if $u \in \text{Var}(C, A)$, where P_C is the metric projection of H onto C.*

Lemma 3. *([22]) Let $\{Y_n\}$ be a sequence of nonnegative real numbers satisfying $Y_{n+1} = (1 - \alpha_n)Y_n + \delta_n$, for all $n \geq 0$,*

where $\{\alpha_n\}$ is a sequence in $(0,1)$ and $\{\delta_n\}$ is a sequence such that

(1) $\sum_{n=1}^{\infty} \alpha_n = +\infty;$

(2) $\limsup_{n \to +\infty} \frac{\delta_n}{\alpha_n} \leq 0$ or $\sum_{n=1}^{+\infty} |\delta_n| < +\infty.$

Then $\lim_{n \to +\infty} Y_n = 0$.

Lemma 4. *([23]) Let $\{Y_n\}$ be a sequence of nonnegative real number satisfying,*

$$Y_{n+1} = (1 - \alpha_n)Y_n + \alpha_n \beta_n, \text{for all } n \geq 0$$

where $\{\alpha_n\}, \{\beta_n\}$ satisfy the conditions

(1) $\{\alpha_n\} \subset [0,1], \sum_{n=1}^{+\infty} \alpha_n = +\infty;$

(2) $\limsup_{n \to +\infty} \beta_n \leq 0$ or $\sum_{n=1}^{+\infty} |\alpha_n \beta_n| < +\infty.$

Then $\lim_{n \to +\infty} Y_n = 0$.

Lemma 5. *For every $i = 1, 2, 3$, let H_i be a real Hilbert spaces and C_i be a nonempty closed convex subset of H_i. Let $B_i : C_i \to H_i$ be β_i-inverse strongly monotonic mappings with $\eta = \min_{i=1,2,3}\{\beta_i\}$ and let $A_2 : H_1 \to H_2, A_3 : H_2 \to H_3$ be bounded linear operators with the adjoint operator A_2^* and A_3^*, respectively. Assume that $\bar{x}_1 \in C_1, A_2\bar{x}_1 = \bar{x}_2, A_3\bar{x}_2 = \bar{x}_3$ and $\Omega \neq \emptyset$. The following are equivalent:*

(i) $\bar{x} \in \Omega$, where $\bar{x} = (\bar{x}_1, \bar{x}_2, \bar{x}_3) \in C_1 \times C_2 \times C_3$.
(ii) $\bar{x}_1 = P_{C_1}(I_1 - \lambda_1 B_1)(\bar{x}_1 - \gamma_2 A_2^*((I_2 - P_{C_2}(I_2 - \lambda_2 B_2))\bar{x}_2 + \gamma_3 A_3^*(I_3 - P_{C_3}(I_3 - \lambda_3 B_3))\bar{x}_3))$,

where $I_i : H_i \to H_i$ is an identity mapping, for all $i = 1, 2, 3$, $\gamma_2(1 + \gamma_3) \leq \dfrac{1}{L}$, $L = \max\{L_1, L_2\} \leq 1$ which L_1, L_2 are spectral radii of $A_2 A_2^*$ and $A_3 A_3^*$, respectively, $\lambda_i \in (0, 2\eta)$, for all $i = 1, 2, 3$ and $\gamma_2, \gamma_3 \geq 0$.

Proof. Let the conditions hold.
$(i) \Rightarrow (ii)$ Let $\bar{x} \in \Omega$ where $\bar{x} = (\bar{x}_1, \bar{x}_2, \bar{x}_3) \in C_1 \times C_2 \times C_3$; we have

$$\bar{x}_i \in Var(C_i, B_i), \text{ for all } i = 1, 2, 3.$$

From Lemma 2, we have

$$\bar{x}_i \in F(P_{C_i}(I_i - \lambda_i B_i)), \text{ for all } i = 1, 2, 3.$$

From determining the definition of $\bar{x}$, we have

$$\bar{x}_1 = P_{C_1}(I_1 - \lambda_1 B_1)(\bar{x}_1 - \gamma_2 A_2^*((I_2 - P_{C_2}(I_2 - \lambda_2 B_2))\bar{x}_2 + \gamma_3 A_3^*(I_3 - P_{C_3}(I_3 - \lambda_3 B_3))\bar{x}_3)).$$

$(ii) \Rightarrow (i)$ Let $\bar{x} = (\bar{x}_1, \bar{x}_2, \bar{x}_3) \in C_1 \times C_2 \times C_3$, where $\bar{x}_2 = A_2 \bar{x}_1, \bar{x}_3 = A_3 \bar{x}_2$ and

$$\bar{x}_1 = P_{C_1}(I_1 - \lambda_1 B_1)(\bar{x}_1 - \gamma_2 A_2^*((I_2 - P_{C_2}(I_2 - \lambda_2 B_2))\bar{x}_2 + \gamma_3 A_3^*(I_3 - P_{C_3}(I_3 - \lambda_3 B_3))\bar{x}_3)).$$

Since B_i is β_i-inverse strongly monotonic with $\lambda_i < 2\eta$, for all $i = 1, 2, 3$, we have $P_{C_i}(I_i - \lambda_i B_i)$ which is a nonexpansive mapping, for all $i = 1, 2, 3$.

Let $w \in \Omega$ where $w = (w_1, w_2, w_3) \in C_1 \times C_2 \times C_3$ where $w_2 = A_2 w_1$, $w_3 = A_3 w_2$. From (i) implies (ii), we have

$$w_1 = P_{C_1}(I - \lambda_1 B_1)(w_1 - \gamma_2 A_2^*((I_2 - P_{C_2}(I_2 - \lambda_2 B_2))w_2 + \gamma_3 A_3^*(I_3 - P_{C_3}(I_3 - \lambda_3 B_3))w_3)).$$

Put $M = (I_2 - P_{C_2}(I_2 - \lambda_2 B_2))\bar{x}_2 + \gamma_3 A_3^*(I_3 - P_{C_3}(I_3 - \lambda_3 B_3))\bar{x}_3$ and $N = (I_2 - P_{C_2}(I_2 - \lambda_2 B_2))w_2 + \gamma_3 A_3^*(I_3 - P_{C_3}(I_3 - \lambda_3 B_3))w_3$.

From determining the definition of $\bar{x}$ and w, we have

$$\begin{aligned}
\|\bar{x}_1 - w_1\|^2 &\leq \|\bar{x}_1 - w_1 - \gamma_2 A_2^*(M-N)\|^2 \\
&= \|\bar{x}_1 - w_1\|^2 - 2\gamma_2 \langle \bar{x}_1 - w_1, A_2^*(M-N)\rangle + \gamma_2^2 \|A_2^*(M-N)\|^2 \\
&\leq \|\bar{x}_1 - w_1\|^2 - 2\gamma_2 \langle \bar{x}_2 - w_2, M-N\rangle + \gamma_2^2 L \|M-N\|^2 \\
&\leq \|\bar{x}_1 - w_1\|^2 - 2\gamma_2 \langle \bar{x}_2 - w_2, (I_2 - P_{C_2}(I_2 - \lambda_2 B_2))\bar{x}_2 + \gamma_3 A_3^*(I_3 - P_{C_3}(I_3 - \lambda_3 B_3))\bar{x}_3\rangle \\
&\quad + \gamma_2^2 L \|(I_2 - P_{C_2}(I_2 - \lambda_2 B_2))\bar{x}_2 + \gamma_3 A_3^*(I_3 - P_{C_3}(I_3 - \lambda_3 B_3))\bar{x}_3\|^2 \\
&= \|\bar{x}_1 - w_1\|^2 - 2\gamma_2 (\langle \bar{x}_2 - w_2, (I_2 - P_{C_2}(I_2 - \lambda_2 B_2))\bar{x}_2\rangle + \gamma_3 \langle \bar{x}_3 - w_3, \\
&\quad (I_3 - P_{C_3}(I_3 - \lambda_3 B_3))\bar{x}_3\rangle) + \gamma_2^2 L \|(I_2 - P_{C_2}(I_2 - \lambda_2 B_2))\bar{x}_2 \\
&\quad + \gamma_3 A_3^*(I_3 - P_{C_3}(I_3 - \lambda_3 B_3))\bar{x}_3\|^2 \\
&= \|\bar{x}_1 - w_1\|^2 + 2\gamma_2 \langle w_2 - \bar{x}_2, (I_2 - P_{C_2}(I_2 - \lambda_2 B_2))\bar{x}_2\rangle + 2\gamma_2 \gamma_3 \langle w_3 - \bar{x}_3, \\
&\quad (I_3 - P_{C_3}(I_3 - \lambda_3 B_3))\bar{x}_3\rangle + \gamma_2^2 L(\|(I_2 - P_{C_2}(I_2 - \lambda_2 B_2))\bar{x}_2\|^2 \\
&\quad + \gamma_3^2 L \|(I_3 - P_{C_3}(I_3 - \lambda_3 B_3))\bar{x}_3\|^2 + 2\gamma_3 \langle (I_2 - P_{C_2}(I_2 - \lambda_2 B_2))\bar{x}_2, \\
&\quad A_3^*(I_3 - P_{C_3}(I_3 - \lambda_3 B_3))\bar{x}_3\rangle) \\
&\leq \|\bar{x}_1 - w_1\|^2 + 2\gamma_2 \langle w_2 - P_{C_2}(I_2 - \lambda_2 B_2)\bar{x}_2 + P_{C_2}(I_2 - \lambda_2 B_2)\bar{x}_2 - \bar{x}_2, \\
&\quad (I_2 - P_{C_2}(I_2 - \lambda_2 B_2))\bar{x}_2\rangle + 2\gamma_2 \gamma_3 \langle w_3 - P_{C_3}(I_3 - \lambda_3 B_3)\bar{x}_3 + P_{C_3}(I_3 - \lambda_3 B_3)\bar{x}_3 - \bar{x}_3, \\
&\quad (I_3 - P_{C_3}(I_3 - \lambda_3 B_3))\bar{x}_3\rangle + \gamma_2^2 L(\|(I_2 - P_{C_2}(I_2 - \lambda_2 B_2))\bar{x}_2\|^2 \\
&\quad + \gamma_3^2 L \|(I_3 - P_{C_3}(I_3 - \lambda_3 B_3))\bar{x}_3\|^2 \\
&\quad + \gamma_3 \|(I_2 - P_{C_2}(I_2 - \lambda_2 B_2))\bar{x}_2\|^2 + \gamma_3 \|A_3^*(I_3 - P_{C_3}(I_3 - \lambda_3 B_3))\bar{x}_3\|^2) \\
&\leq \|\bar{x}_1 - w_1\|^2 + 2\gamma_2(\frac{1}{2}\|(I_2 - P_{C_2}(I_2 - \lambda_2 B_2))\bar{x}_2\|^2 - \|(I_2 - P_{C_2}(I_2 - \lambda_2 B_2))\bar{x}_2\|^2) \\
&\quad + 2\gamma_2 \gamma_3(\frac{1}{2}\|(I_3 - P_{C_3}(I_3 - \lambda_3 B_3))\bar{x}_3\|^2 - \|(I_3 - P_{C_3}(I_3 - \lambda_3 B_3))\bar{x}_3\|^2) \\
&\quad + \gamma_2^2 L(\|((I_2 - P_{C_2}(I_2 - \lambda_2 B_2))\bar{x}_2\|^2 + \gamma_3^2 L \|(I_3 - P_{C_3}(I_3 - \lambda_3 B_3))\bar{x}_3\|^2 \\
&\quad + \gamma_3 \|(I_2 - P_{C_2}(I_2 - \lambda_2 B_2))\bar{x}_2\|^2 + \gamma_3 L \|(I_3 - P_{C_3}(I_3 - \lambda_3 B_3))\bar{x}_3\|^2) \\
&= \|\bar{x}_1 - w_1\|^2 - \gamma_2(1 - \gamma_2 L(1+\gamma_3))\|(I_2 - P_{C_2}(I_2 - \lambda_2 B_2))\bar{x}_2\|^2 \\
&\quad - \gamma_2 \gamma_3(1 - \gamma_2 L^2(1+\gamma_3))\|(I_3 - P_{C_3}(I_3 - \lambda_3 B_3))\bar{x}_3\|^2.
\end{aligned} \tag{8}$$

By applying above equation and Lemma 2, we have

$$\bar{x}_2 \in F\left(P_{C_2}(I - \lambda_2 B_2)\right) = Var(C_2, B_2) \text{ and } \bar{x}_3 \in F\left(P_{C_3}(I - \lambda_3 B_3)\right) = Var(C_3, B_3). \tag{9}$$

From determining the definitions of $\bar{x}$ and (9), we have

$$\bar{x}_1 \in F\left(P_{C_1}(I - \lambda_1 B_1)\right) = Var(C_1, B_1).$$

Hence $\bar{x} \in \Omega$. □

Lemma 6. *Let C be a nonempty closed convex subset of Hilbert space H. Let $\vartheta : C \to C$ be a nonspreading mapping and $\varrho : C \to C$ be κ-pseudo-nonspreding mapping with $F(\vartheta) \cap F(\varrho) \neq \emptyset$. Then $F(P_C(I - \gamma(a(I-\vartheta) + (1-a)(I-\varrho)))) = F(\vartheta) \cap F(\varrho)$ for all $a \in (0,1)$ and $\gamma > 0$. Moreover, if $\gamma < 1-\kappa$, then*

$$\|I - \gamma(a(I-\vartheta) + (1-a)(I-\varrho)))x - \omega^*\| \leq \|\omega^* - x\|,$$

for all $x \in C$ and $\omega^* \in F(\varrho) \cap F(\vartheta)$.

Proof. Let $\omega_0 \in F(\varrho) \cap F(\vartheta)$, we have

$$P_C(I - \gamma(a(I - \vartheta) + (1 + a)(I - \varrho)))\omega_0 = \omega_0.$$

It follows that $\omega_0 \in F(P_C(I - \gamma(a(I - \vartheta) + (1 + a)(I - \varrho))))$. Therefore

$$F(\vartheta) \cap F(\varrho) \subseteq F(P_C(I - \gamma(a(I - \vartheta) + (1 + a)(I - \varrho))))$$

Let $\omega_0 \in F(P_C(I - \gamma(a(I - \vartheta) + (1 + a)(I - \varrho))))$ and $\omega^* \in F(\vartheta) \cap F(\varrho)$. From Lemma 2, we have

$$\langle y - \omega_0, a(I - \vartheta)\omega_0 + (1 - a)(I - \varrho)\omega_0 \rangle \geq 0,$$

for all $y \in C$.

From determining the definition of ϱ, we have

$$\begin{aligned}
\|\omega_0 - \omega^*\|^2 + \kappa \|(I - \varrho)\omega_0\|^2 &\geq \|\varrho\omega_0 - \omega^*\|^2 \\
&= \|(I - \varrho)\omega_0 - (\omega_0 - \omega^*)\|^2 \\
&= \|(I - \varrho)\omega_0\|^2 - 2\langle (I - \varrho)\omega_0, \omega_0 - \omega^* \rangle + \|\omega_0 - \omega^*\|^2
\end{aligned} \quad (10)$$

From the result of the calculation from the inequality (10), we get

$$\langle (I - \varrho)\omega_0, \omega_0 - \omega^* \rangle \geq \left(\frac{1 - \kappa}{2}\right) \|(I - \varrho)\omega_0\|^2 \quad (11)$$

Assume that $\omega_0 \neq \vartheta\omega_0$; then we have $\|(I - \vartheta)\omega_0\| > 0$. Using the same method as (11) and definitions of ϑ, we get

$$\langle (I - \vartheta)\omega_0, \omega_0 - \omega^* \rangle \geq \frac{1}{2} \|(I - \vartheta)\omega_0\|^2 \quad (12)$$

From (11) and $a \in (0, 1)$, we obtain

$$\begin{aligned}
\langle \omega^* - \omega_0, a(I - \vartheta)\omega_0 \rangle &= \langle \omega^* - \omega_0, a(I - \vartheta)\omega_0 + (1 - a)(I - \varrho)\omega_0 \rangle \\
&\quad - (1 - a)\langle \omega^* - \omega_0, (I - \varrho)\omega_0 \rangle \\
&\geq (1 - a)\langle \omega_0 - \omega^*, (I - \varrho)\omega_0 \rangle.
\end{aligned} \quad (13)$$

From (13), we have

$$\langle \omega^* - \omega_0, (I - \vartheta)\omega_0 \rangle \geq 0.$$

From above and (12), we have

$$0 \leq \langle \omega^* - \omega_0, (I - \vartheta)\omega_0 \rangle \leq -\frac{1}{2}\|(I - \vartheta)\omega_0\|^2.$$

Thus, $\|(I - \vartheta)\omega_0\| \leq 0$. This is a contradiction.

Thus, we have $\omega_0 = \vartheta\omega_0$ and it implies that

$$\omega_0 \in F(\vartheta). \quad (14)$$

Similarly, by using the same technique as (14), we have

$$\omega_0 \in F(\varrho). \tag{15}$$

From (14) and (15), we have

$$F(P_C(I - \gamma(a(I - \vartheta) + (1+a)(I - \varrho)))) \subseteq F(\varrho) \cap F(\vartheta).$$

Let $\omega^* \in F(\varrho) \cap F(\vartheta)$ and $x \in C$; we have

$$\begin{aligned}
\|(I - \gamma(a(I - \vartheta) + (1-a)(I - \varrho)))x - \omega^*\|^2 &= \|(I - \gamma(a(I - \vartheta) + (1-a)(I - \varrho)))x \\
&\quad - (I - \gamma(a(I - \vartheta) + (1-a)(I - \varrho)))\omega^*\|^2 \\
&= \|x - \omega^* - \gamma(a((I - \vartheta)x - (I - \vartheta)\omega^*) \\
&\quad + (1-a)((I - \varrho)x - (I - \varrho)\omega^*))\|^2 \\
&= \|x - \omega^*\|^2 - 2\gamma\langle a((I - \vartheta)x - (I - \vartheta)\omega^*) \\
&\quad + (1-a)((I - \varrho)x - (I - \varrho)\omega^*), x - \omega^*\rangle \\
&\quad + \gamma^2 \|a((I - \vartheta)x - (I - \vartheta)\omega^*) \\
&\quad + (1-a)((I - \varrho)x - (I - \varrho)\omega^*)\|^2 \\
&\leq \|x - \omega^*\|^2 - 2\gamma a\langle(I - \vartheta)x - (I - \vartheta)\omega^*, x - \omega^*\rangle \\
&\quad - 2\gamma(1-a)\langle(I - \varrho)x - (I - \varrho)\omega^*, x - \omega^*\rangle \\
&\quad + \gamma^2 a\|(I - \vartheta)x - (I - \vartheta)\omega^*\|^2 \\
&\quad + (1-a)\gamma^2 \|(I - \varrho)x - (I - \varrho)\omega^*\|^2 \\
&\leq \|x - \omega^*\|^2 - 2\gamma a \frac{\|(I - T)x\|^2}{2} \\
&\quad - 2\gamma(1-a)(1-\kappa)\frac{\|(I - \varrho)x\|^2}{2} \\
&\quad + \gamma^2 a\|(I - \vartheta)x\|^2 + (1-a)\gamma^2 \|(I - \varrho)x\|^2 \\
&\leq \|x - \omega^*\|^2.
\end{aligned}$$

□

2. The Split Various Variational Inequality Theorem

Theorem 7. *For every $i = 1, 2, 3$, let C_i be a closed convex subset of a real Hilbert space H_i. Let $B_i : C_i \to H_i$ be β_i-inverse strongly monotonic mappings with $\eta = \min_{i=1,2,3} \{\beta_i\}$ and let $A_2 : H_1 \to H_2, A_3 : H_2 \to H_3$ be a bounded linear operator with the adjoint operator A_2^* and A_3^*, respectively. Assume that $\bar{x}_1 \in H_1, \bar{x}_2 = A_2 \bar{x}_1, \bar{x}_3 = A_3 \bar{x}_2$ and $\Omega \neq \emptyset$. Let $\vartheta, \varrho : C \to C$ be nonspreading and κ-strictly pseudo-nonspreading mappings, respectively. Assume that $\Omega \cap F(\vartheta) \cap F(\varrho) \neq \emptyset$ and let the sequence $\{x_n\}$ generated by $u, x_1 \in C$, and*

$$\begin{aligned}
x_{n+1} &= \alpha_n u + \beta_n x_n + \gamma_n P_{C_1}(I_1 - \lambda_n(a(I_1 - \vartheta) + (1-a)(I_1 - \varrho)))x_n \\
&\quad + \delta_n P_{C_1}(I_1 - \lambda_1 B_1)(x_n^1 - \gamma_2 A_2^*((I_2 - P_{C_2}(I - \lambda_2 B_2))x_n^2 + \gamma_3 A_3^*(I_3 - P_{C_3}(I - \lambda_3 B_3))x_n^3)),
\end{aligned}$$

for all $n \geq 1$ and $a \in (0,1)$, $I_i : H_i \to H_i$ are identity mappings, for all $i = 1, 2, 3$, where $\{\alpha_n\}, \{\beta_n\}, \{\gamma_n\}, \{\delta_n\} \subseteq [0,1]$ and $\alpha_n + \beta_n + \gamma_n + \delta_n = 1$ and $x_n = x_n^1, x_n^2 = A_2 x_n^1, x_n^3 = A_3 x_n^2$, for all $n \in \mathbb{N}, 0 < \lambda_i < 2\eta$, for all $i = 1, 2, 3$ and $\gamma_j > 0$, for all $j = 2, 3$. Suppose that the conditions (i)–(v) are true;

(i) $\lim_{n\to+\infty} \alpha_n = 0, \sum_{n=1}^{+\infty} \alpha_n = +\infty$;

(ii) $\gamma_2(1+\gamma_3) \leq \dfrac{1}{L}$, where $L = \max\{L_{A_1}, L_{A_2}\} \leq 1$ where $L_{A_1} L_{A_2}$ are spectral radius of $A_2 A_2^*, A_3 A_3^*$, respectively;

(iii) $0 < a \leq \beta_n, \gamma_n, \delta_n \leq b < 1$, for some $a, b > 0$, for all $n \in \mathbb{N}$;

(iv) $\sum_{n=1}^{+\infty} \lambda_n < +\infty$ and $0 < \lambda_n < 1 - \kappa$, for all $n \in \mathbb{N}$;

(v) $\sum_{n=1}^{+\infty} |\alpha_{n+i} - \alpha_n|, \sum_{n=1}^{+\infty} |\beta_{n+i} - \beta_n|, \sum_{n=1}^{+\infty} |\gamma_{n+1} - \gamma_n| < +\infty$.

Then the sequence $\{x_n\}$ converges strongly to $z_0 = P_{\Omega \cap F(\vartheta) \cap F(\varrho)} u$.

Proof. Put $M = a(I - \vartheta) + (1-a)(I - \varrho)$ and $u_n = P_{C_1}(I - \lambda_1 B_1)(x_n^1 - \gamma_2 A_2^*((I_2 - P_{C_2}(I - \lambda_2 B_2))x_n^2 + \gamma_3 A_3^*(I_3 - P_{C_3}(I - \lambda_n B_3))x_n^3))$. Thus, we can rewrite x_n as follows:

$$x_{n+1} = \alpha_n u + \beta_n x_n + \gamma_n P_{C_1}(I - \lambda_n M) x_n + \delta_n u_n, \tag{16}$$

for all $n \geq 1$.

From determining the definition of u_n put $w_n = (I_2 - P_{C_2}(I - \lambda_2 B_2))x_n^2 + \gamma_3 A_3^*(I_3 - P_{C_3}(I - \lambda_3 B_3))x_n^3$ and $z_n = (I_3 - P_{C_3}(I - \lambda_3 B_3))x_n^3$, we have

$$u_n = P_{C_1}(I_1 - \lambda_1 B_1)(x_n - \gamma_2 A_2^* w_n).$$

For every $n \in \mathbb{N}$, we have

$$\begin{aligned}
\|u_n - u_{n-1}\|^2 &\leq \|x_n - \gamma_2 A_2^* w_n - x_{n-1} + \gamma_2 A_2^* w_{n-1}\|^2 \\
&= \|x_n - x_{n-1} - \gamma_2 A_2^*(w_n - w_{n-1})\|^2 \\
&= \|x_n - x_{n-1}\|^2 - 2\gamma_2 \langle A_2 x_n - A_2 x_{n-1}, w_n - w_{n-1}\rangle + \gamma_2^2 \|A_2^*(w_n - w_{n-1})\|^2 \\
&= \|x_n - x_{n-1}\|^2 - 2\gamma_2 \langle x_n^2 - x_{n-1}^2, (I_2 - P_{C_2}(I - \lambda_2 B_2))x_n^2 \\
&\quad + \gamma_3 A_3^* z_n - (I_2 - P_{C_2}(I - \lambda_2 B_2))x_{n-1}^2 - \gamma_3 A_3^* z_{n-1}\rangle \\
&\quad + \gamma_2^2 \|A_2^*(w_n - w_{n-1})\|^2 \\
&= \|x_n - x_{n-1}\|^2 + 2\gamma_2 \langle x_{n-1}^2 - x_n^2, (I_2 - P_{C_2}(I - \lambda_2 B_2))x_n^2 \\
&\quad - (I_2 - P_{C_2}(I - \lambda_2 B_2))x_{n-1}^2\rangle + 2\gamma_2 \gamma_3 \langle x_{n-1}^3 - x_n^3, z_n \\
&\quad - z_{n-1}\rangle + \gamma_2^2 \|A_2^*(w_n - w_{n-1})\|^2 \\
&\leq \|x_n - x_{n-1}\|^2 + 2\gamma_2 \langle x_{n-1}^2 - x_n^2, (I_2 - P_{C_2}(I - \lambda_2 B_2))x_n^2 \\
&\quad - (I_2 - P_{C_2}(I - \lambda_2 B_2))x_{n-1}^2\rangle + 2\gamma_2 \gamma_3 \langle x_{n-1}^3 - x_n^3, z_n - z_{n-1}\rangle \\
&\quad + \gamma_2^2 L \left\| (I_2 - P_{C_2}(I_2 - \lambda_2 B_2))x_n^2 - (I_2 - P_{C_2}(I_2 - \lambda_2 B_2))x_{n-1}^2 + \gamma_3 A_3^*(z_n - z_{n-1})\right\|^2 \\
&\leq \|x_n - x_{n-1}\|^2 + 2\gamma_2 \langle x_{n-1}^2 - x_n^2, (I_2 - P_{C_2}(I - \lambda_2 B_2))x_n^2 \\
&\quad - (I_2 - P_{C_2}(I - \lambda_2 B_2))x_{n-1}^2\rangle + 2\gamma_2 \gamma_3 \langle x_{n-1}^3 - x_n^3, z_n - z_{n-1}\rangle \\
&\quad + \gamma_2^2 L (\|(I_2 - P_{C_2}(I - \lambda_2 B_2))x_n^2 - (I_2 - P_{C_2}(I - \lambda_2 B_2))x_{n-1}^2\|^2 \\
&\quad + \gamma_3^2 L \|z_n - z_{n-1}\|^2 + 2\gamma_3 \langle (I_2 - P_{C_2}(I - \lambda_2 B_2))x_n^2 \\
&\quad - (I_2 - P_{C_2}(I - \lambda_2 B_2))x_{n-1}^2, A^*(z_n - z_{n-1})\rangle)
\end{aligned}$$

$$
\begin{aligned}
&\leq \|x_n - x_{n-1}\|^2 + 2\gamma_2 \langle x_{n-1}^2 - x_n^2, (I_2 - P_{C_2}(I - \lambda_2 B_2))x_n^2 - (I_2 - P_{C_2}(I - \lambda_2 B_2))x_{n-1}^2 \rangle \\
&\quad + 2\gamma_2\gamma_3 \langle x_{n-1}^3 - x_n^3, (I_3 - P_{C_3}(I - \lambda_3 B_3))x_n^3 - (I_3 - P_{C_3}(I - \lambda_3 B_3))x_{n-1}^3 \rangle \\
&\quad + \gamma_2^2 L(\|(I_2 - P_{C_2}(I - \lambda_2 B_2))x_n^2 - (I_2 - P_{C_2}(I - \lambda_2 B_2))x_{n-1}^2\|^2 + \gamma_3^2 L \|z_n - z_{n-1}\|^2 \\
&\quad + \gamma_3 \|(I_2 - P_{C_2}(I - \lambda_2 B_2))x_n^2 - (I_2 - P_{C_2}(I - \lambda_2 B_2))x_{n-1}^2\|^2 + \gamma_3 L \|z_n - z_{n-1}\|^2) \\
&\leq \|x_n - x_{n-1}\|^2 + 2\gamma_2 (-\left\|(I_2 - P_{C_2}(I - \lambda_2 B_2))x_n^2 - (I_2 - P_{C_2}(I - \lambda_2 B_2))x_{n-1}^2\right\|^2 \\
&\quad + \frac{1}{2} \left\|(I_2 - P_{C_2}(I - \lambda_2 B_2))x_n^2 - (I_2 - P_{C_2}(I - \lambda_2 B_2))x_{n-1}^2\right\|^2) \\
&\quad + 2\gamma_2\gamma_3 (-\|z_n - z_{n-1}\|^2 + \frac{1}{2} \|z_n - z_{n-1}\|^2) \\
&\quad + \gamma_2^2 L(\|(I_2 - P_{C_2}(I - \lambda_2 B_2))x_n^2 - (I_2 - P_{C_2}(I - \lambda_2 B_2))x_{n-1}^2\|^2 + \gamma_3^2 L \|z_n - z_{n-1}\|^2 \\
&\quad + \gamma_3 \|(I_2 - P_{C_2}(I - \lambda_2 B_2))x_n^2 - (I_2 - P_{C_2}(I - \lambda_2 B_2))x_{n-1}^2\|^2 + \gamma_3 L \|z_n - z_{n-1}\|) \\
&= \|x_n - x_{n-1}\|^2 - \gamma_2 \left\|(I_2 - P_{C_2}(I - \lambda_2 B_2))x_n^2 - (I_2 - P_{C_2}(I - \lambda_2 B_2))x_{n-1}^2\right\|^2 \\
&\quad - \gamma_2\gamma_3 \|z_n - z_{n-1}\|^2 + \gamma_2^2 L \|(I_2 - P_{C_2}(I - \lambda_2 B_2))x_n^2 - (I_2 - P_{C_2}(I - \lambda_2 B_2))x_{n-1}^2\|^2 \\
&\quad + \gamma_2^2 \gamma_3^2 L^2 \|z_n - z_{n-1}\|^2 + \gamma_2^2 \gamma_3 L \|(I_2 - P_{C_2}(I - \lambda_2 B_2))x_n^2 - (I_2 - P_{C_2}(I - \lambda_2 B_2))x_{n-1}^2\|^2 \\
&\quad + \gamma_2^2 \gamma_3 L^2 \|z_n - z_{n-1}\| \\
&= \|x_n - x_{n-1}\|^2 - \gamma_2(1 - \gamma_2 L(1 + \gamma_3)) \|(I_2 - P_{C_2}(I - \lambda_2 B_2))x_n^2 \\
&\quad - (I_2 - P_{C_2}(I - \lambda_2 B_2))x_{n-1}^2\|^2 - \gamma_2 \gamma_3 (1 - \gamma_2 L^2 (1 + \gamma_3)) \|z_n - z_{n-1}\|^2 \\
&\leq \|x_n - x_{n-1}\|^2.
\end{aligned}
\tag{17}
$$

Let $\varpi^* \in \Omega \cap F(\varrho) \cap F(\vartheta)$. From Lemma 6 and utilization of (8), we have

$$
\begin{aligned}
\|x_{n+1} - \varpi^*\| &\leq \alpha_n \|u - \varpi^*\| + \beta_n \|x_n - \varpi^*\| + \gamma_n \|P_{C_1}(I - \lambda_n M)x_n - \varpi^*\| \\
&\quad + \delta_n \|u_n - \varpi^*\| \\
&\leq \alpha_n \|u - \varpi^*\| + (1 - \alpha_n) \|x_n - \varpi^*\| \\
&\leq \widetilde{M},
\end{aligned}
\tag{18}
$$

where $\widetilde{M} = \max\{\|u - \varpi^*\|, \|x_1 - \varpi^*\|\}$. By induction we can conclude that the sequence $\{x_n\}$ is bounded and so are $\{u_n\}$ and $\{P_{C_1}(I - \lambda_n M)x_n\}$.

From determining the definition of x_n and (17), we have

$$\begin{aligned}
\|x_{n+1} - x_n\| &= \|\alpha_n u + \beta_n x_n + \gamma_n P_{C1}(I - \lambda_n M)x_n + \delta_n u_n \\
&\quad - \alpha_{n-1} u - \beta_{n-1} x_{n-1} - \gamma_{n-1} P_{C1}(I - \lambda_{n-1} M)x_{n-1} - \delta_{n-1} u_{n-1}\| \\
&\leq |\alpha_n - \alpha_{n-1}|\|u\| + \beta_n \|x_n - x_{n-1}\| + |\beta_n - \beta_{n-1}|\|x_{n-1}\| + \delta_n \|u_n - u_{n-1}\| \\
&\quad + |\delta_n - \delta_{n-1}|\|u_{n-1}\| + \gamma_n \|P_{C_1}(I - \lambda_n M)x_n - P_{C_1}(I - \lambda_{n-1}M)x_{n-1}\| \\
&\quad + |\gamma_n - \gamma_{n-1}|\|P_{C_1}(I_1 - \lambda_{n-1}M)x_{n-1}\| \\
&\leq |\alpha_n - \alpha_{n-1}|\|u\| + \beta_n \|x_n - x_{n-1}\| + |\beta_n - \beta_{n-1}|\|x_{n-1}\| + \delta_n \|u_n - u_{n-1}\| \\
&\quad + |\delta_n - \delta_{n-1}|\|u_{n-1}\| + \gamma_n \|x_n - x_{n-1}\| + \|\lambda_{n-1} M x_{n-1} - \lambda_n M x_n\| \\
&\quad + |\gamma_n - \gamma_{n-1}|\|P_{C_1}(I_1 - \lambda_{n-1}M)x_{n-1}\| \\
&\leq |\alpha_n - \alpha_{n-1}|\|u\| + \beta_n \|x_n - x_{n-1}\| + |\beta_n - \beta_{n-1}|\|x_{n-1}\| + \delta_n \|x_n - x_{n-1}\| \\
&\quad + |\delta_n - \delta_{n-1}|\|u_{n-1}\| + \gamma_n \|x_n - x_{n-1}\| + \lambda_{n-1}\|M x_{n-1}\| + \lambda_n\|M x_n\| \\
&\quad + |\gamma_n - \gamma_{n-1}|\|P_{C_1}(I_1 - \lambda_{n-1}M)x_{n-1}\| \\
&= (1 - \alpha_n)\|x_n - x_{n-1}\| + |\alpha_n - \alpha_{n-1}|\|u\| + |\beta_n - \beta_{n-1}|\|x_{n-1}\| \\
&\quad + |\delta_n - \delta_{n-1}|\|u_{n-1}\| + \lambda_{n-1}\|M x_{n-1}\| + \lambda_n\|M x_n\| \\
&\quad + |\gamma_n - \gamma_{n-1}|\|P_{C_1}(I_1 - \lambda_{n-1}M)x_{n-1}\|.
\end{aligned}$$

From the conditions $(i), (iv), (v)$ and Lemma 3, we have

$$\lim_{n \to +\infty} \|x_{n+1} - x_n\| = 0 \tag{19}$$

Applying (8) and the definition of x_n, we have

$$\begin{aligned}
\|x_{n+1} - \omega^*\|^2 &\leq \alpha_n \|u - \omega^*\|^2 + \beta_n \|x_n - \omega^*\|^2 + \gamma_n \|P_{C_1}(I - \lambda_n M)x_n - \omega^*\|^2 \\
&\quad + \delta_n \|u_n - \omega^*\|^2 - \gamma_n \beta_n \|P_{C_1}(I - \lambda_n M)x_n - x_n\|^2 - \delta_n \beta_n \|u_n - x_n\|^2 \\
&\leq \alpha_n \|u - \omega^*\|^2 + \|x_n - \omega^*\|^2 - \gamma_n \beta_n \|P_{C1}(I - \lambda_n M)x_n - x_n\|^2 \\
&\quad - \delta_n \beta_n \|u_n - x_n\|^2,
\end{aligned}$$

which implies that

$$\gamma_n \beta_n \|P_{C_1}(I - \lambda_n M)x_n - x_n\|^2 + \delta_n \beta_n \|u_n - x_n\|^2 \leq (\|x_{n+1} - \omega^*\| + \|x_n - \omega^*\|)\|x_{n+1} - x_n\| \\
+ \alpha_n \|u - \omega^*\|^2.$$

From the conditions $(i), (iii)$ and (19) we can conclude the following results

$$\lim_{n \to +\infty} \|u_n - x_n\| = \lim_{n \to +\infty} \|P_{C1}(I - \lambda_n M)x_n - x_n\| = 0. \tag{20}$$

Next, we show that

$$\limsup_{n \to +\infty} \langle u - z_0, z_0 - x_n \rangle \leq 0, \tag{21}$$

where $z_0 = P_{\Omega \cap F(\varrho) \cap F(\vartheta)} u$. In order to prove this we may assume that

$$\limsup_{n \to +\infty} \langle u - z_0, x_n - z_0 \rangle = \lim_{k \to +\infty} \langle u - z_0, x_{n_k} - z_0 \rangle, \tag{22}$$

where $\{x_{n_k}\}$ is a subsequence of $\{x_n\}$. Since $\{x_n\}$ is bounded, we may assume that $x_{n_k} \rightharpoonup q$ as $k \to +\infty$. Assume that $q \notin F(\varrho) \cap F(\vartheta)$. From Lemma 6, we have $q \notin F\left(P_{C_1}\left(I - \lambda_{n_k} M\right)\right)$. By using properties of Opial's condition and (20), we have

$$\begin{aligned}
\liminf_{k \to +\infty} \|x_{n_k} - q\| &< \liminf_{k \to +\infty} \|x_{n_k} - P_{C_1}\left(I - \lambda_{n_k} M\right) q\| \\
&\leq \liminf_{k \to +\infty} (\|x_{n_k} - P_{C_1}\left(I - \lambda_{n_k} M\right) x_{n_k}\| \\
&\quad + \|P_{C_1}\left(I - \lambda_{n_k} M\right) x_{n_k} - P_{C_1}\left(I - \lambda_{n_k} M\right) q\|) \\
&\leq \liminf_{k \to +\infty} (\|x_{n_k} - q\| + \lambda_{n_k} \|M x_{n_k} - M q\|) \\
&\leq \liminf_{k \to +\infty} \|x_{n_k} - q\|.
\end{aligned}$$

This is a contradiction. Therefore $q \in F(\varrho) \cap F(\vartheta)$.
Assume $q \notin \Omega$. From Lemma 5, we have

$$q \neq P_{C_1}\left(I - \lambda_1 B_1\right)\left(q - \gamma_2 A_2^*((I_2 - P_{C_2}(I - \lambda_2 B_2)) A_2 q + \gamma_3 A_3^*(I_3 - P_{C_3}(I - \lambda_3 B_3)) A_3 A_2 q)\right).$$

By using properties of Opial's condition, and the definitions of u_n and (20), we have

$$\begin{aligned}
\liminf_{k \to +\infty} \|x_{n_k} - q\| &< \liminf_{k \to +\infty} \|x_{n_k} - P_{C_1}\left(I - \lambda_1 B_1\right)(q - \gamma_2 A_2^*((I_2 - P_{C_2}(I - \lambda_2 B_2)) A_2 q \\
&\quad + \gamma_3 A_3^*(I_3 - P_{C_3}(I - \lambda_3 B_3)) A_3 A_2 q))\| \\
&\leq \liminf_{k \to +\infty} (\|x_{n_k} - u_{n_k}\| \\
&\quad + \|u_{n_k} - P_{C_1}\left(I - \lambda_1 B_1\right)(q - \gamma_2 A_2^*((I_2 - P_{C_2}(I - \lambda_2 B_2)) A_2 q \\
&\quad + \gamma_3 A_3^*(I_3 - P_{C_3}(I - \lambda_3 B_3)) A_3 A_2 q))\|) \\
&\leq \liminf_{k \to +\infty} \|x_{n_k} - q\|.
\end{aligned}$$

This is a contradiction. Then $q \in \Omega$. Therefore $q \in \Omega \cap F(\varrho) \cap F(\vartheta)$.
From (22) and the well-known properties of metric projection, we have

$$\limsup_{n \to +\infty} \langle u - z_0, x_n - z_0 \rangle \leq 0.$$

From determining the definition of x_n, we can conclude that

$$\|x_{n+1} - z_0\|^2 \leq (1 - \alpha_n) \|x_n - z_0\|^2 + 2\alpha_n \langle u - z_0, x_{n+1} - z_0 \rangle,$$

where $z_0 = P_{\Omega \cap F(\varrho) \cap F(\vartheta)} u$. From Lemma 4, we can conclude that the sequence $\{x_n\}$ converses strongly to $z_0 = P_{\Omega \cap F(\varrho) \cap F(\vartheta)} u$. □

The following results were obtained directly from the main theorem.

Corollary 8. *For every* $i = 1, 2, 3$, *let* C_i *be a closed convex subset of a real Hilbert space* H_i. *Let* $B_i : C_i \to H_i$ *be* β_i-*inverse strongly monotonic mappings with* $\eta = \min_{i=1,2,3}\{\beta_i\}$ *and let* $A_2 : H_1 \to H_2$, $A_3 : H_2 \to lH_3$ *be a bounded linear operator with the adjoint operator* A_2^* *and* A_3^*, *respectively. Assume that* $\bar{x}_1 \in H_1$, $\bar{x}_2 = A_2\bar{x}_1$, $\bar{x}_3 = A_3\bar{x}_2$ *and* $\Omega \neq \emptyset$. *Let* $\vartheta, \varrho : C \to C$ *be nonspreading mappings, respectively. Assume that* $\Omega \cap F(\vartheta) \cap F(\varrho) \neq \emptyset$ *and let the sequence* $\{x_n\}$ *generated by* $u, x_1 \in C$, *and*

$$x_{n+1} = \alpha_n u + \beta_n x_n + \gamma_n P_{C_1}(I_1 - \lambda_n(a(I_1 - \vartheta) + (1-a)(I_1 - \varrho)))x_n$$
$$+ \delta_n P_{C_1}(I_1 - \lambda_1 B_1)(x_n^1 - \gamma_2 A_2^*((I_2 - P_{C_2}(I - \lambda_2 B_2))x_n^2 + \gamma_3 A_3^*(I_3 - P_{C_3}(I - \lambda_3 B_3))x_n^3)),$$

for all $n \geq 1$ *and* $a \in (0,1)$, $I_i : H_i \to H_i$ *are identity mappings, for all* $i = 1, 2, 3$, *where* $\{\alpha_n\}, \{\beta_n\}, \{\gamma_n\}, \{\delta_n\} \subseteq [0,1]$ *and* $\alpha_n + \beta_n + \gamma_n + \delta_n = 1$ *and* $x_n = x_n^1, x_n^2 = A_2 x_n^1, x_n^3 = A_3 x_n^2$, *for all* $n \in \mathbb{N}$, $0 < \lambda_i < 2\eta$, *for all* $i = 1, 2, 3$ *and* $\gamma_j > 0$, *for all* $j = 2, 3$. *Suppose that the conditions (i) to (v) are true;*

(i) $\lim_{n \to +\infty} \alpha_n = 0$, $\sum_{n=1}^{+\infty} \alpha_n = +\infty$;

(ii) $\gamma_2(1 + \gamma_3) \leq \dfrac{1}{L}$, *where* $L = \max\{L_{A_1}, L_{A_2}\} \leq 1$, *where* L_{A_1}, L_{A_2} *are spectral radius of* $A_2 A_2^*$, $A_3 A_3^*$, *respectively;*

(iii) $0 < a \leq \beta_n, \gamma_n, \delta_n \leq b < 1$, *for some* $a, b > 0$, *for all* $n \in \mathbb{N}$;

(iv) $\sum_{n=1}^{+\infty} \lambda_n < +\infty$ *and* $0 < \lambda_n < 1 - \kappa$, *for all* $n \in \mathbb{N}$;

(v) $\sum_{n=1}^{+\infty} |\alpha_{n+i} - \alpha_n|$, $\sum_{n=1}^{+\infty} |\beta_{n+i} - \beta_n|$, $\sum_{n=1}^{+\infty} |\gamma_{n+1} - \gamma_n| < +\infty$.

Then the sequence $\{x_n\}$ *converses strongly to* $z_0 = P_{\Omega \cap F(\vartheta) \cap F(\varrho)} u$.

3. Application

We have applied the problem (7) for the various fixed point problems in three Hilbert spaces as follows:

For every $i = 1, 2, 3$, let H_i be a real Hilbert space and C_i be a nonempty closed convex subset of H_i. Let $\vartheta_i : C_i \to C_i$ be a mapping, for all $i = 1, 2, 3$ and let $A_2 : H_1 \to H_2$ and $A_3 : H_2 \to H_3$. The fixed points problem in three Hilbert spaces is meant to find the point

$$\begin{cases} \omega_1^* \in C_1, \text{ such that } \omega_1^* \in F(\vartheta_1) \text{ and} \\ \omega_2^* = A_2 \omega_1^* \in C_2, \text{ such that } \omega_2^* \in F(\vartheta_2) \text{ and} \\ \omega_3^* = A_3 \omega_2^* \in C_3, \text{ such that } \omega_3^* \in F(\vartheta_3). \end{cases} \quad (23)$$

The set of the solutions of (23) is denoted by $\Omega = \{\omega^* = (\omega_1^*, \omega_2^*, \omega_3^*) \in C_1 \times C_2 \times C_3 : \omega_i^* \in F(\vartheta_i)$, for all $i = 1, 2, 3\}$. It is clear that $Var(C, I - T) = F(\vartheta)$, where $\vartheta : C \to C$ is a nonexpansive mapping with $F(\vartheta) \neq \emptyset$. By leveraging Lemma 5 and such knowledge, we have the following results:

Lemma 9. *For every* $i = 1, 2, 3$, *let* H_i *be a real Hilbert spaces and* C_i *be a nonempty closed convex subset of* H_i. *Let* $\vartheta_i : C_i \to C_i$ *be nonexpansive mappings and let* $A_2 : H_1 \to H_2$, $A_3 : H_2 \to H_3$ *be a bounded linear operator with the adjoint operator* A_2^* *and* A_3^*, *respectively. Assume that* $\bar{x}_1 \in C_1$, $A_2 \bar{x}_1 = \bar{x}_2$, $A_3 \bar{x}_2 = \bar{x}_3$ *and* $\Omega \neq \emptyset$. *The following are equivalent:*

(i) $\bar{x} \in \Omega$, *where* $\bar{x} = (\bar{x}_1, \bar{x}_2, \bar{x}_3) \in C_1 \times C_2 \times C_3$.

(ii) $\bar{x}_1 = P_{C_1}(I_1 - \lambda_1(I_1 - \vartheta_1))(\bar{x}_1 - \gamma_2 A_2^*((I_2 - P_{C_2}(I_2 - \lambda_2(I_2 - \vartheta_2)))\bar{x}_2$
$+ \gamma_3 A_3^*(I_3 - P_{C_3}(I - \lambda_3(I_3 - \vartheta_3)))\bar{x}_3)),$

where $I_i : H_i \to H_i$ *is an identity mapping, for all* $i = 1, 2, 3$, $\gamma_2(1 + \gamma_3) \leq \dfrac{1}{L}$, $L = \max\{L_1, L_2\} \leq 1$ *which* L_1, L_2 *are spectral radii of* $A_2 A_2^*$ *and* $A_3 A_3^*$, *respectively,* $\lambda_i \in (0,1)$, *for all* $i = 1, 2, 3$ *and* $\gamma_2, \gamma_3 \geq 0$

Proof. Given $F(\vartheta_i) = Var(C, I_i - \vartheta_i)$ for all $i = 1, 2, 3$; $(I_i - \vartheta_i)$—a $\frac{1}{2}$-inverse strongly monotonic; and Lemma 5, we can summarize the result of Lemma 9. □

As the direct benefits of Lemma 9, we get Corollary 10.

Corollary 10. *For every $i = 1, 2, 3$, let C_i be a closed convex subset of a real Hilbert space H_i. Let $\vartheta_i : C_i \to C_i$ be a nonexpansive mapping and let $A_2 : H_1 \to H_2$, $A_3 : H_2 \to H_3$ be a bounded linear operator with the adjoint operator A_2^* and A_3^*, respectively. Assume that $\overline{x}_1 \in H_1, \overline{x}_2 = A_2 \overline{x}_1, \overline{x}_3 = A_3 \overline{x}_2$. Let $\vartheta, \varrho : C \to C$ be nonspreading and κ-strictly pseudo-nonspreading mappings, respectively. Assume that $\underline{\Omega} \cap F(\vartheta) \cap F(\varrho) \neq \emptyset$ and let the sequence $\{x_n\}$ generated by $u, x_1 \in C$, and*

$$\begin{aligned} x_{n+1} &= \alpha_n u + \beta_n x_n + \gamma_n P_{C_1} \left(I_1 - \lambda_n \left(a(I_1 - \vartheta) + (1-a)(I_1 - \varrho) \right) \right) x_n \\ &+ \delta_n P_{C_1}(I_1 - \lambda_1(I_1 - \vartheta_1))(x_n^1 - \gamma_2 A_2^*((I_2 - P_{C_2}(I - \lambda_2(I_2 - \vartheta_2)))x_n^2 \\ &+ \gamma_3 A_3^*(I_3 - P_{C_3}(I - \lambda_3(I_3 - \vartheta_3)))x_n^3)), \end{aligned}$$

for all $n \geq 1$ and $a \in (0,1)$, $I_i : H_i \to H_i$ is an identity mapping, for all $i = 1, 2, 3$, where $\{\alpha_n\}, \{\beta_n\}, \{\gamma_n\}, \{\delta_n\} \subseteq [0,1]$ and $\alpha_n + \beta_n + \gamma_n + \delta_n = 1$ and $x_n = x_n^1, x_n^2 = A_2 x_n^1, x_n^3 = A_3 x_n^2$, for all $n \in \mathbb{N}, 0 < \lambda_i < 1$, for all $i = 1, 2, 3$ and $\gamma_j > 0$, for all $j = 2, 3$. Suppose that the conditions (i) to (v) are true;

(i) $\lim_{n \to +\infty} \alpha_n = 0, \sum_{n=1}^{+\infty} \alpha_n = +\infty$;

(ii) $\gamma_2(1 + \gamma_3) \leq \frac{1}{L}$, *where* $L = \max\{L_{A_1}, L_{A_2}\} \leq 1$, *where* L_{A_1}, L_{A_2} *are spectral radii of* $A_2 A_2^*, A_3 A_3^*$, *respectively;*

(iii) $0 < a \leq \beta_n, \gamma_n, \delta_n \leq b < 1$, *for some* $a, b > 0$, *for all* $n \in \mathbb{N}$;

(iv) $\sum_{n=1}^{+\infty} \lambda_n < +\infty$ *and* $0 < \lambda_n < 1 - \kappa$, *for all* $n \in \mathbb{N}$;

(v) $\sum_{n=1}^{+\infty} |\alpha_{n+i} - \alpha_n|, \sum_{n=1}^{+\infty} |\beta_{n+i} - \beta_n|, \sum_{n=1}^{+\infty} |\gamma_{n+1} - \gamma_n| < +\infty$.

Then the sequence $\{x_n\}$ converses strongly to $z_0 = P_{\underline{\Omega} \cap F(T) \cap F(S)} u$.

4. Conclusions

We have proposed a new split variational inequality in three Hilbert spaces. The convergence theorem for finding a common element of the set of solutions of such problems and the sets of fixed-points of discontinuous mappings are proved.

Author Contributions: Investigation, writing and editing by A.K. and C.C. All authors have read and agreed to the published version of the manuscript.

Funding: This research received no external funding.

Acknowledgments: This paper is supported by the Faculty of Science, King Mongkut's Institute of Technology Ladkrabang, Bangkok, Thailand.

Conflicts of Interest: The authors declare no conflict of interest.

References

1. Kinderlehrer, D.; Stampaccia, G. *An Iteration to Variational Inequalities and Their Applications*; Academic Press: New York, NY, USA, 1990.
2. Lions, J.L.; Stampacchia, G. Variational inequalities. *Comm. Pure Appl. Math.* **1967**, *20*, 493–517. [CrossRef]

3. Kangtunyakarn, A. The variational inequality problem in Hilbert spaces endowed with graphs. *J. Fixed Point Theory Appl.* **2020**, 22. [CrossRef]
4. Censor, Y.; Gibali, A.; Reich, S. Algorithms for the split variational inequality problem. *Numer. Algorithms* **2012**, *59*, 301–323. [CrossRef]
5. Dušan Đ.; Paunovic, L.; Radenovic, S. Convergence of iterates with errors of uniformly quasi-Lipschitzian mappings in cone metric spaces. *Kragujev. J. Math.* **2011**, *35*, 399–410.
6. Kazmi, K.R.; Rizvi, S.H. An iterative method for split variational inclusion problem and fixed point problem for a nonexpansive mapping. *Optim. Lett.* **2013**. [CrossRef]
7. Chuang, C.-S. Strong convergence theorems for the split variational inclusion problem in Hilbert spaces. *Fixed Point Theory Appl.* **2013**, *2013*, 350. [CrossRef]
8. Ćirić, L.; Rafiq, A.; Radenović, S.; Rajović, M.; Ume, J.S. On Mann implicit iterations for strongly accreative and strongly pseudo-contractive mappings. *Appl. Math. And Comput.* **2008**, *198*, 128–137.
9. Kohsaka, F.; Takahashi, W. Fixed point theorems for a class of nonlinear mappings related to maximal monotone operators in Banach spaces. *Arch. Math.* **2008**, *91*, 166–177. [CrossRef]
10. Browder, F.E.; Petryshyn, W.V. Construction of fixed points of nonlinear mappings in Hilbert spaces. *J. Math. Anal. Appl.* **1967**, *20*, 197–228. [CrossRef]
11. Osilike, M.O.; Isiogugu, F.O. Weak and strong convergence theorems for nonspreading-type mappings in Hilbert spaces. *Non. Anal.* **2011**, *74*, 1814–1822. [CrossRef]
12. Bnouhachem, A. A modified projection method for a common solution of a system of variational inequalities, a split equilibrium problem and a hierarchical fixed-point problem. *Fixed Point Theory Appl.* **2014**, 22. [CrossRef]
13. Chidume, C.E.; Nnakwe, M.O. Iterative algorithms for split variational Inequalities and generalized split feasibility problems with applications. *J. Nonlinear Var. Anal.* **2019**, *3*, 127–140.
14. Cheng, Q.; Srivastava, R.; Yuan, Q. Hybrid iterative methods for multiple sets split feasibility problems. *Appl. Set-Valued Anal. Optim.* **2019**, *1*, 135–147.
15. Moudafi, A. A relaxed alternating CQ algorithm for convex feasibility problems. *Nonlinear Anal.* **2013**, *79*, 117–121. [CrossRef]
16. Moudafi, A. Split monotone variational inclusions. *J. Optim. Theory Appl.* **2010**, *150*, 275–283. [CrossRef]
17. Ceng, L.C. Approximation of common solutions of a split inclusion problem and a fixed-point problem. *J. Appl. Numer. Optim.* **2019**, *1*, 1–12.
18. Tian, D.; Jiang, L.; Shi, L.; Chen, R. Split equality fixed point problems of quasinonexpansive operators in Hilbert spaces. *J. Nonlinear Funct. Anal.* **2019**, *2019*, 11.
19. Todorčević, V. *Harmonic Quasiconformal Mappings and Hyperbolic Type Metrics*; Springer Nature Switzerland AG: Cham, Switzerland, 2019.
20. Takahashi, W. *Nonlinear Functional Analysis*; Yokohama Publisher: Yokohama, Japan, 2000.
21. Zhou, H. *Fixed Points of Nonlinear Operators Iterative Methods*; De Gruyter: Berlin, Germany, 2020.
22. Xu, H.K. An iterative approach to quadratic optimization. *J. Optim. Theory Appl.* **2003**, *116*, 659–678. [CrossRef]
23. Xu, H.K. Iterative algorithms for nonlinear opearators. *J. Lond. Math. Soc.* **2002**, *66*, 240–256. [CrossRef]

© 2020 by the authors. Licensee MDPI, Basel, Switzerland. This article is an open access article distributed under the terms and conditions of the Creative Commons Attribution (CC BY) license (http://creativecommons.org/licenses/by/4.0/).

Article

A General Inertial Projection-Type Algorithm for Solving Equilibrium Problem in Hilbert Spaces with Applications in Fixed-Point Problems

Nopparat Wairojjana [1], Habib ur Rehman [2], Manuel De la Sen [3,*] and Nuttapol Pakkaranang [2]

1. Applied Mathematics Program, Faculty of Science and Technology,
Valaya Alongkorn Rajabhat University under the Royal Patronage (VRU), 1 Moo 20 Phaholyothin Road, Klong Neung, Klong Luang, Pathumthani 13180, Thailand; nopparat@vru.ac.th
2. KMUTTFixed Point Research Laboratory, KMUTT-Fixed Point Theory and Applications Research Group, SCL 802 Fixed Point Laboratory, Department of Mathematics, Faculty of Science,
King Mongkut's University of Technology Thonburi (KMUTT), 126 Pracha-Uthit Road, Bang Mod, Thrung Khru, Bangkok 10140, Thailand; habib.rehman@mail.kmutt.ac.th (H.u.R.);
nuttapol.pak@mail.kmutt.ac.th (N.P.)
3. Institute of Research and Development of Processes IIDP, University of the Basque Country, Leioa 48940, Spain
* Correspondence: manuel.delasen@ehu.eus; Tel.:+34-94-601-2548

Received: 30 July 2020; Accepted: 28 August 2020; Published: 31 August 2020

Abstract: A plethora of applications from mathematical programming, such as minimax, and mathematical programming, penalization, fixed point to mention a few can be framed as equilibrium problems. Most of the techniques for solving such problems involve iterative methods that is why, in this paper, we introduced a new extragradient-like method to solve equilibrium problems in real Hilbert spaces with a Lipschitz-type condition on a bifunction. The advantage of a method is a variable stepsize formula that is updated on each iteration based on the previous iterations. The method also operates without the previous information of the Lipschitz-type constants. The weak convergence of the method is established by taking mild conditions on a bifunction. For application, fixed-point theorems that involve strict pseudocontraction and results for pseudomonotone variational inequalities are studied. We have reported various numerical results to show the numerical behaviour of the proposed method and correlate it with existing ones.

Keywords: convex optimization; pseudomonotone bifunction; equilibrium problems; variational inequality problems; weak convergence; fixed point problems

1. Introduction

For a nonempty, closed and convex subset $\mathcal{K}$ of a real Hilbert space $\mathcal{E}$ and $f : \mathcal{E} \times \mathcal{E} \to \mathcal{R}$ is a bifunction with $f(p_1, p_1) = 0$, for each $p_1 \in \mathcal{K}$. A equilibrium problem [1,2] for f on the set $\mathcal{K}$ is defined in the following way:

$$\text{Find } \wp^* \in \mathcal{K} \text{ such that } f(\wp^*, p_1) \geq 0, \ \forall p_1 \in \mathcal{K}. \tag{1}$$

The problem (1) is very general, it includes many problems, such as fixed point problems, variational inequalities problems, the optimization problems, the Nash equilibrium of non-cooperative games, the complementarity problems, the saddle point problems, and the vector optimization problem (for further details see [1,3,4]). The equilibrium problem is also considered as the famous Ky Fan inequality [2]. This above-defined particular format of an equilibrium problem (1) is initiated by Muu and Oettli [5] in 1992 and further investigation on its theoretical properties studied by Blum and

Oettli [1]. The construction of new optimization-based methods and the modification and extension of existing methods, as well as the examination of their convergence analysis, is an important research direction in equilibrium problem theory. Many methods have been developed over the last few years to numerically solve the equilibrium problems in both finite and infinite dimensional Hilbert spaces, i.e., the extragradient algorithms [6–14] subgradient algorithms [15–21] inertial methods [22–25], and others in [26–34].

In particular, a proximal method [35] is an efficient way to solve equilibrium problems that are equivalent to solving minimization problems on each step. This approach is also considered as the two-step extragradient-like method in [6], because of the early contribution of the Korpelevich [36] extragradient method to solve the saddle point problems. More precisely, Tran et al. introduced a method in [6], in which an iterative sequence $\{u_{n+1}\}$ was generated in the following manner:

$$\begin{cases} u_n \in \mathcal{K}, \\ v_n = \arg\min\{\xi f(u_n, y) + \frac{1}{2}\|u_n - y\|^2 : y \in \mathcal{K}\}, \\ u_{n+1} = \arg\min\{\xi f(v_n, y) + \frac{1}{2}\|u_n - y\|^2 : y \in \mathcal{K}\}, \end{cases}$$

where $0 < \xi < \min\left\{\frac{1}{2k_1}, \frac{1}{2k_2}\right\}$ and k_1, k_2 are Lipschitz constants. Moreover, $\arg\min_{y \in \mathcal{K}} f(x)$ is the value of x in set $\mathcal{K}$ for which $f(x)$ attains it's minimum. The iterative sequence generated from the above-described method provides a weak convergent iterative sequence and in order to operate it, previous knowledge of the Lipschitz-like constants are required. These Lipschitz-type constants are normally unknown or hard to evaluate. In order to overcome this situation, Hieu et al. [12] introduced an extension of the method in [37] to solve the problems of equilibrium in the following manner: let $[t]_+ := \max\{t, 0\}$ and choose $u_0 \in \mathcal{K}$, $\mu \in (0, 1)$ with $\xi_0 > 0$, such that

$$\begin{cases} v_n = \arg\min\{\xi_n f(u_n, y) + \frac{1}{2}\|u_n - y\|^2 : y \in \mathcal{K}\}, \\ u_{n+1} = \arg\min\{\xi_n f(v_n, y) + \frac{1}{2}\|u_n - y\|^2 : y \in \mathcal{K}\}, \end{cases}$$

where the stepsize sequence $\{\xi_n\}$ is updated in the following way:

$$\xi_{n+1} = \min\left\{\xi_n, \frac{\mu(\|u_n - v_n\|^2 + \|u_{n+1} - v_n\|^2)}{2[f(u_n, u_{n+1}) - f(u_n, v_n) - f(v_n, u_{n+1})]_+}\right\}.$$

Recently, Vinh and Muu proposed an inertial iterative algorithm in [38] to solve a pseudomonotone equilibrium problem. The key contribution is an inertial factor in the method that used to enhance the convergence speed of the iterative sequence. The iterative sequence $\{u_n\}$ was defined in the following manner:

(i) Choose $u_{-1}, u_0 \in \mathcal{K}$, $\theta \in [0, 1)$, $0 < \xi < \min\{\frac{1}{2k_1}, \frac{1}{2k_2}\}$ where a sequence $\{\rho_n\} \subset [0, +\infty)$ is satisfies the following conditions:

$$\sum_{n=0}^{+\infty} \rho_n < +\infty. \tag{2}$$

(ii) Choose θ_n satisfying $0 \leq \theta_n \leq \bar{\theta}_n$ and

$$\bar{\theta}_n = \begin{cases} \min\left\{\theta, \frac{\rho_n}{\|u_n - u_{n-1}\|}\right\} & \text{if } u_n \neq u_{n-1}, \\ \theta & \text{else.} \end{cases} \tag{3}$$

(iii) Compute
$$\begin{cases} \varrho_n = u_n + \theta_n(u_n - u_{n-1}), \\ v_n = \arg\min\{\xi f(\varrho_n, y) + \frac{1}{2}\|\varrho_n - y\|^2 : y \in \mathcal{K}\}, \\ u_{n+1} = \arg\min\{\xi f(v_n, y) + \frac{1}{2}\|\varrho_n - y\|^2 : y \in \mathcal{K}\}. \end{cases}$$

Recently, another efficient inertial algorithm proposed by Hieu et al. in [39] as follows: let $u_{n-1}, u_n, v_n \in \mathcal{K}$, $\theta \in [0,1)$, $0 < \xi \leq \frac{1}{2k_2 + 8k_1}$ and the sequence $\{u_n\}$ was defined in the following manner:

$$\begin{cases} \varrho_n = u_n + \theta(u_n - u_{n-1}), \\ u_{n+1} = \arg\min\{\xi f(v_n, y) + \frac{1}{2}\|\varrho_n - y\|^2 : y \in \mathcal{K}\}, \\ \varrho_{n+1} = u_{n+1} + \theta(u_{n+1} - u_n), \\ v_{n+1} = \arg\min\{\xi f(v_n, y) + \frac{1}{2}\|\varrho_{n+1} - y\|^2 : y \in \mathcal{K}\}. \end{cases}$$

In this article, we concentrates on projection methods that are normally well-established and easy to execute due to their efficient numerical computation. Motivated by the works of [12,38], we formulate an inertial explicit subgradient extragradient method to solve the pseudomonotone equilibrium problem. These results can be seen as the modification of the methods appeared in [6,12,38,39]. Under certain mild conditions, a weak convergence theorem is proved regarding the iterative sequence of the algorithm. Moreover, experimental studies have documented that the designed method tends to be more efficient when compared to the existing methods that are presented in [38,39].

The remainder of the paper has been arranged, as follows: Section 2 contains the elementary results used in this paper. Section 3 contains our main algorithm and proves their convergence. Sections 4 and 5 incorporate the applications of our main results. Section 6 carries out the numerical results that prove the computational effectiveness of our suggested method.

2. Preliminaries

Assume that $h : \mathcal{K} \to \mathcal{R}$ be a convex function on a nonempty, closed and convex subset $\mathcal{K}$ of a real Hilbert space $\mathcal{E}$ and subdifferential of a function h at $p_1 \in \mathcal{K}$ is defined by

$$\partial h(p_1) = \{p_3 \in \mathcal{E} : h(p_2) - h(p_1) \geq \langle p_3, p_2 - p_1 \rangle, \forall p_2 \in \mathcal{K}\}.$$

Assume that $\mathcal{K}$ be a nonempty, closed and convex subset of a real Hilbert space $\mathcal{E}$ and Normal cone of $\mathcal{K}$ at $p_1 \in \mathcal{K}$ is defined by

$$N_\mathcal{K}(p_1) = \{p_3 \in \mathcal{E} : \langle p_3, p_2 - p_1 \rangle \leq 0, \forall p_2 \in \mathcal{K}\}.$$

A metric projection $P_\mathcal{K}(p_1)$ for $p_1 \in \mathcal{E}$ onto a closed and convex subset $\mathcal{K}$ of $\mathcal{E}$ is defined by

$$P_\mathcal{K}(p_1) = \arg\min\{\|p_2 - p_1\| : p_2 \in \mathcal{K}\}.$$

Now, consider the following definitions of monotonicity a bifunction (see for details [1,40]). Assume that $f : \mathcal{E} \times \mathcal{E} \to \mathcal{R}$ on $\mathcal{K}$ for $\gamma > 0$ is said to be

(1) γ-strongly monotone if
$$f(p_1, p_2) + f(p_2, p_1) \leq -\gamma\|p_1 - p_2\|^2, \forall p_1, p_2 \in \mathcal{K};$$

(2) monotone if
$$f(p_1, p_2) + f(p_2, p_1) \leq 0, \forall p_1, p_2 \in \mathcal{K};$$

(3)   *γ-strongly pseudomonotone* if

$$f(p_1, p_2) \geq 0 \implies f(p_2, p_1) \leq -\gamma\|p_1 - p_2\|^2, \ \forall p_1, p_2 \in \mathcal{K};$$

(4)   *pseudomonotone* if

$$f(p_1, p_2) \geq 0 \implies f(p_2, p_1) \leq 0, \ \forall p_1, p_2 \in \mathcal{K}.$$

We have the following implications from the above definitions:

$$(1) \longrightarrow (2) \implies (4) \text{ and } (1) \implies (3) \implies (4).$$

In general, the converses are not true. Suppose that $f : \mathcal{E} \times \mathcal{E} \to \mathcal{R}$ satisfy the Lipschitz-type condition [41] on a set $\mathcal{K}$ if there exist two constants $k_1, k_2 > 0$, such that

$$f(p_1, p_2) + f(p_2, p_3) + k_1\|p_1 - p_2\|^2 + k_2\|p_2 - p_3\|^2 \geq f(p_1, p_3), \ \forall p_1, p_2, p_3 \in \mathcal{K}.$$

Lemma 1 ([42]). *Suppose $\mathcal{K}$ be a nonempty, closed and convex subset of $\mathcal{E}$ and $P_\mathcal{K} : \mathcal{E} \to \mathcal{K}$ is metric projection from $\mathcal{E}$ onto $\mathcal{K}$.*

(i)   *Let $p_1 \in \mathcal{K}$ and $p_2 \in \mathcal{E}$, we have*

$$\|p_1 - P_\mathcal{K}(p_2)\|^2 + \|P_\mathcal{K}(p_2) - p_2\|^2 \leq \|p_1 - p_2\|^2.$$

(ii) $p_3 = P_\mathcal{K}(p_1)$ *if and only if*

$$\langle p_1 - p_3, p_2 - p_3 \rangle \leq 0, \ \forall p_2 \in \mathcal{K}.$$

(iii)   *For any $p_2 \in \mathcal{K}$ and $p_1 \in \mathcal{E}$*

$$\|p_1 - P_\mathcal{K}(p_1)\| \leq \|p_1 - p_2\|.$$

Lemma 2 ([43,44]). *Assume that $h : \mathcal{K} \to \mathcal{R}$ be a convex, lower semicontinuous and subdifferentiable function on $\mathcal{K}$, where $\mathcal{K}$ is a nonempty, convex and closed subset of a Hilbert space $\mathcal{E}$. Subsequently, $p_1 \in \mathcal{K}$ is minimizer of a function h if and only if $0 \in \partial h(p_1) + N_\mathcal{K}(p_1)$, where $\partial h(p_1)$ and $N_\mathcal{K}(p_1)$ denotes the subdifferential of h at $p_1 \in \mathcal{K}$ and the normal cone of $\mathcal{K}$ at p_1, respectively.*

Lemma 3 ([45]). *Let $\{u_n\}$ be a sequence in $\mathcal{E}$ and $\mathcal{K} \subset \mathcal{E}$, such that the following conditions are satisfied:*

(i)   *for every $u \in \mathcal{K}$, the $\lim_{n \to \infty} \|u_n - u\|$ exists;*
(ii)   *each sequentially weak cluster limit point of the sequence $\{u_n\}$ belongs to $\mathcal{K}$.*

Then, $\{u_n\}$ weakly converge to some element in $\mathcal{K}$.

Lemma 4 ([46]). *Let $\{q_n\}$ and $\{p_n\}$ be sequences of non-negative real numbers satisfying $q_{n+1} \leq q_n + p_n$, for each $n \in \mathcal{N}$. If $\sum p_n < \infty$, then $\lim_{n \to \infty} q_n$ exists.*

Lemma 5 ([47]). *For every $p_1, p_2 \in \mathcal{E}$ and $\zeta \in \mathcal{R}$, then*

$$\|\zeta p_1 + (1 - \zeta)p_2\|^2 = \zeta\|p_1\|^2 + (1 - \zeta)\|p_2\|^2 - \zeta(1 - \zeta)\|p_1 - p_2\|^2.$$

Suppose that bifunction f satisfies the following conditions:

(f1) f is pseudomonotone on $\mathcal{K}$ and $f(p_2, p_2) = 0$, for every $p_2 \in \mathcal{K}$;
(f2) f satisfies the Lipschitz-type condition on $\mathcal{E}$ with constants $k_1 > 0$ and $k_2 > 0$;
(f3) $\limsup\limits_{n \to \infty} f(p_n, v) \leq f(p^*, v)$ for every $v \in \mathcal{K}$ and $\{p_n\} \subset \mathcal{K}$ satisfying $p_n \rightharpoonup p^*$;
(f4) $f(p_1, .)$ needs to be convex and subdifferentiable on $\mathcal{E}$ for all $p_1 \in \mathcal{E}$.

3. The Modified Extragradient Algorithm for the Problem (1) and Its Convergence Analysis

We provide a method consisting of two strongly convex minimization problems with an inertial term and an explicit stepsize formula that are being used to enhance the convergence rate of the iterative sequence and to make the algorithm independent of the Lipschitz constants. For the sake of simplicity in the presentation, we will use the notation $[t]_+ = \max\{0, t\}$ and follow the conventions $\frac{0}{0} = +\infty$ and $\frac{a}{0} = +\infty$ ($a \neq 0$). The detailed method is provided below (Algorithm 1):

Algorithm 1 (Modified Extragradient Algorithm for the Problem (1))

Initialization: Choose $u_{-1}, u_0 \in \mathcal{K}$, $\mu \in (0,1)$, $\beta_n \in (0,1]$, $\theta \in [0,1)$ and $\{\rho_n\} \subset [0, +\infty)$ satisfying

$$\sum_{n=0}^{+\infty} \rho_n < +\infty. \quad (4)$$

Iterative steps: Choose θ_n satisfying $0 \leq \theta_n \leq \bar{\theta}_n$ and

$$\bar{\theta}_n = \begin{cases} \min\left\{\theta, \frac{\rho_n}{\|u_n - u_{n-1}\|}\right\} & \text{if } u_n \neq u_{n-1}, \\ \theta & \text{else.} \end{cases} \quad (5)$$

Step 1: Compute

$$v_n = \arg\min_{y \in \mathcal{K}} \{\zeta_n f(\varrho_n, y) + \frac{1}{2}\|\varrho_n - y\|^2\},$$

where $\varrho_n = u_n + \theta_n(u_n - u_{n-1})$. If $\varrho_n = v_n$; STOP. Else, go to next step.

Step 2: Compute $u_{n+1} = (1 - \beta_n)\varrho_n + \beta_n z_n$, where

$$z_n = \arg\min_{y \in \mathcal{K}} \{\zeta_n f(v_n, y) + \frac{1}{2}\|\varrho_n - y\|^2\}.$$

Step 3: Update the stepsize in the following manner:

$$\zeta_{n+1} = \min\left\{\zeta_n, \frac{\mu\|\varrho_n - v_n\|^2 + \mu\|z_n - v_n\|^2}{2[f(\varrho_n, z_n) - f(\varrho_n, v_n) - f(v_n, z_n)]_+}\right\}.$$

Put $n := n + 1$ and return to **Iterative steps**.

Lemma 6. *The sequence $\{\zeta_n\}$ is monotonically decreasing with a lower bound $\min\{\frac{\mu}{2\max\{k_1,k_2\}}, \zeta_0\}$ and it converges to $\zeta > 0$.*

Proof. From the definition of sequence $\{\zeta_n\}$ implies that sequence $\{\zeta_n\}$ decreasing monotonically. It is given that f satisfy the Lipschitz-type condition with k_1 and k_2. Let $f(\varrho_n, z_n) - f(\varrho_n, v_n) - f(v_n, z_n) > 0$, such that

$$\frac{\mu(\|\varrho_n - v_n\|^2 + \|z_n - v_n\|^2)}{2[f(\varrho_n, z_n) - f(\varrho_n, v_n) - f(v_n, z_n)]} \geq \frac{\mu(\|\varrho_n - v_n\|^2 + \|z_n - v_n\|^2)}{2[k_1\|\varrho_n - v_n\|^2 + k_2\|z_n - v_n\|^2]}$$

$$\geq \frac{\mu}{2\max\{k_1, k_2\}}. \quad (6)$$

The above implies that $\{\zeta_n\}$ has a lower bound $\min\{\frac{\mu}{2\max\{k_1,k_2\}}, \zeta_0\}$. Moreover, there exists a fixed real number $\zeta > 0$, such that $\lim_{n\to\infty} \zeta_n = \zeta$. □

Remark 1. *Because of the summability of $\sum_{n=0}^{+\infty} \rho_n$ and the expression (5) implies that*

$$\sum_{n=1}^{\infty} \theta_n \|u_n - u_{n-1}\| \leq \sum_{n=1}^{\infty} \bar{\theta}_n \|u_n - u_{n-1}\| \leq \sum_{n=1}^{\infty} \theta \|u_n - u_{n-1}\| < \infty, \tag{7}$$

that implies

$$\lim_{n \to \infty} \theta \|u_n - u_{n-1}\| = 0. \tag{8}$$

Lemma 7. *Suppose that $f : \mathcal{E} \times \mathcal{E} \to \mathcal{R}$ be a bifunction satisfies the conditions (f1)–(f4). For each $\wp^* \in EP(f, \mathcal{K}) \neq \emptyset$, we have*

$$\|z_n - \wp^*\|^2 \leq \|\varrho_n - \wp^*\|^2 - \left(1 - \frac{\mu \xi_n}{\xi_{n+1}}\right) \|\varrho_n - v_n\|^2 - \left(1 - \frac{\mu \xi_n}{\xi_{n+1}}\right) \|z_n - v_n\|^2.$$

Proof. From the value of z_n, we have

$$0 \in \partial_2 \left\{ \xi_n f(v_n, y) + \frac{1}{2} \|\varrho_n - y\|^2 \right\}(z_n) + N_{\mathcal{K}}(z_n).$$

For some $\omega \in \partial f(v_n, z_n)$, there exists $\overline{\omega} \in N_{\mathcal{K}}(z_n)$, such that

$$\xi_n \omega + z_n - \varrho_n + \overline{\omega} = 0.$$

The above expression implies that

$$\langle \varrho_n - z_n, y - z_n \rangle = \xi_n \langle \omega, y - z_n \rangle + \langle \overline{\omega}, y - z_n \rangle, \ \forall y \in \mathcal{K}.$$

For given $\overline{\omega} \in N_{\mathcal{K}}(z_n)$, imply that $\langle \overline{\omega}, y - z_n \rangle \leq 0, \forall y \in \mathcal{K}$. It provides that

$$\langle \varrho_n - z_n, y - z_n \rangle \leq \xi_n \langle \omega, y - z_n \rangle, \ \forall y \in \mathcal{K}. \tag{9}$$

From $\omega \in \partial f(v_n, z_n)$, we have

$$f(v_n, y) - f(v_n, z_n) \geq \langle \omega, y - z_n \rangle, \ \forall y \in \mathcal{E}. \tag{10}$$

Combining expressions (9) and (10) we obtain

$$\xi_n f(v_n, y) - \xi_n f(v_n, z_n) \geq \langle \varrho_n - z_n, y - z_n \rangle, \ \forall y \in \mathcal{K}. \tag{11}$$

By substituting $y = \wp^*$ in (11), gives that

$$\xi_n f(v_n, \wp^*) - \xi_n f(v_n, z_n) \geq \langle \varrho_n - z_n, \wp^* - z_n \rangle. \tag{12}$$

Because $f(\wp^*, v_n) \geq 0$, then $f(v_n, \wp^*) \leq 0$, provides that

$$\langle \varrho_n - z_n, z_n - \wp^* \rangle \geq \xi_n f(v_n, z_n). \tag{13}$$

From the formula of ξ_{n+1}, we obtain

$$f(\varrho_n, z_n) - f(\varrho_n, v_n) - f(v_n, z_n) \leq \frac{\mu \|\varrho_n - v_n\|^2 + \mu \|z_n - v_n\|^2}{2 \xi_{n+1}} \tag{14}$$

From the expressions (13) and (14), we have

$$\langle \varrho_n - z_n, z_n - \wp^* \rangle \geq \xi_n \{ f(\varrho_n, z_n) - f(\varrho_n, v_n) \}$$
$$- \frac{\mu \tilde{\xi}_n}{2 \xi_{n+1}} \| \varrho_n - v_n \|^2 - \frac{\mu \tilde{\xi}_n}{2 \xi_{n+1}} \| z_n - v_n \|^2. \tag{15}$$

Similar to expression (11), the value of v_n gives that

$$\xi_n f(\varrho_n, y) - \xi_n f(\varrho_n, v_n) \geq \langle \varrho_n - v_n, y - v_n \rangle, \ \forall y \in \mathcal{K}. \tag{16}$$

By substituting $y = z_n$ in the above expression, we have

$$\xi_n \{ f(\varrho_n, z_n) - f(\varrho_n, v_n) \} \geq \langle \varrho_n - v_n, z_n - v_n \rangle. \tag{17}$$

Combining the expressions (15) and (17), we obtain

$$\langle \varrho_n - z_n, z_n - \wp^* \rangle \geq \langle \varrho_n - v_n, z_n - v_n \rangle$$
$$- \frac{\mu \tilde{\xi}_n}{2 \xi_{n+1}} \| \varrho_n - v_n \|^2 - \frac{\mu \tilde{\xi}_n}{2 \xi_{n+1}} \| z_n - v_n \|^2. \tag{18}$$

We have the given formulas:

$$-2 \langle \varrho_n - z_n, z_n - \wp^* \rangle = - \| \varrho_n - \wp^* \|^2 + \| z_n - \varrho_n \|^2 + \| z_n - \wp^* \|^2.$$

$$2 \langle v_n - \varrho_n, v_n - z_n \rangle = \| \varrho_n - v_n \|^2 + \| z_n - v_n \|^2 - \| \varrho_n - z_n \|^2.$$

The above expressions with (18), we have

$$\| z_n - \wp^* \|^2 \leq \| \varrho_n - \wp^* \|^2 - \left(1 - \frac{\mu \tilde{\xi}_n}{\xi_{n+1}} \right) \| \varrho_n - v_n \|^2 - \left(1 - \frac{\mu \tilde{\xi}_n}{\xi_{n+1}} \right) \| z_n - v_n \|^2.$$

□

Theorem 1. *Assume that $f : \mathcal{E} \times \mathcal{E} \to \mathcal{R}$ be a bifunction satisfies the conditions* (f1)–(f4) *and $\wp^*$ belongs to solution set $EP(f, \mathcal{K})$. Subsequently, the sequences $\{\varrho_n\}, \{v_n\}, \{z_n\}$ and $\{u_n\}$ generated through Algorithm 1 weakly converges to $\wp^*$. In addition, $\lim_{n \to \infty} P_{EP(f, \mathcal{K})}(u_n) = \wp^*$.*

Proof. By value of u_{n+1} through Lemma 5, we obtain

$$\| u_{n+1} - \wp^* \|^2 = \| (1 - \beta_n) \varrho_n + \beta_n z_n - \wp^* \|^2$$
$$= \| (1 - \beta_n)(\varrho_n - \wp^*) + \beta_n (z_n - \wp^*) \|^2$$
$$= (1 - \beta_n) \| \varrho_n - \wp^* \|^2 + \beta_n \| z_n - \wp^* \|^2 - \beta_n (1 - \beta_n) \| \varrho_n - z_n \|^2$$
$$\leq (1 - \beta_n) \| \varrho_n - \wp^* \|^2 + \beta_n \| z_n - \wp^* \|^2. \tag{19}$$

By Lemma 7 and expression (19), we obtain

$$\| u_{n+1} - \wp^* \|^2 \leq \| \varrho_n - \wp^* \|^2$$
$$- \beta_n \left(1 - \frac{\mu \tilde{\xi}_n}{\xi_{n+1}} \right) \| \varrho_n - v_n \|^2 - \beta_n \left(1 - \frac{\mu \tilde{\xi}_n}{\xi_{n+1}} \right) \| z_n - v_n \|^2. \tag{20}$$

Because $\tilde{\xi}_n \to \xi$, then there exists a fixed number $\epsilon \in (0, 1 - \mu)$, such that

$$\lim_{n \to \infty} \left(1 - \frac{\mu \tilde{\xi}_n}{\xi_{n+1}} \right) = 1 - \mu > \epsilon > 0.$$

Subsequently, there exist a fixed real number $N_1 \in \mathcal{N}$ such that

$$\left(1 - \frac{\mu \tilde{\zeta}_n}{\tilde{\zeta}_{n+1}}\right) > \epsilon > 0, \ \forall n \geq N_1. \tag{21}$$

Combining the expressions (20) and (21), we obtain

$$\|u_{n+1} - \wp^*\|^2 \leq \|\varrho_n - \wp^*\|^2, \ \forall n \geq N_1. \tag{22}$$

By definition of the ϱ_n, we have

$$\|\varrho_n - \wp^*\| = \|u_n + \theta_n(u_n - u_{n-1}) - \wp^*\| \leq \|u_n - \wp^*\| + \theta_n \|u_n - u_{n-1}\|. \tag{23}$$

From the definition of ϱ_n in Algorithm 1, we obtain

$$\|\varrho_n - \wp^*\|^2 = \|u_n + \theta_n(u_n - u_{n-1}) - \wp^*\|^2$$
$$= \|(1 + \theta_n)(u_n - \wp^*) - \theta_n(u_{n-1} - \wp^*)\|^2$$
$$= (1 + \theta_n)\|u_n - \wp^*\|^2 - \theta_n\|u_{n-1} - \wp^*\|^2 + \theta_n(1 + \theta_n)\|u_n - u_{n-1}\|^2 \tag{24}$$
$$\leq (1 + \theta_n)\|u_n - \wp^*\|^2 - \theta_n\|u_{n-1} - \wp^*\|^2 + 2\theta\|u_n - u_{n-1}\|^2. \tag{25}$$

The expression (22) can also be written as

$$\|u_{n+1} - \wp^*\| \leq \|u_n - \wp^*\| + \theta\|u_n - u_{n-1}\|, \ \forall n \geq N_1. \tag{26}$$

By using Lemma 4 with expressions (7) and (26), we have

$$\lim_{n \to \infty} \|u_n - \wp^*\| = l, \text{ for some finite } l \geq 0. \tag{27}$$

The equality (8) implies that

$$\lim_{n \to \infty} \|u_n - u_{n-1}\| = 0. \tag{28}$$

By letting $n \to \infty$ in (24) implies that

$$\lim_{n \to \infty} \|\varrho_n - \wp^*\| = l. \tag{29}$$

From the expression (20) and (25), we have

$$\|u_{n+1} - \wp^*\|^2$$
$$\leq (1 + \theta_n)\|u_n - \wp^*\|^2 - \theta_n\|u_{n-1} - \wp^*\|^2 + 2\theta\|u_n - u_{n-1}\|^2$$
$$- \beta_n\left(1 - \frac{\mu \tilde{\zeta}_n}{\tilde{\zeta}_{n+1}}\right)\|\varrho_n - v_n\|^2 - \beta_n\left(1 - \frac{\mu \tilde{\zeta}_n}{\tilde{\zeta}_{n+1}}\right)\|z_n - v_n\|^2, \tag{30}$$

which further implies that (for $n \geq N_1$)

$$\epsilon \beta \|\varrho_n - v_n\|^2 + \epsilon \beta \|v_n - z_n\|^2$$
$$\leq \|u_n - \wp^*\|^2 - \|u_{n+1} - \wp^*\|^2 + \theta_n\left(\|u_n - \wp^*\|^2 - \|u_{n-1} - \wp^*\|^2\right) + 2\theta\|u_n - u_{n-1}\|^2. \tag{31}$$

By letting $n \to \infty$ in (31), we obtain

$$\lim_{n \to \infty} \|\varrho_n - v_n\| = \lim_{n \to \infty} \|v_n - z_n\| = 0. \tag{32}$$

By using the Cauchy inequality and expression (32), we obtain

$$\lim_{n\to\infty} \|\varrho_n - z_n\| \leq \lim_{n\to\infty} \|\varrho_n - v_n\| + \lim_{n\to\infty} \|z_n - v_n\| = 0. \tag{33}$$

The expressions (29) and (32) imply that

$$\lim_{n\to\infty} \|v_n - \wp^*\| = \lim_{n\to\infty} \|z_n - \wp^*\| = l. \tag{34}$$

It follows from the expressions (27), (29) and (34) that the sequences $\{\varrho_n\}$, $\{u_n\}$, $\{v_n\}$ and $\{z_n\}$ are bounded. Now, we need to use Lemma 3, for this it is compulsory to show that any weak sequential limit points of $\{u_n\}$ lies in the set $EP(f, \mathcal{K})$. Consider z to be a weak limit point of $\{u_n\}$ i.e., there is a $\{u_{n_k}\}$ of $\{u_n\}$ that is weakly converges to z. Because $\|u_n - v_n\| \to 0$, then $\{v_{n_k}\}$ also weakly converge to z and so $z \in \mathcal{K}$. Now, it is renaming to show that $z \in EP(f, \mathcal{K})$. From relation (11), due to ξ_{n+1} and (17), we have

$$\begin{aligned}
\xi_{n_k} f(v_{n_k}, y) &\geq \xi_{n_k} f(v_{n_k}, z_{n_k}) + \langle \varrho_{n_k} - z_{n_k}, y - z_{n_k} \rangle \\
&\geq \xi_{n_k} f(\varrho_{n_k}, z_{n_k}) - \xi_{n_k} f(\varrho_{n_k}, v_{n_k}) - \frac{\mu \xi_{n_k}}{2\xi_{n_k+1}} \|\varrho_{n_k} - v_{n_k}\|^2 \\
&\quad - \frac{\mu \xi_{n_k}}{2\xi_{n_k+1}} \|v_{n_k} - z_{n_k}\|^2 + \langle \varrho_{n_k} - z_{n_k}, y - z_{n_k} \rangle \\
&\geq \langle \varrho_{n_k} - v_{n_k}, z_{n_k} - v_{n_k} \rangle - \frac{\mu \xi_{n_k}}{2\xi_{n_k+1}} \|\varrho_{n_k} - v_{n_k}\|^2 \\
&\quad - \frac{\mu \xi_{n_k}}{2\xi_{n_k+1}} \|v_{n_k} - z_{n_k}\|^2 + \langle \varrho_{n_k} - z_{n_k}, y - z_{n_k} \rangle,
\end{aligned} \tag{35}$$

where $y \in \mathcal{K}$. It follows from (28), (32), (33) and the boundedness of $\{u_n\}$ right hand side tend to zero. Due to $\xi_{n_k} > 0$, condition (f3) and $v_{n_k} \rightharpoonup z$, implies

$$0 \leq \limsup_{k\to\infty} f(v_{n_k}, y) \leq f(z, y), \quad \forall y \in \mathcal{K}. \tag{36}$$

Because $z \in \mathcal{K}$ imply that $f(z, y) \geq 0, \forall y \in \mathcal{K}$. It is prove that $z \in EP(f, \mathcal{K})$. By Lemma 3, provides that $\{\varrho_n\}$, $\{v_n\}$, $\{z_n\}$ and $\{u_n\}$ weakly converges to $\wp^*$ as $n \to \infty$.

Finally, to prove that $\lim_{n\to\infty} P_{EP(f, \mathcal{K})}(u_n) = \wp^*$. Let $q_n := P_{EP(f, \mathcal{K})}(u_n)$, $\forall n \in \mathcal{N}$. For any $\wp^* \in EP(f, \mathcal{K})$, we have

$$\|q_n\| \leq \|q_n - u_n\| + \|u_n\| \leq \|\wp^* - u_n\| + \|u_n\|. \tag{37}$$

Clearly, the above implies that sequence $\{q_n\}$ is bounded. Next, we need to show that $\{q_n\}$ is a Cauchy sequence. By using Lemma 1(iii) and (23), we have

$$\|u_{n+1} - q_{n+1}\| \leq \|u_{n+1} - q_n\| \leq \|u_n - q_n\| + \theta\|u_n - u_{n-1}\|, \quad \forall n \geq N_1. \tag{38}$$

Thus, Lemma 4 provides the existence of $\lim_{n\to\infty} \|u_n - q_n\|$. Next, take (23) $\forall \, m > n \geq N_1$, we have

$$\begin{aligned}
\|q_n - u_m\| &\leq \|q_n - u_{m-1}\| + \theta\|u_n - u_{n-1}\| \\
&\leq \cdots \leq \|q_n - u_n\| + \theta \sum_{k=n}^{m-1} \|u_n - u_{n-1}\|.
\end{aligned} \tag{39}$$

Suppose that $q_m, q_n \in EP(f, \mathcal{K})$ for $m > n \geq N_1$, through Lemma 1(i) and (39), we have

$$
\begin{aligned}
\|q_n - q_m\|^2 &\leq \|q_n - u_m\|^2 - \|q_m - u_m\|^2 \\
&\leq \|q_n - u_n\|^2 + \left(\theta \sum_{k=n}^{m-1} \|u_n - u_{n-1}\|\right)^2 + 2\theta \|q_n - u_n\| \sum_{k=n}^{m-1} \|u_n - u_{n-1}\| - \|q_m - u_m\|^2.
\end{aligned}
\tag{40}
$$

The existence of $\lim_{n \to \infty} \|u_n - q_n\|$ and the summability of the series $\sum_n \|u_n - u_{n-1}\| < +\infty$, imply $\lim_{n \to \infty} \|q_n - q_m\| = 0$, $\forall m > n$. As a result, $\{q_n\}$ is a Cauchy sequence and due the closeness of the set $EP(f, \mathcal{K})$ the sequence $\{q_n\}$ strongly converges to $q^* \in EP(f, \mathcal{K})$. Next, remaining to show that $q^* = \wp^*$. From Lemma 1(ii) and $\wp^*, q^* \in EP(f, \mathcal{K})$, we have

$$
\langle u_n - q_n, \wp^* - q_n \rangle \leq 0.
\tag{41}
$$

Because of $q_n \to q^*$ and $u_n \rightharpoonup \wp^*$, we obtain

$$
\langle \wp^* - q^*, \wp^* - q^* \rangle \leq 0,
$$

implies that $\wp^* = q^* = \lim_{n \to \infty} P_{EP(f, \mathcal{K})}(u_n)$. □

4. Applications to Solve Fixed Point Problems

Now, consider the applications of our results that are discussed in Section 3 to solve fixed-point problems involving κ-strict pseudo-contraction. Let $T : \mathcal{K} \to \mathcal{K}$ be a mapping and the fixed point problem is formulated in the following manner:

$$
\text{Find } \wp^* \in \mathcal{K} \text{ such as } T(\wp^*) = \wp^*.
$$

Let a mapping $T : \mathcal{K} \to \mathcal{K}$ is said to be

(i) sequentially weakly continuous on $\mathcal{K}$ if

$$
T(p_n) \rightharpoonup T(p) \text{ for every sequence in } \mathcal{K} \text{ satisfying } p_n \rightharpoonup p \text{ (weakly converges);}
$$

(ii) κ-strict pseudo-contraction [48] on $\mathcal{K}$ if

$$
\|Tp_1 - Tp_2\|^2 \leq \|p_1 - p_2\|^2 + \kappa\|(p_1 - Tp_1) - (p_2 - Tp_2)\|^2, \quad \forall p_1, p_2 \in \mathcal{K};
\tag{42}
$$

that is equivalent to

$$
\langle Tp_1 - Tp_2, p_1 - p_2 \rangle \leq \|p_1 - p_2\|^2 - \frac{1-\kappa}{2}\|(p_1 - Tp_1) - (p_2 - Tp_2)\|^2, \quad \forall p_1, p_2 \in \mathcal{K}.
\tag{43}
$$

Note: if we define $f(p_1, p_2) = \langle p_1 - Tp_1, p_2 - p_1 \rangle$, $\forall p_1, p_2 \in \mathcal{K}$. Then, the problem (1) convert into the fixed point problem with $2k_1 = 2k_2 = \frac{3-2\kappa}{1-\kappa}$. The value of v_n in Algorithm 1 convert into followings:

$$\begin{aligned}
v_n &= \arg\min_{y \in \mathcal{K}} \{\xi_n f(\varrho_n, y) + \frac{1}{2}\|\varrho_n - y\|^2\} \\
&= \arg\min_{y \in \mathcal{K}} \{\xi_n \langle \varrho_n - T(\varrho_n), y - \varrho_n \rangle + \frac{1}{2}\|\varrho_n - y\|^2\} \\
&= \arg\min_{y \in \mathcal{K}} \{\xi_n \langle \varrho_n - T(\varrho_n), y - \varrho_n \rangle + \frac{1}{2}\|\varrho_n - y\|^2 + \frac{\xi_n^2}{2}\|\varrho_n - T(\varrho_n)\|^2 - \frac{\xi_n^2}{2}\|\varrho_n - T(\varrho_n)\|^2\} \\
&= \arg\min_{y \in \mathcal{K}} \{\frac{1}{2}\|y - \varrho_n + \xi_n(\varrho_n - T(\varrho_n))\|^2\} \\
&= P_\mathcal{K}[\varrho_n - \xi_n(\varrho_n - T(\varrho_n))] = P_\mathcal{K}[(1-\xi_n)\varrho_n + \xi_n T(\varrho_n)]. \quad (44)
\end{aligned}$$

In the similar way to the expression (44), we obtain

$$z_n = P_\mathcal{K}[\varrho_n - \xi_n(v_n - T(v_n))]. \quad (45)$$

As a consequence of the results in Section 3, we have the following fixed point theorem:

Corollary 1. *Assume that $T : \mathcal{K} \to \mathcal{K}$ to be a weakly continuous and κ-strict pseudocontraction with $\text{Fix}(T) \neq \emptyset$. The sequences ϱ_n, v_n, z_n and u_n be generated in the following way:*

(i) *Choose $u_{-1}, u_0 \in \mathcal{K}$, $\mu \in (0,1)$, $\beta_n \in (0,1]$, $\theta \in [0,1)$ and $\{\rho_n\} \subset [0, +\infty)$ satisfies the following condition:*

$$\sum_{n=0}^{+\infty} \rho_n < +\infty. \quad (46)$$

(ii) *Choose θ_n satisfies $0 \leq \theta_n \leq \bar{\theta}_n$, such that*

$$\bar{\theta}_n = \begin{cases} \min\left\{\theta, \frac{\rho_n}{\|u_n - u_{n-1}\|}\right\} & \text{if } u_n \neq u_{n-1}, \\ \theta & \text{else.} \end{cases} \quad (47)$$

(iii) *Compute $u_{n+1} = (1-\beta_n)\varrho_n + \beta_n z_n$, where*

$$\begin{cases} \varrho_n = u_n + \theta_n(u_n - u_{n-1}), \\ v_n = P_\mathcal{K}[\varrho_n - \xi_n(\varrho_n - T(\varrho_n))], \\ z_n = P_\mathcal{K}[\varrho_n - \xi_n(v_n - T(v_n))]. \end{cases} \quad (48)$$

(iv) *Revised the stepsize ξ_{n+1} in the following way:*

$$\xi_{n+1} = \min\left\{\xi_n, \frac{\mu\|\varrho_n - v_n\|^2 + \mu\|z_n - v_n\|^2}{2[\langle(\varrho_n - v_n) - (T(\varrho_n) - T(v_n)), z_n - v_n\rangle]_+}\right\}$$

Subsequently, $\{\varrho_n\}, \{v_n\}, \{z_n\}$ and $\{u_n\}$ be the sequences converges weakly to $\wp^* \in \text{Fix}(T)$.

5. Application to Solve Variational Inequality Problems

Now, consider the applications of our results that are discussed in Section 3 in order to solve variational inequality problems involving pseudomonotone and Lipschitz-type continuous operator. Let a operator $L: \mathcal{K} \to \mathcal{K}$ and the variational inequality problem is formulated as follows:

$$\text{Find } \wp^* \in \mathcal{K} \text{ such that } \langle L(\wp^*), y - \wp^* \rangle \geq 0, \ \forall y \in \mathcal{K}.$$

A mapping $L: \mathcal{E} \to \mathcal{E}$ is said to be

(i) *L-Lipschitz continuous* on $\mathcal{K}$ if

$$\|L(p_1) - L(p_2)\| \leq L\|p_1 - p_2\|, \ \forall p_1, p_2 \in \mathcal{K};$$

(ii) *monotone* on $\mathcal{K}$ if

$$\langle L(p_1) - L(p_2), p_1 - p_2 \rangle \geq 0, \ \forall p_1, p_2 \in \mathcal{K};$$

(iii) *pseudomonotone* on $\mathcal{K}$ if

$$\langle L(p_1), p_2 - p_1 \rangle \geq 0 \implies \langle L(p_2), p_1 - p_2 \rangle \leq 0, \ \forall p_1, p_2 \in \mathcal{K}.$$

Note: let $f(p_1, p_2) := \langle L(p_1), p_2 - p_1 \rangle, \forall p_1, p_2 \in \mathcal{K}$. Thus, problem (1) translates into the problem (VIP) with $L = 2k_1 = 2k_2$. From the value of v_n, we have

$$v_n = \underset{y \in \mathcal{K}}{\arg\min} \left\{ \xi_n f(\varrho_n, y) + \frac{1}{2}\|\varrho_n - y\|^2 \right\}$$

$$= \underset{y \in \mathcal{K}}{\arg\min} \left\{ \xi_n \langle L(\varrho_n), y - \varrho_n \rangle + \frac{1}{2}\|\varrho_n - y\|^2 + \frac{\xi_n^2}{2}\|L(\varrho_n)\|^2 - \frac{\xi_n^2}{2}\|L(\varrho_n)\|^2 \right\}$$

$$= \underset{y \in \mathcal{K}}{\arg\min} \left\{ \frac{1}{2}\|y - (\varrho_n - \xi_n L(\varrho_n))\|^2 \right\}$$

$$= P_\mathcal{K}[\varrho_n - \xi_n L(\varrho_n)]. \tag{49}$$

In similar way to the expression (49), we obtain

$$z_n = P_\mathcal{K}[\varrho_n - \xi_n L(v_n)].$$

Suppose that a mapping L satisfies the following conditions:

(L1) L is monotone on $\mathcal{K}$ with $VI(L, \mathcal{K}) \neq \varnothing$;
(L2) L is L-Lipschitz continuous on $\mathcal{K}$ with $L > 0$;
(L3) L is pseudomonotone on $\mathcal{K}$ with $VI(L, \mathcal{K}) \neq \varnothing$; and,
(L4) $\limsup_{n \to \infty} \langle L(p_n), p - p_n \rangle \leq \langle L(p), y - p \rangle, \forall y \in \mathcal{K}$ and $\{p_n\} \subset \mathcal{K}$ satisfying $p_n \rightharpoonup p$.

Next, let L to be monotone and (L4) can be removed. The condition (L4) is used to defined $f(u, v) = \langle L(u), v - u \rangle$ and satisfy the conditions (L4). The condition (f3) is required to show $z \in EP(f, \mathcal{K})$ see (36). The condition (L4) is required to show $z \in VI(L, \mathcal{K})$. Further, to show that $z \in VI(L, \mathcal{K})$. By letting the monotonicity of operator L, we have

$$\langle L(y), y - v_n \rangle \geq \langle L(v_n), y - v_n \rangle, \ \forall y \in \mathcal{K}. \tag{50}$$

By letting $f(u, v) = \langle L(u), v - u \rangle$ with expression (35), implies that

$$\limsup_{k \to \infty} \langle L(v_{n_k}), y - v_{n_k} \rangle \geq 0, \ \forall y \in \mathcal{K}. \tag{51}$$

Combining (50) with (51), we deduce that

$$\limsup_{k\to\infty}\langle L(y), y - v_{n_k}\rangle \geq 0, \ \forall y \in \mathcal{K}. \tag{52}$$

Therefore, $v_{n_k} \rightharpoonup z \in \mathcal{K}$, provides $\langle L(y), y - z\rangle \geq 0, \ \forall y \in \mathcal{K}$. Let $v_t = (1-t)z + ty, \ \forall t \in [0,1]$. Since $v_t \in \mathcal{K}$ for $t \in (0,1)$, we have

$$0 \leq \langle L(v_t), v_t - z\rangle = t\langle L(v_t), y - z\rangle. \tag{53}$$

That is $\langle L(v_t), y - z\rangle \geq 0$ every $t \in (0,1)$. Due to $v_t \to z$, while $t \to 0$, we have $\langle L(z), y - z\rangle \geq 0$, for all $y \in \mathcal{K}$, consequently $z \in VI(L, \mathcal{K})$.

Corollary 2. *Let $L : \mathcal{K} \to \mathcal{E}$ be a mapping and satisfying the conditions (L1)–(L2). Assume that the sequences $\{\varrho_n\}, \{v_n\}, \{z_n\}$ and $\{u_n\}$ generated in the following manner:*

(i) *Choose $u_{-1}, u_0 \in \mathcal{K}, \mu \in (0,1), \beta_n \in (0,1], \theta \in [0,1)$ and $\{\rho_n\} \subset [0,+\infty)$, such that*

$$\sum_{n=0}^{+\infty} \rho_n < +\infty. \tag{54}$$

(ii) *Let θ_n satisfies $0 \leq \theta_n \leq \bar{\theta}_n$ and*

$$\bar{\theta}_n = \begin{cases} \min\left\{\theta, \frac{\rho_n}{\|u_n - u_{n-1}\|}\right\} & \text{if } u_n \neq u_{n-1}, \\ \theta & \text{otherwise.} \end{cases} \tag{55}$$

(iii) *Compute $u_{n+1} = (1 - \beta_n)\varrho_n + \beta_n z_n$, where*

$$\begin{cases} \varrho_n = u_n + \theta_n(u_n - u_{n-1}), \\ v_n = P_\mathcal{K}[\varrho_n - \xi_n L(\varrho_n)], \\ z_n = P_\mathcal{K}[\varrho_n - \xi_n L(v_n)]. \end{cases} \tag{56}$$

(iv) *Stepsize ξ_{n+1} is revised in the following way:*

$$\xi_{n+1} = \min\left\{\xi_n, \frac{\mu\|\varrho_n - v_n\|^2 + \mu\|z_n - v_n\|^2}{2[\langle L(\varrho_n) - L(v_n), z_n - v_n\rangle]_+}\right\}$$

Subsequently, the sequences $\{\varrho_n\}, \{v_n\}, \{z_n\}$ and $\{z_n\}$ converge weakly to $\wp^* \in VI(L, \mathcal{K})$.

Corollary 3. *Let $L : \mathcal{K} \to \mathcal{E}$ be a mapping and satisfying the conditions (L2)–(L4). Assume that the sequences $\{\varrho_n\}, \{v_n\}, \{z_n\}$ and $\{u_n\}$ generated in the following manner:*

(i) *Choose $u_{-1}, u_0 \in \mathcal{K}, \mu \in (0,1), \beta_n \in (0,1], \theta \in [0,1)$ and $\{\rho_n\} \subset [0,+\infty)$, such that*

$$\sum_{n=0}^{+\infty} \rho_n < +\infty. \tag{57}$$

(ii) *Choose θ_n satisfying $0 \leq \theta_n \leq \bar{\theta}_n$, such that*

$$\bar{\theta}_n = \begin{cases} \min\left\{\theta, \frac{\rho_n}{\|u_n - u_{n-1}\|}\right\} & \text{if } u_n \neq u_{n-1}, \\ \theta & \text{else.} \end{cases} \tag{58}$$

(iii) Compute $u_{n+1} = (1 - \beta_n)\varrho_n + \beta_n z_n$, where

$$\begin{cases} \varrho_n = u_n + \theta_n(u_n - u_{n-1}), \\ v_n = P_{\mathcal{K}}[\varrho_n - \xi_n L(\varrho_n)], \\ z_n = P_{\mathcal{K}}[\varrho_n - \xi_n L(v_n)]. \end{cases} \quad (59)$$

(iv) The stepsize ξ_{n+1} is updated in the following way:

$$\xi_{n+1} = \min\left\{\xi_n, \frac{\mu\|\varrho_n - v_n\|^2 + \mu\|z_n - v_n\|^2}{2[\langle L(\varrho_n) - L(v_n), z_n - v_n\rangle]_+}\right\}$$

Subsequently, the sequences $\{\varrho_n\}$, $\{v_n\}$, $\{z_n\}$ and $\{z_n\}$ converge weakly to $\wp^* \in VI(L, \mathcal{K})$.

6. Numerical Experiments

The computational results present this section to prove the effectiveness of Algorithm 1 when compared to Algorithm 3.1 in [39] and Algorithm 1 in [38].

(i) For Algorithm 3.1 (Alg3.1) in [39]:

$$\xi = \frac{1}{10\max\{k_1, k_2\}}, \; \theta = \frac{1}{2}, \; \text{Error term } (D_n) = \max\{\|u_{n+1} - v_n\|^2, \|u_{n+1} - \varrho_n\|^2\}.$$

(ii) For Algorithm 1 (Alg1) in [38]:

$$\xi = \frac{1}{4\max\{k_1, k_2\}}, \; \theta = \frac{1}{2}, \; \rho_n = \frac{1}{n^2}, \; \text{Error term } (D_n) = \|\varrho_n - v_n\|^2.$$

(iii) For Algorithm 1 (mAlg1):

$$\xi = \frac{1}{2}, \; \theta = \frac{1}{2}, \; \mu = \frac{1}{3}, \; \rho_n = \frac{1}{n^2}, \; \beta_n = \frac{8}{10}, \; \text{Error term } (D_n) = \|\varrho_n - v_n\|^2.$$

Example 1. *Let take the Nash–Cournot Equilibrium Model that found in the paper [6]. A bifunction f consider into the following form:*

$$f(p_1, p_2) = \langle Pp_1 + Qp_2 + q, p_2 - p_1\rangle,$$

where $q \in \mathcal{R}^m$ with matrices P, Q of order m and Lipschitz constants are $k_1 = k_2 = \frac{1}{2}\|P - Q\|$ (see for more details [6]). In our case, P, Q are taken at random (choose diagonal matrices A_1 and A_2 randomly entries from $[0, 2]$ and $[-2, 0]$, respectively. Two random orthogonal matrices B_1 and B_2 provide positive semidefinite matrix $M_1 = B_1 A_1 B_1^T$ and negative semidefinite matrix $M_2 = B_2 A_2 B_2^T$. Finally, set $Q = M_1 + M_1^T$, $S = M_2 + M_2^T$ and $P = Q - S$.) and elements of q are taken arbitrary form $[-1, 1]$. A set $\mathcal{K} \subset \mathcal{R}^m$ is taken as

$$\mathcal{K} := \{u \in \mathcal{R}^m : -10 \leq u_i \leq 10\}.$$

Tables 1 and 2 and Figures 1–8 presented the numerical results by taking $u_{-1} = u_0 = v_0 = (1, \cdots, 1)$ and $D_n \leq 10^{-9}$.

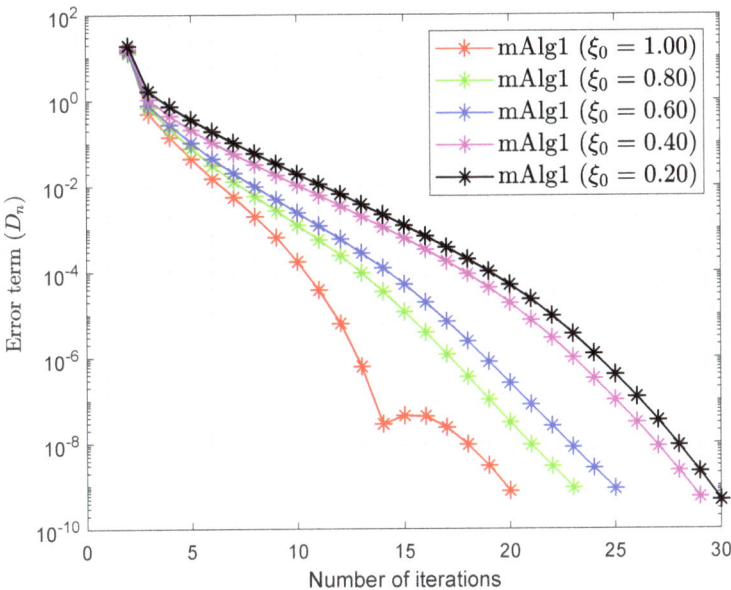

Figure 1. Example 1: numerical behaviour of Algorithm 1 by letting different options for ξ_0, while m = 10.

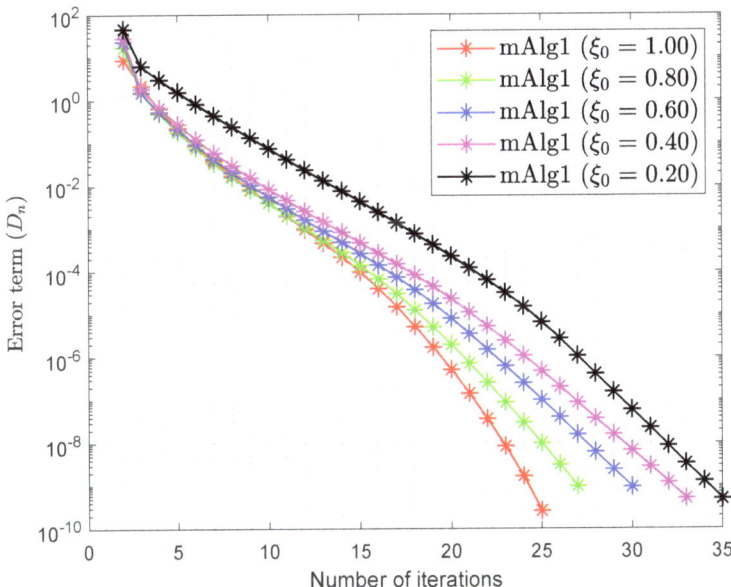

Figure 2. Example 1: numerical behaviour of Algorithm 1 by letting different options for ξ_0, while m = 20.

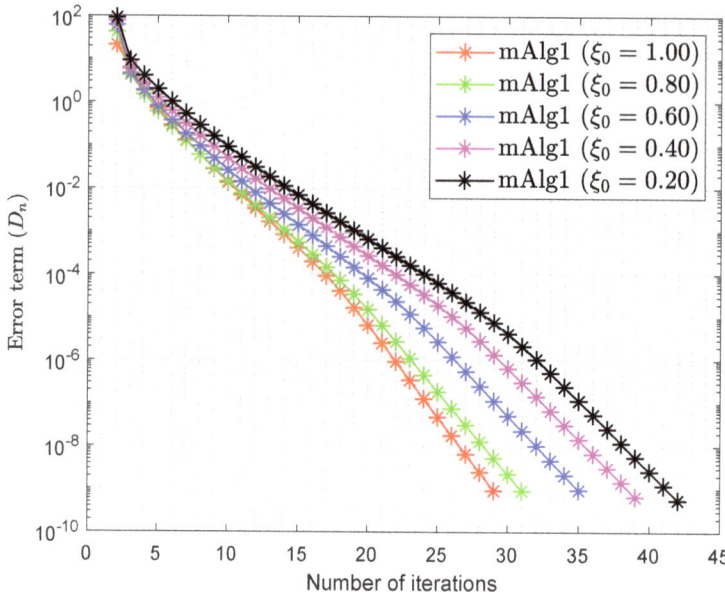

Figure 3. Example 1: numerical behaviour of Algorithm 1 by letting different options for ξ_0 while m = 50.

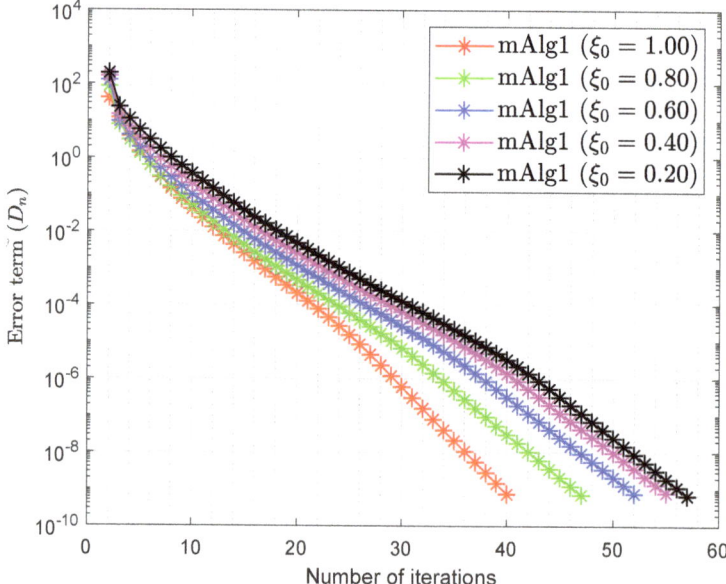

Figure 4. Example 1: numerical behaviour of Algorithm 1 by letting different options for ξ_0 while m = 100.

Table 1. Example 1: Algorithm 1 numerical behaviour by letting different options for ζ_0 and m.

ζ_0	m = 10		m = 20		m = 50		m = 100	
	iter.	time	iter.	time	iter.	time	iter.	time
1.00	20	0.1701	25	0.2153	29	0.2726	40	0.5570
0.80	23	0.1945	27	0.2326	31	0.2788	47	0.5469
0.60	25	0.1995	30	0.2634	35	0.3285	52	0.6228
0.40	29	0.1467	33	0.2979	39	0.3549	55	0.6542
0.20	30	0.2632	35	0.2868	42	0.3849	57	0.6662

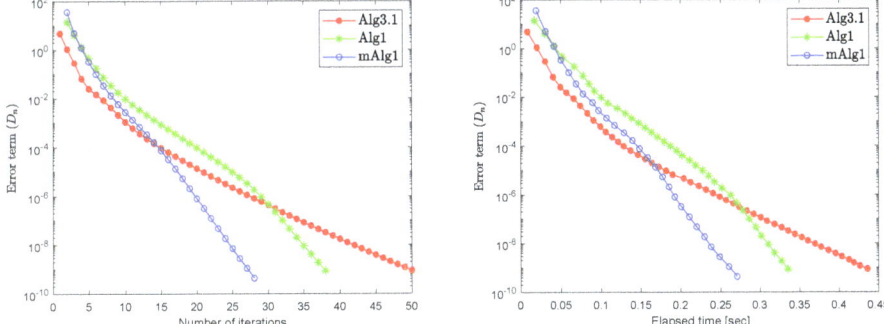

Figure 5. Example 1: Algorithm 1 (mAlg1) numerical comparison with Algorithm 3.1 (Alg3.1) in [39] and Algorithm 1 (Alg1) in [38] while m = 60.

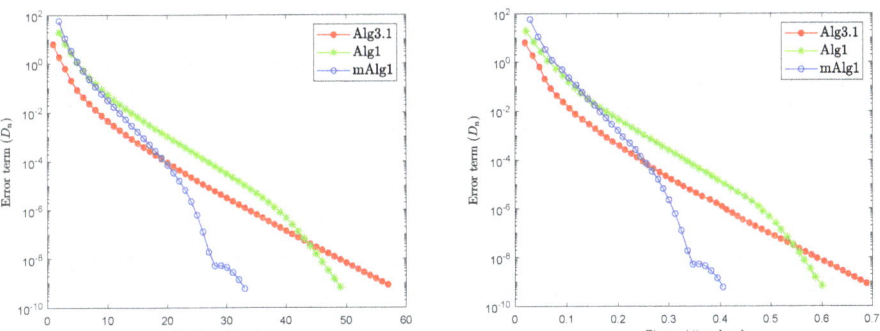

Figure 6. Example 1: Algorithm 1 (mAlg1) numerical comparison with Algorithm 3.1 (Alg3.1) in [39] and Algorithm 1 (Alg1) in [38] while m = 120.

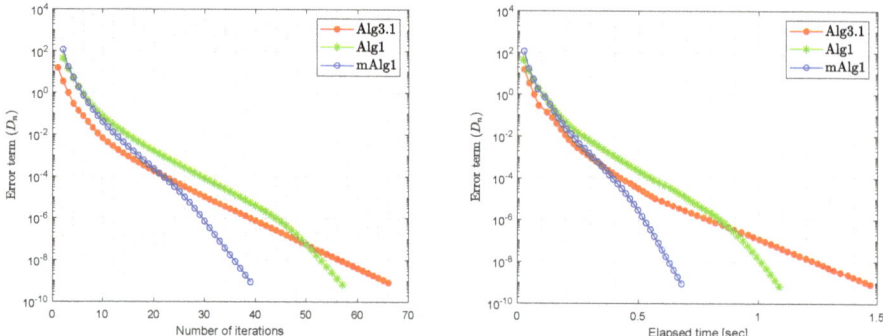

Figure 7. Example 1: Algorithm 1 (mAlg1) numerical comparison with Algorithm 3.1 (Alg3.1) in [39] and Algorithm 1 (Alg1) in [38] while m = 200.

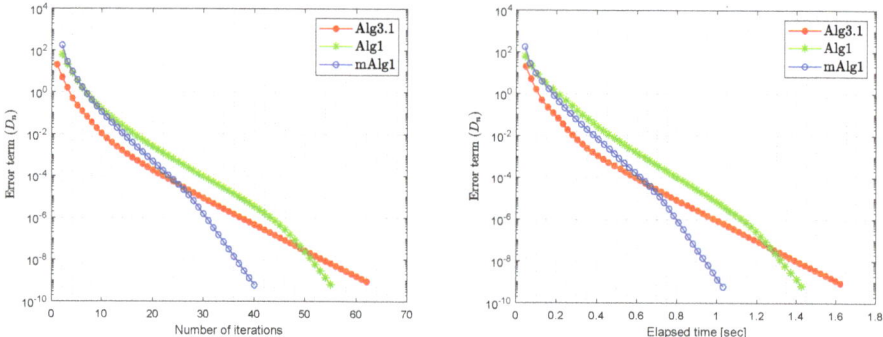

Figure 8. Example 1: Algorithm 1 (mAlg1) numerical comparison with Algorithm 3.1 (Alg3.1) in [39] and Algorithm 1 (Alg1) in [38] while m = 300.

Table 2. Example 1: Algorithm 1 (mAlg1) numerical comparison with Algorithm 3.1 (Alg3.1) in [39] and Algorithm 1 (Alg1) in [38].

	Number of Iterations			Execution Time in Seconds		
m	Alg3.1	Alg1	mAlg1	Alg3.1	Alg1	mAlg1
60	50	38	28	0.4362	0.3352	0.2705
120	57	49	33	0.6888	0.6000	0.4047
200	66	57	39	1.4708	1.0881	0.6794
300	62	55	40	1.6213	1.4251	1.0303

Example 2. *Suppose that $f : \mathcal{K} \times \mathcal{K} \to \mathcal{R}$ be a bifunction defined in the following way*

$$f(p,q) = \sum_{i=2}^{5}(q_i - p_i)\|p\|, \ \forall p,q \in \mathcal{R}^5,$$

where $\mathcal{K} = \{(p_1, \cdots, p_5) : p_1 \geq -1, p_i \geq 1, i = 2, \cdots, 5\}$. A bifunction f is Lipschitz-type continuous with constants $k_1 = k_2 = 2$ and satisfy the conditions (f1)–(f4). *In order to evaluate the best possible value of the control parameters, a numerical test is performed taking the variation of the inertial factor θ. The numerical comparison results are shown in the Table 3 by using $u_{-1} = u_0 = v_0 = (2,3,2,5,5)$ and $D_n \leq 10^{-6}$.*

Table 3. Example 2: Algorithm 1 (mAlg1) numerical comparison with Algorithm 3.1 (Alg3.1) in [39] and Algorithm 1 (Alg1) in [38].

	Number of Iterations			Execution Time in Seconds		
θ	Alg3.1	Alg1	mAlg1	Alg3.1	Alg1	mAlg1
0.90	67	56	47	2.8674	2.5324	1.6734
0.70	63	53	45	2.7813	2.6423	1.5026
0.50	57	47	41	2.0912	2.4212	1.4991
0.30	61	48	44	2.4115	2.3567	1.5092
0.10	69	60	47	2.9229	2.2881	1.5098

Example 3. Let $\mathcal{E} = L^2([0,1])$ be a Hilbert space with an inner product $\langle p,q \rangle = \int_0^1 p(r)q(r)dr$, and the induced norm $\|p\| = \sqrt{\int_0^1 p^2(r)dr}$, $\forall p,q \in \mathcal{E}$. The set $\mathcal{K} := \{p \in L^2([0,1]) : \int_0^1 rp(r)dr = 2\}$. Suppose that $f : \mathcal{E} \times \mathcal{E} \to \mathcal{R}$ is defined by

$$f(p,q) = \langle L(p), q - p \rangle,$$

where $L(p(r)) = \int_0^r p(s)ds$, for every $p \in L^2([0,1])$ and $r \in [0,1]$. The projection on set $\mathcal{K}$ is computed in the following way:

$$P_{\mathcal{K}}(p)(r) := p(r) - \frac{\int_0^1 rp(r)dr - 2}{\int_0^1 r^2 dr} r, \quad r \in [0,1].$$

Table 4 reports the numerical results by using stopping criterion $D_n \leq 10^{-6}$ and letting $u_{-1} = u_0 = v_0$.

Table 4. Example 3: Algorithm 1 (mAlg1) numerical comparison with Algorithm 3.1 (Alg3.1) in [39] and Algorithm 1 (Alg1) in [38].

	Number of Iterations			Execution time in Seconds		
u_0	Alg3.1	Alg1	mAlg1	Alg3.1	Alg1	mAlg1
$3t$	33	28	19	4.7654	3.9782	2.9342
$3t^2$	38	31	20	5.2598	4.1458	3.0987
$3\sin(t)$	41	33	22	5.9876	5.3976	4.4298
$3\cos(t)$	47	39	22	6.9921	5.4765	4.4611
$3\exp(t)^2$	58	43	31	8.4691	5.8329	5.0321

Example 4. Assume that a bifunction f is defined by

$$f(p,q) = \langle L(p), q - p \rangle \text{ and } L(p) = G(p) + H(p),$$

where

$$G(p) = (g_1(p), g_2(p), \cdots, g_m(p)), \quad H(p) = Ep + c, \quad c = (-1, -1, \cdots, -1),$$

and

$$g_i(p) = p_{i-1}^2 + p_i^2 + p_{i-1}p_i + p_i p_{i+1}, \quad i = 1, 2, \ldots, m, \quad p_0 = p_{m+1} = 0.$$

Let the matrix E of order m are consider in the following way:

$$e_{i,j} = \begin{cases} 4 & j = i \\ 1 & i - j = 1 \\ -2 & i - j = -1 \\ 0 & \text{otherwise,} \end{cases}$$

where $\mathcal{K} = \{(u_1, \cdots, u_m) \in \mathcal{R}^m : u_i \geq 1, i = 2, \cdots, m\}$. Figures 9–13 and Table 5 report the numerical results by taking $u_{-1} = u_0 = v_0 = (1, \cdots, 1)$ and $D_n \leq 10^{-6}$.

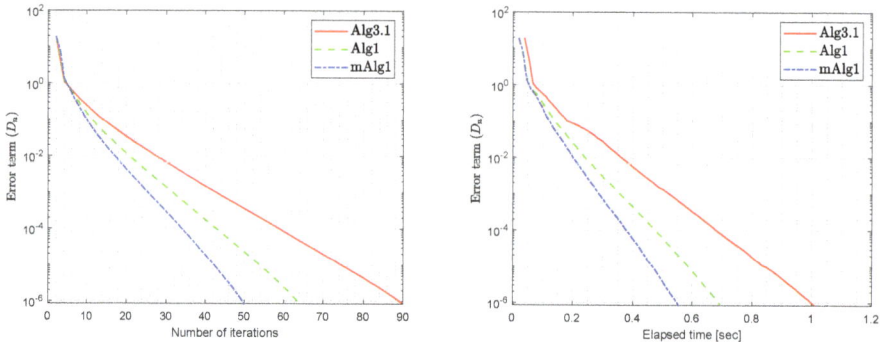

Figure 9. Example 4: Algorithm 1 (mAlg1) numerical comparison with Algorithm 3.1 (Alg3.1) in [39] and Algorithm 1 (Alg1) in [38] while m = 20.

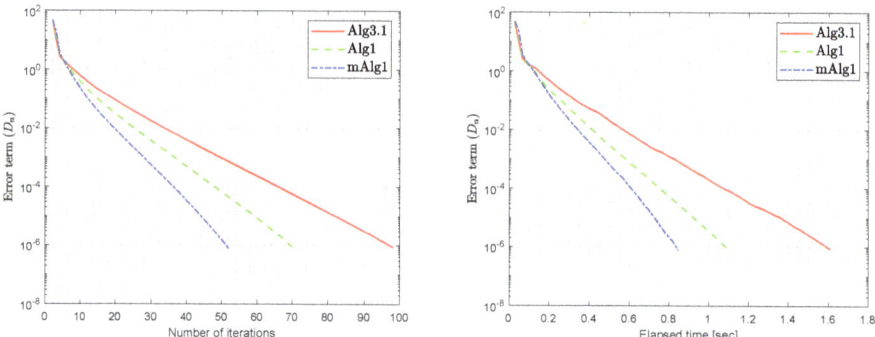

Figure 10. Example 4: Algorithm 1 (mAlg1) numerical comparison with Algorithm 3.1 (Alg3.1) in [39] and Algorithm 1 (Alg1) in [38] while m = 50.

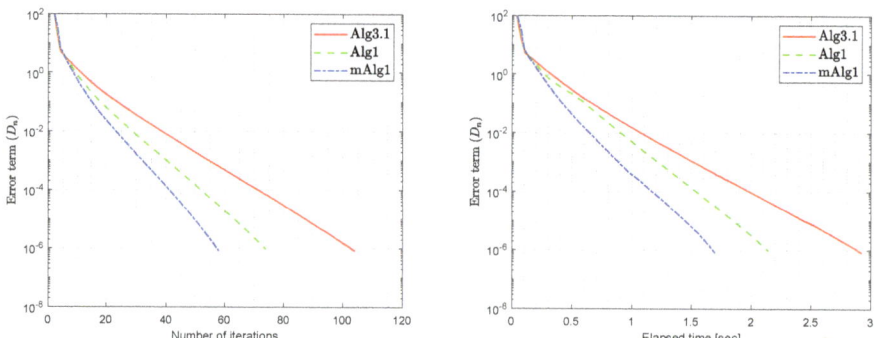

Figure 11. Example 4: Algorithm 1 (mAlg1) numerical comparison with Algorithm 3.1 (Alg3.1) in [39] and Algorithm 1 (Alg1) in [38] while m = 100.

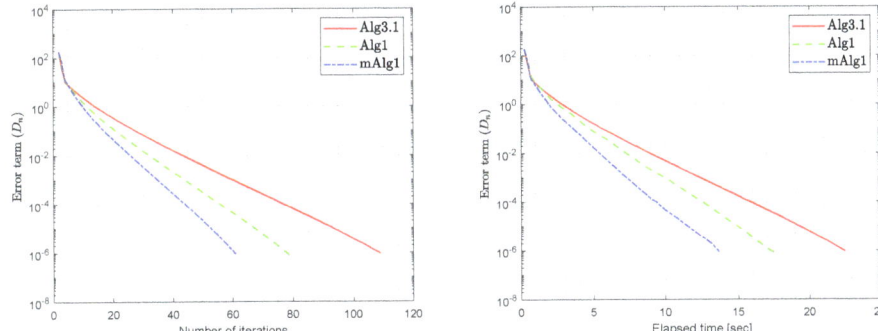

Figure 12. Example 4: Algorithm 1 (mAlg1) numerical comparison with Algorithm 3.1 (Alg3.1) in [39] and Algorithm 1 (Alg1) in [38] while m = 200.

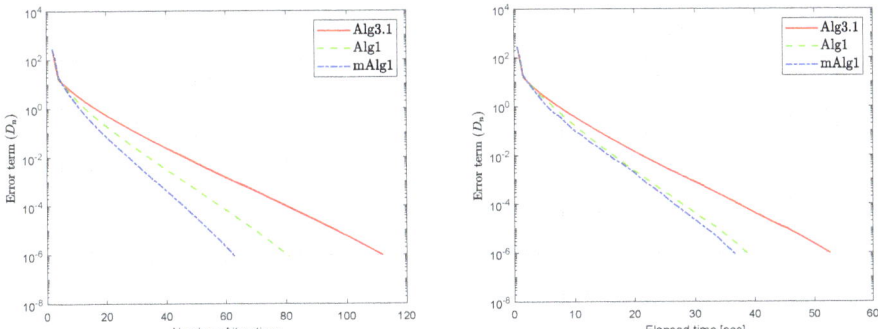

Figure 13. Example 4: Algorithm 1 (mAlg1) numerical comparison with Algorithm 3.1 (Alg3.1) in [39] and Algorithm 1 (Alg1) in [38] while m = 300.

Table 5. Example 4: Algorithm 1 (mAlg1) numerical comparison with Algorithm 3.1 (Alg3.1) in [39] and Algorithm 1 (Alg1) in [38].

	Number of Iterations			Execution Time in Seconds		
m	Alg3.1	Alg1	mAlg1	Alg3.1	Alg1	mAlg1
20	90	64	50	1.0089	0.6923	0.5541
50	98	70	52	1.6089	1.9092	0.8464
100	104	74	58	2.9231	2.1456	1.6970
200	109	79	61	22.5299	17.6267	13.6542
300	112	81	63	52.6776	39.0018	36.6305

Remark 2.

(i) It is also significant that the value of ξ_0 is crucial and performs best when it is nearer to 1.

(ii) It is observed that the selection of the value ϑ is often significant and roughly the value $\vartheta \in (3,6)$ performs better than most other values.

7. Conclusions

In this paper, we consider the convergence result for pseudomonotone equilibrium problems that involve Lipschitz-type continuous bifunction but the Lipschitz-type constants are unknown. We modify the extragradient methods with an inertial term and new step size formula. Weak convergence theorem is proved for sequences generated by the algorithm. Several numerical experiments confirm the effectiveness of the proposed algorithms.

Author Contributions: Conceptualization, H.u.R., N.P. and M.D.l.S.; Writing-Original Draft Preparation, N.W., N.P. and H.u.R.; Writing-Review & Editing, N.W., N.P., H.u.R. and M.D.l.S.; Methodology, N.P. and H.u.R.; Visualization, N.W. and N.P.; Software, H.u.R.; Funding Acquisition, M.D.l.S.; Supervision, M.D.l.S. and H.u.R.; Project Administration; M.D.l.S.; Resources; M.D.l.S. and H.u.R. All authors have read and agreed to the published version of this manuscript.

Funding: This research work was financially supported by Spanish Government for Grant RTI2018-094336-B-I00 (MCIU/AEI/FEDER, UE) and to the Basque Government for Grant IT1207-19.

Acknowledgments: We are very grateful to the Editor and the anonymous referees for their valuable and useful comments, which helped improve the quality of this work. Nopparat Wairojjana was partially supported by Valaya Alongkorn Rajabhat University under the Royal Patronage, Thailand. The corresponding author are grateful to the Spanish Government for Grant RTI2018-094336-B-I00 (MCIU/AEI/FEDER, UE) and to the Basque Government for Grant IT1207-19.

Conflicts of Interest: The authors declare no conflict of interest.

References

1. Blum, E. From optimization and variational inequalities to equilibrium problems. *Math. Stud.* **1994**, *63*, 123–145.
2. Fan, K. *A Minimax Inequality and Applications, Inequalities III*; Shisha, O., Ed.; Academic Press: New York, NY, USA, 1972.
3. Facchinei, F.; Pang, J.S. *Finite-Dimensional Variational Inequalities and Complementarity Problems*; Springer Science & Business Media: Berlin, Germany, 2007.
4. Konnov, I. *Equilibrium Models and Variational Inequalities*; Elsevier: Amsterdam, The Netherlands, 2007; Volume 210.
5. Muu, L.D.; Oettli, W. Convergence of an adaptive penalty scheme for finding constrained equilibria. *Nonlinear Anal. Theory Methods Appl.* **1992**, *18*, 1159–1166. [CrossRef]
6. Quoc, T.D.; Le Dung, M.N.V.H. Extragradient algorithms extended to equilibrium problems. *Optimization* **2008**, *57*, 749–776. [CrossRef]
7. Quoc, T.D.; Anh, P.N.; Muu, L.D. Dual extragradient algorithms extended to equilibrium problems. *J. Glob. Optim.* **2011**, *52*, 139–159. [CrossRef]
8. Lyashko, S.I.; Semenov, V.V. A New Two-Step Proximal Algorithm of Solving the Problem of Equilibrium Programming. In *Optimization and Its Applications in Control and Data Sciences*; Springer International Publishing: Berlin/Heidelberg, Germany, 2016; pp. 315–325. [CrossRef]
9. Takahashi, S.; Takahashi, W. Viscosity approximation methods for equilibrium problems and fixed point problems in Hilbert spaces. *J. Math. Anal. Appl.* **2007**, *331*, 506–515. [CrossRef]
10. ur Rehman, H.; Kumam, P.; Cho, Y.J.; Yordsorn, P. Weak convergence of explicit extragradient algorithms for solving equilibirum problems. *J. Inequalities Appl.* **2019**, *2019*. [CrossRef]
11. Anh, P.N.; Hai, T.N.; Tuan, P.M. On ergodic algorithms for equilibrium problems. *J. Glob. Optim.* **2015**, *64*, 179–195. [CrossRef]
12. Hieu, D.V.; Quy, P.K.; Vy, L.V. Explicit iterative algorithms for solving equilibrium problems. *Calcolo* **2019**, *56*. [CrossRef]
13. Hieu, D.V. New extragradient method for a class of equilibrium problems in Hilbert spaces. *Appl. Anal.* **2017**, *97*, 811–824. [CrossRef]
14. ur Rehman, H.; Kumam, P.; Je Cho, Y.; Suleiman, Y.I.; Kumam, W. Modified Popov's explicit iterative algorithms for solving pseudomonotone equilibrium problems. *Optim. Methods Softw.* **2020**, pp. 1–32. [CrossRef]
15. ur Rehman, H.; Kumam, P.; Abubakar, A.B.; Cho, Y.J. The extragradient algorithm with inertial effects extended to equilibrium problems. *Comput. Appl. Math.* **2020**, *39*. [CrossRef]
16. Santos, P.; Scheimberg, S. An inexact subgradient algorithm for equilibrium problems. *Comput. Appl. Math.* **2011**, *30*, 91–107.

17. Hieu, D.V. Halpern subgradient extragradient method extended to equilibrium problems. *Revista de la Real Academia de Ciencias Exactas, Físicas y Naturales Serie A Matemáticas* **2016**, *111*, 823–840. [CrossRef]
18. ur Rehman, H.; Kumam, P.; Kumam, W.; Shutaywi, M.; Jirakitpuwapat, W. The Inertial Sub-Gradient Extra-Gradient Method for a Class of Pseudo-Monotone Equilibrium Problems. *Symmetry* **2020**, *12*, 463. [CrossRef]
19. Anh, P.N.; An, L.T.H. The subgradient extragradient method extended to equilibrium problems. *Optimization* **2012**, *64*, 225–248. [CrossRef]
20. Muu, L.D.; Quoc, T.D. Regularization Algorithms for Solving Monotone Ky Fan Inequalities with Application to a Nash-Cournot Equilibrium Model. *J. Optim. Theory Appl.* **2009**, *142*, 185–204. [CrossRef]
21. ur Rehman, H.; Kumam, P.; Argyros, I.K.; Deebani, W.; Kumam, W. Inertial Extra-Gradient Method for Solving a Family of Strongly Pseudomonotone Equilibrium Problems in Real Hilbert Spaces with Application in Variational Inequality Problem. *Symmetry* **2020**, *12*, 503. [CrossRef]
22. ur Rehman, H.; Kumam, P.; Argyros, I.K.; Alreshidi, N.A.; Kumam, W.; Jirakitpuwapat, W. A Self-Adaptive Extra-Gradient Methods for a Family of Pseudomonotone Equilibrium Programming with Application in Different Classes of Variational Inequality Problems. *Symmetry* **2020**, *12*, 523. [CrossRef]
23. ur Rehman, H.; Kumam, P.; Argyros, I.K.; Shutaywi, M.; Shah, Z. Optimization Based Methods for Solving the Equilibrium Problems with Applications in Variational Inequality Problems and Solution of Nash Equilibrium Models. *Mathematics* **2020**, *8*, 822. [CrossRef]
24. Yordsorn, P.; Kumam, P.; ur Rehman, H.; Ibrahim, A.H. A Weak Convergence Self-Adaptive Method for Solving Pseudomonotone Equilibrium Problems in a Real Hilbert Space. *Mathematics* **2020**, *8*, 1165. [CrossRef]
25. Yordsorn, P.; Kumam, P.; Rehman, H.U. Modified two-step extragradient method for solving the pseudomonotone equilibrium programming in a real Hilbert space. *Carpathian J. Math.* **2020**, *36*, 313–330.
26. La Sen, M.D.; Agarwal, R.P.; Ibeas, A.; Alonso-Quesada, S. On the Existence of Equilibrium Points, Boundedness, Oscillating Behavior and Positivity of a SVEIRS Epidemic Model under Constant and Impulsive Vaccination. *Adv. Differ. Equ.* **2011**, *2011*, 1–32. [CrossRef]
27. La Sen, M.D.; Agarwal, R.P. Some fixed point-type results for a class of extended cyclic self-mappings with a more general contractive condition. *Fixed Point Theory Appl.* **2011**, *2011*. [CrossRef]
28. Wairojjana, N.; ur Rehman, H.; Argyros, I.K.; Pakkaranang, N. An Accelerated Extragradient Method for Solving Pseudomonotone Equilibrium Problems with Applications. *Axioms* **2020**, *9*, 99. [CrossRef]
29. La Sen, M.D. On Best Proximity Point Theorems and Fixed Point Theorems for -Cyclic Hybrid Self-Mappings in Banach Spaces. *Abstr. Appl. Anal.* **2013**, *2013*, 1–14. [CrossRef]
30. ur Rehman, H.; Kumam, P.; Shutaywi, M.; Alreshidi, N.A.; Kumam, W. Inertial Optimization Based Two-Step Methods for Solving Equilibrium Problems with Applications in Variational Inequality Problems and Growth Control Equilibrium Models. *Energies* **2020**, *13*, 3292. [CrossRef]
31. Rehman, H.U.; Kumam, P.; Dong, Q.L.; Peng, Y.; Deebani, W. A new Popov's subgradient extragradient method for two classes of equilibrium programming in a real Hilbert space. *Optimization* **2020**, 1–36. [CrossRef]
32. Wang, L.; Yu, L.; Li, T. Parallel extragradient algorithms for a family of pseudomonotone equilibrium problems and fixed point problems of nonself-nonexpansive mappings in Hilbert space. *J. Nonlinear Funct. Anal.* **2020**, *2020*, 13.
33. Shahzad, N.; Zegeye, H. Convergence theorems of common solutions for fixed point, variational inequality and equilibrium problems, J. *Nonlinear Var. Anal.* **2019**, *3*, 189–203.
34. Farid, M. The subgradient extragradient method for solving mixed equilibrium problems and fixed point problems in Hilbert spaces. *J. Appl. Numer. Optim.* **2019**, *1*, 335–345.
35. Flåm, S.D.; Antipin, A.S. Equilibrium programming using proximal-like algorithms. *Math. Program.* **1996**, *78*, 29–41. [CrossRef]
36. Korpelevich, G. The extragradient method for finding saddle points and other problems. *Matecon* **1976**, *12*, 747–756.
37. Yang, J.; Liu, H.; Liu, Z. Modified subgradient extragradient algorithms for solving monotone variational inequalities. *Optimization* **2018**, *67*, 2247–2258. [CrossRef]
38. Vinh, N.T.; Muu, L.D. Inertial Extragradient Algorithms for Solving Equilibrium Problems. *Acta Math. Vietnam.* **2019**, *44*, 639–663. [CrossRef]

39. Hieu, D.V.; Cho, Y.J.; bin Xiao, Y. Modified extragradient algorithms for solving equilibrium problems. *Optimization* **2018**, *67*, 2003–2029. [CrossRef]
40. Bianchi, M.; Schaible, S. Generalized monotone bifunctions and equilibrium problems. *J. Optim. Theory Appl.* **1996**, *90*, 31–43. [CrossRef]
41. Mastroeni, G. On Auxiliary Principle for Equilibrium Problems. In *Nonconvex Optimization and Its Applications*; Springer: New York, NY, USA, 2003; pp. 289–298. [CrossRef]
42. Kreyszig, E. *Introductory Functional Analysis with Applications*, 1st ed.; Wiley Classics Library, Wiley: Hoboken, NJ, USA, 1989.
43. Tiel, J.V. *Convex Analysis: An Introductory Text*, 1st ed.; Wiley: New York, NY, USA, 1984.
44. Ioffe, A.D.; Tihomirov, V.M. (Eds.) Theory of Extremal Problems. In *Studies in Mathematics and Its Applications 6*; North-Holland, Elsevier: Amsterdam, thte Netherlands; New York, NY, USA, 1979.
45. Opial, Z. Weak convergence of the sequence of successive approximations for nonexpansive mappings. *Bull. Amer. Math. Soc.* **1967**, *73*, 591–598. [CrossRef]
46. Tan, K.; Xu, H. Approximating Fixed Points of Nonexpansive Mappings by the Ishikawa Iteration Process. *J. Math. Anal. Appl.* **1993**, *178*, 301–308. [CrossRef]
47. Bauschke, H.H.; Combettes, P.L. *Convex Analysis and Monotone Operator Theory in Hilbert Spaces*; Springer: New York, NY, USA, 2011; Volume 408.
48. Browder, F.; Petryshyn, W. Construction of fixed points of nonlinear mappings in Hilbert space. *J. Math. Anal. Appl.* **1967**, *20*, 197–228. [CrossRef]

© 2020 by the authors. Licensee MDPI, Basel, Switzerland. This article is an open access article distributed under the terms and conditions of the Creative Commons Attribution (CC BY) license (http://creativecommons.org/licenses/by/4.0/).

Article

Inertial Subgradient Extragradient Methods for Solving Variational Inequality Problems and Fixed Point Problems

Godwin Amechi Okeke [1,2,*], Mujahid Abbas [3,4] and Manuel de la Sen [5]

1. Department of Mathematics, School of Physical Sciences, Federal University of Technology, Owerri, P.M.B. 1526 Owerri, Imo State, Nigeria
2. Abdus Salam School of Mathematical Sciences, Government College University, Lahore 54600, Pakistan
3. Department of Mathematics, Government College University, Lahore 54000, Pakistan; abbas.mujahid@gmail.com or abbas.mujahid@gcu.edu.pk
4. Department of Mathematics and Applied Mathematics, University of Pretoria (Hatfield Campus), Lynnwood Road, Pretoria 0002, South Africa
5. Institute of Research and Development of Processes, University of the Basque Country, Campus of Leioa (Bizkaia), P.O. Box 644-Bilbao, Barrio Sarriena, 48940 Leioa, Spain; manuel.delasen@ehu.eus
* Correspondence: godwin.okeke@futo.edu.ng or gaokeke1@yahoo.co.uk

Received: 16 March 2020; Accepted: 7 May 2020; Published: 11 May 2020

Abstract: We propose two new iterative algorithms for solving K-pseudomonotone variational inequality problems in the framework of real Hilbert spaces. These newly proposed methods are obtained by combining the viscosity approximation algorithm, the Picard Mann algorithm and the inertial subgradient extragradient method. We establish some strong convergence theorems for our newly developed methods under certain restriction. Our results extend and improve several recently announced results. Furthermore, we give several numerical experiments to show that our proposed algorithms performs better in comparison with several existing methods.

Keywords: K-pseudomonotone; inertial iterative algorithms; variational inequality problems; Hilbert spaces; strong convergence

MSC: 47H10; 47H05; 68W10; 65K15; 65Y05

1. Introduction

In this paper, the set C denotes a nonempty closed convex subset of a real Hilbert space H. The inner product of H is denoted by $\langle .,. \rangle$ and the induced norm by $\|.\|$. Suppose $A : H \to H$ is an operator. The variational inequality problem (VIP) for the operator A on $C \subset H$ is to find a point $x^* \in C$ such that

$$\langle Ax^*, x - x^* \rangle \geq 0 \text{ for each } x \in C. \tag{1}$$

In this study, we denote the solution set of (VIP) (1) by Γ. The theory of variational inequalities was introduced by Stampacchia [1]. It is known that the (VIP) problem arise in various models involving problems in many fields of study such as mathematics, physics, sciences, social sciences, management sciences, engineering and so on. The ideas and methods of the variational inequalities have been highly applied innovatively in diverse areas of sciences and engineering and have proved very effective in solving certain problems. The theory of (VIP) provides a natural, simple and unified setting for a comprehensive treatment of unrelated problems (see, e.g., [2]). Several authors have developed efficient numerical methods in solving the (VIP) problem. These methods includes the projection methods and its variants (see, e.g., [3–13]). The fundamental objective involves extending the well-known projected gradient algorithm, which is useful in solving the minimization problem $f(x)$ subject to $x \in C$. This method is given as follows:

$$x_{n+1} = P_C(x_n - \alpha_n \nabla f(x_n)), \quad n \geq 0, \tag{2}$$

where the real sequence $\{\alpha_n\}$ satisfy certain conditions and P_C is the well-known metric projection onto $C \subset H$. The interested reader may consult [14] for convergence analysis of this algorithm for the special case in which the mapping $f : H \to \mathbb{R}$ is convex and differentiable. Equation (2) has been extended to the (VIP) problem and it is known as the projected gradient method for optimization problems. This is done by replacing the gradient with the operator A thereby generating a sequence $\{x_n\}$ as follows:

$$x_{n+1} = P_C(x_n - \alpha_n A x_n), \quad n \geq 0. \tag{3}$$

However, the major drawback of this method is the restrictive condition that the operator A is strongly monotone or inverse strongly monotone (see, e.g., [15]) to guarantee the convergence of this method. In 1976, Korpelevich [16] removed this strong condition by introducing the extragradient method for solving saddle point problems. The extragradient method was extended to solving variational inequality problems in both Hilbert and Euclidean spaces. The only required restriction for the extragradient algorithm to converge is that the operator A is monotone and L-Lipschitz continuous. The extragradient method is given as follows:

$$\begin{cases} y_n = P_C(x_n - \tau A x_n) \\ x_{n+1} = P_C(x_n - \tau A y_n), \end{cases} \tag{4}$$

where $\tau \in (0, \frac{1}{L})$ and the metric projection from H onto C is denoted by P_C. If the solution set of the (VIP) denoted by Γ is nonempty, then the sequence $\{x_n\}$ generated by iterative algorithm (4) converges weakly to an element in Γ.

Observe that by using the extragradient method, we need to calculate two projections onto the set $C \subset H$ in every iteration. It is known that the projection onto a closed convex set $C \subset H$ has a close relationship with the minimun distance problem. Let C be a closed and convex set, this method may require a prohibitive amount of computation time. In view of this drawback, in 2011 Censor et al. [5] introduced the subgradient extragradient method by modifying iterative algorithm in Equation (4) above. They replaced the two projections in the extragradient method in Equation (4) onto the set C by only one projection onto the set $C \subset H$ and one onto a half-space. It has been established that the projection onto a given half-space is easier to calculate. Next, we give the subgradient extragradient method of Censor et al. [5] as follows:

$$\begin{cases} y_n = P_C(x_n - \tau A x_n) \\ T_n = \{x \in H | \langle x_n - \tau A x_n - y_n, x - y_n \rangle \leq 0\} \\ x_{n+1} = P_{T_n}(x_n - \tau A y_n), \end{cases} \tag{5}$$

where $\tau \in (0, \frac{1}{L})$. Several authors have studied the subgradient extragradient method and obtained some interesting and applicable results (see, e.g., [11]) and the references therein.

The theory of pseudomonotone operators is very crucial in studies in nonlinear analysis, variational inequalities and optimization problems (see, e.g., [17–20]). One important class of pseudomonotone operators was introduced in 1976 by Karamardian [21] and have been utilized in solving problems in variational inequalities, optimization and economics (see, e.g., [17,20]). In this paper, we shall call the class of pseudomonotone in the sense of Karamardian K-pseudomonotone. Yao [20] utilized K-pseudomonotone in solving some variational inequalities problems in Banach spaces. He established some new existence results which extend many known results in infinite-dimensional spaces under some weak assumptions. He also proved some uniqueness results for the complementarity problem with K-pseudomonotone operators in Banach spaces. It is our purpose in the present paper to introduce two new inertial subgradient extragradient iterative algorithms for solving K-pseudomonotone variational inequality problems in the framework of real Hilbert spaces.

The inertial type iterative algorithms are based on a discrete version of a second order dissipative dynamical system (see, [22,23]). These kind of algorithms can be seen as a process of accelerating the convergence properties of a given method (see, e.g., [24–26]). Alvarez and Attouch [24] in 2001 used the inertial method to derive a proximal algorithm for solving the problem of finding zero of a maximal monotone operator. Their method is given as follows: given $x_{n-1}, x_n \in H$ and any two parameters $\theta_n \in [0,1)$, $\lambda_n > 0$, obtain $x_{n+1} \in H$ such that

$$0 \in \lambda_n A(x_{n+1}) + x_{n+1} - x_n - \theta_n(x_n - x_{n-1}). \tag{6}$$

This algorithm can be written equivalently as follows:

$$x_{n+1} = J^A_{\lambda_n}(x_n + \theta_n(x_n - x_{n-1})), \tag{7}$$

where $J^A_{\lambda_n}$ is the resolvent of the operator A with the given parameter λ_n and the inertial is induced by the term $\theta_n(x_n - x_{n-1})$.

Several researchers have developed some fast iterative algorithms by using inertial methods. These methods includes the inertial Douglas–Rachford splitting method (see, e.g., [27]), inertial forward–backward splitting methods (see, e.g., [28]), inertial ADMM (see, e.g., [29]), inertial proximal–extragradient method (see, e.g., [30]), inertial forward–backward–forward method (see, e.g., [31]), inertial contraction method (see, e.g., [32]), inertial Tseng method (see, e.g., [33]) and inertial Mann method (see, e.g., [11]).

Inspired by the results above, we propose two inertial subgradient extragradient methods for finding a solution of K-pseudomonotone and Lipschitz continuous (VIP). Our first proposed iterative algorithm is a hybrid of the inertial subgradient extragradient method [11], the viscosity method [34] and the Picard Mann method [35]. Our second method combines the inertial subgradient extragradient method [11] and the Picard Mann method [35].

This paper is organized as follows. In Section 2, we give some preliminary definitions of concepts and results that will be crucial in this study. In Section 3, we present our proposed iterative algorithms and prove some convergence results for them. In Section 4, we present some numerical experiments to support the convergence of our proposed iterative algorithms. In Section 5, we give the concluding remarks of the study.

2. Preliminaries

In this paper, the set C denotes a nonempty closed convex subset of a real Hilbert space H. The inner product of H is denoted by $\langle .,. \rangle$ and the induced norm by $\|.\|$.

We denote the weak convergence of the sequence $\{x_n\}$ to x by $x_n \rightharpoonup x$ as $n \to \infty$, we denote the strong convergence of $\{x_n\}$ to x by $x_n \to x$ as $n \to \infty$.

For each $x, y \in H$ and $\alpha \in \mathbb{R}$, we recall the following inequalities in Hilbert spaces:

$$\|\alpha x + (1-\alpha)y\|^2 = \alpha \|x\|^2 + (1-\alpha)\|y\|^2 - \alpha(1-\alpha)\|x-y\|^2. \tag{8}$$

$$\|x+y\|^2 \leq \|x\|^2 + 2\langle y, x+y\rangle. \tag{9}$$

$$\|x+y\|^2 = \|x\|^2 + 2\langle x, y\rangle + \|y\|^2. \tag{10}$$

A mapping $A : H \to H$ is said to be nonexpansive if for each $x, y \in H$, we have

$$\|Ax - Ay\| \leq \|x - y\|.$$

For each $x \in H$, we can find a unique nearest point in $C \subset H$, denoted by $P_C x$ such that we have

$$\|x - P_C x\| \leq \|x - y\| \tag{11}$$

for each $y \in C$. Then P_C is known as the *metric projection* of H onto $C \subset H$. It has been proved that the mapping P_C is nonexpansive.

Lemma 1 ([36]). *Suppose that C is a closed convex subset of a real Hilbert space H and for each $x \in H$. Then the following holds:*

(i) $\|P_C x - P_C y\|^2 \leq \langle P_C x - P_C y, x - y\rangle$ *for all $y \in H$.*
(ii) $\|P_C x - y\|^2 \leq \|x - y\|^2 - \|x - P_C x\|^2$ *for all $y \in H$.*

(iii) *Given $x \in H$ and $z \in C$. Then we have*

$$z = P_C x \iff \langle x - z, z - y\rangle \geq 0$$

for all $y \in C$.

For more of the metric projection P_C, the interested reader should see Section 3 of [36].

The fixed point problem involves finding the fixed point of an operator $A : H \to H$. The set of fixed point of the operator A is denoted by $F(A)$ and we assume that it is nonempty, that is $F(A) \neq \emptyset$. The fixed point problem (FP) is then formulated as follows:

$$\text{find } x \in H \text{ such that } x = A(x). \tag{12}$$

In this paper, our problem of interest is to find a point $x \in H$ such that

$$x \in \Gamma \cap F(A). \tag{13}$$

Definition 1. *Let $A : H \to H$ be a mapping. Then for all $x, y \in H$*

(i) *A is said to be L-Lipschitz continuous with $L > 0$ if*

$$\|Ax - Ay\| \leq L\|x - y\|. \tag{14}$$

If $L \in [0, 1)$ then A is called a contraction mapping.
(ii) *A is said to be monotone if*

$$\langle Ax - Ay, x - y\rangle \geq 0. \tag{15}$$

(iii) *The mapping $A : H \to H$ is said to be pseudomonotone in the sense of Karamardian [21] or K-pseudomonotone for short, if for all $x, y \in H$*

$$\langle Ay, x - y\rangle \geq 0 \implies \langle Ax, x - y\rangle \geq 0. \tag{16}$$

The following lemmas will be needed in this paper.

Lemma 2. ([37]) *Suppose $\{x_n\}$ is a real sequence of nonnegative numbers such that there is a subsequence $\{x_{n_j}\}$ of $\{x_n\}$ such that $x_{n_j} < x_{n_{j+1}}$ for any $j \in \mathbb{N}$. Then there is a nondecreasing sequence $\{m_k\}$ of $\mathbb{N}$ such that $\lim_{k\to\infty} m_k = \infty$ and the following properties are fulfilled: for each (sufficiently large) number $k \in \mathbb{N}$,*

$$x_{m_k} \leq x_{m_k+1}, \quad x_k \leq x_{m_k+1}.$$

In fact, m_k is the largest number n in the set $\{1, 2, \cdots, k\}$ such that $x_n < x_{n+1}$.

Lemma 3. ([38]) *Let $\{a_n\}$ be a sequence of nonnegative real numbers such that*

$$a_{n+1} \leq (1 - \alpha_n)a_n + \alpha_n b_n$$

for all $n \geq 0$, where $\{\alpha_n\} \subset (0, 1)$ and $\{b_n\}$ is a sequence such that

(a) $\sum_{n=0}^{\infty} \alpha_n = \infty$;
(b) $\limsup_{n\to\infty} b_n \leq 0$.

Then $\lim_{n\to\infty} a_n = 0$.

3. Main Results

The following condition will be needed in this study.

Condition 3.1

The operator $A : H \to H$ is K-pseudomonotone and L-Lipschitz continuous on the real Hilbert space H, with the solution set of the (VIP) (1.1) $\Gamma \neq \emptyset$ and the contraction mapping $f : H \to H$ with the contraction parameter $k \in [0, 1)$. The feasible set $C \subset H$ is non-empty, closed and convex.

3.1. The Viscosity Inertial Subgradient Extragradient Algorithm

We propose the following algorithm

Algorithm 3.1

Step 0: Given $\tau \in (0, \frac{1}{L})$. $\{\alpha_n\} \subset [0, \alpha)$ for some $\alpha > 0$ and $\{\beta_n\} \subset (0, 1)$ satisfying the following conditions:

$$\lim_{n\to\infty} \beta_n = 0, \quad \sum_{n=1}^{\infty} \beta_n = \infty. \tag{17}$$

Choose initial $x_0, x_1 \in C$ and set $n := 1$.
Step 1: Compute

$$w_n = x_n + \alpha_n(x_n - x_{n-1}), \tag{18}$$

$$y_n = P_C(w_n - \tau A w_n). \tag{19}$$

If $y_n = w_n$, then stop, y_n is a solution to the (**VIP**) problem. Otherwise, go to **Step 2**.
Step 2: Construct the half-space

$$T_n := \{z \in H : \langle w_n - \tau A w_n - y_n, z - y_n \rangle \leq 0\} \tag{20}$$

and compute

$$z_n = P_{T_n}(w_n - \tau A y_n). \tag{21}$$

Step 3: Calculate

$$h_n = (1 - \beta_n)z_n + \beta_n f(z_n), \tag{22}$$

and compute

$$x_{n+1} = f(h_n). \tag{23}$$

Let $n := n + 1$ and return to **Step 1**.

Next, we prove the following results which will be useful in this study.

Lemma 4. *Let $\{x_n\}$ be a sequence generated by Algorithm 3.1. Then*

$$\|x_{n+1} - p\|^2 \leq \|w_n - p\|^2 - (1 - \tau L)\|y_n - x_{n+1}\|^2 - (1 - \tau L)\|y_n - w_n\|^2, \tag{24}$$

for all $p \in \Gamma$.

Proof. Since $p \in \Gamma \subset C \subset T_n$, then by Equation (10) and Lemma 2 (i) we have the following

$$\begin{aligned}
\|x_{n+1} - p\|^2 &= \|P_{T_n}(w_n - \tau A y_n) - P_{T_n} p\|^2 \\
&\leq \langle x_{n+1} - p, w_n - \tau A y_n - p \rangle \\
&= \tfrac{1}{2}\|x_{n+1} - p\|^2 + \tfrac{1}{2}\|w_n - \tau A y_n - p\|^2 - \tfrac{1}{2}\|x_{n+1} - w_n + \tau A y_n\|^2 \\
&= \tfrac{1}{2}\|x_{n+1} - p\|^2 + \tfrac{1}{2}\|w_n - p\|^2 + \tfrac{1}{2}\tau^2\|A y_n\|^2 - \langle w_n - p, \tau A y_n \rangle - \tfrac{1}{2}\|x_{n+1} - w_n\|^2 \\
&\quad - \tfrac{1}{2}\tau^2\|A y_n\|^2 - \langle x_{n+1} - w_n, \tau A y_n \rangle \\
&= \tfrac{1}{2}\|x_{n+1} - p\|^2 + \tfrac{1}{2}\|w_n - p\|^2 - \tfrac{1}{2}\|x_{n+1} - w_n\|^2 - \langle x_{n+1} - p, \tau A y_n \rangle.
\end{aligned} \tag{25}$$

Hence, from Equation (25) we obtain

$$\|x_{n+1} - p\|^2 \leq \|w_n - p\|^2 - \|x_{n+1} - w_n\|^2 - 2\langle x_{n+1} - p, \tau A y_n \rangle. \tag{26}$$

Using the condition that A is K-pseudomonotone, we have that $2\tau \langle A y_n, y_n - p \rangle \geq 0$. We now add this to the right hand side of inequality (25) to obtain the following

$$\begin{aligned}
\|x_{n+1} - p\|^2 &\leq \|w_n - p\|^2 - \|x_{n+1} - w_n\|^2 - 2\langle x_{n+1} - p, \tau A y_n \rangle + 2\tau \langle A y_n, y_n - p \rangle \\
&= \|w_n - p\|^2 - \|x_{n+1} - w_n\|^2 - 2\tau \langle x_{n+1} - p, A y_n \rangle + 2\tau \langle y_n - p, A y_n \rangle \\
&= \|w_n - p\|^2 - \|x_{n+1} - w_n\|^2 - 2\tau \langle x_{n+1} - y_n, A y_n \rangle - 2\tau \langle y_n - p, A y_n \rangle + \\
&\quad 2\tau \langle y_n - p, A y_n \rangle \\
&= \|w_n - p\|^2 - \|x_{n+1} - w_n\|^2 - 2\tau \langle x_{n+1} - y_n, A y_n - A w_n \rangle - 2\tau \langle x_{n+1} - y_n, A w_n \rangle \\
&= \|w_n - p\|^2 - \|x_{n+1} - w_n\|^2 + 2\tau \langle y_n - x_{n+1}, A y_n - A w_n \rangle + 2\tau \langle y_n - x_{n+1}, A w_n \rangle.
\end{aligned} \tag{27}$$

Next, we have the following estimates using the condition that A is L-Lipschitz continuous

$$\begin{aligned}
2\tau \langle y_n - x_{n+1}, A y_n - A w_n \rangle &\leq 2\tau \|y_n - x_{n+1}\| \|A y_n - A w_n\| \\
&\leq 2\tau L \|y_n - x_{n+1}\| \|y_n - w_n\| \\
&\leq \tau L \|y_n - x_{n+1}\|^2 + \tau L \|y_n - w_n\|^2.
\end{aligned} \tag{28}$$

Since $y_n = P_{T_n}(w_n - \tau A y_n)$ and $x_{n+1} \in T_n$, we obtain $\langle w_n - \tau A w_n - y_n, x_{n+1} - y_n \rangle \leq 0$. This implies that

$$\begin{aligned}
2\tau \langle y_n - x_{n+1}, A w_n \rangle &\leq 2\langle y_n - w_n, x_{n+1} - y_n \rangle \\
&= \|x_{n+1} - w_n\|^2 - \|y_n - w_n\|^2 - \|x_{n+1} - y_n\|^2.
\end{aligned} \tag{29}$$

Using Equations (28) and (29) in Equation (27), we obtain:

$$\begin{aligned}\|x_{n+1}-p\|^2 &\leq \|w_n-p\|^2 - \|x_{n+1}-w_n\|^2 + \tau L\|y_n-x_{n+1}\|^2 + \tau L\|y_n-w_n\|^2 + \|x_{n+1}-w_n\|^2 \\ &\quad -\|y_n-w_n\|^2 - \|x_{n+1}-y_n\|^2 \\ &= \|w_n-p\|^2 - (1-\tau L)\|y_n-x_{n+1}\|^2 - (1-\tau L)\|y_n-w_n\|^2.\end{aligned} \quad (30)$$

The proof of Lemma 4 is completed. □

Next, we prove the following results for Algorithm 3.1.

Theorem 1. *Assume that the sequence $\{\alpha_n\}$ is chosen such that*

$$\lim_{n\to\infty}\frac{\alpha_n}{\beta_n}\|x_n-x_{n-1}\|=0. \quad (31)$$

Suppose that $\{x_n\}$ is a sequence generated by our Algorithm 3.1, then $\{x_n\}$ converges strongly to an element $p \in \Gamma$, where we have that $p = P_\Gamma \circ f(p)$.

Proof. Claim I
We need to prove that the sequence $\{x_n\}$ is bounded, for each $p = P_\Gamma \circ f(p)$. By Lemma 4 we have

$$\|z_n-p\|^2 \leq \|w_n-p\|^2 - (1-\tau L)\|y_n-x_{n+1}\|^2 - (1-\tau L)\|y_n-w_n\|^2. \quad (32)$$

This implies that

$$\|z_n-p\| \leq \|w_n-p\|. \quad (33)$$

Using Equation (18), we have

$$\begin{aligned}\|w_n-p\| &= \|x_n + \alpha_n(x_n-x_{n-1})-p\| \\ &\leq \|x_n-p\| + \alpha_n\|x_n-x_{n-1}\| \\ &= \|x_n-p\| + \beta_n \cdot \frac{\alpha_n}{\beta_n}\|x_n-x_{n-1}\|.\end{aligned} \quad (34)$$

Using the condition that $\lim_{n\to\infty}\frac{\alpha_n}{\beta_n}\|x_n-x_{n-1}\|=0$; it follows that there exist a constant $\ell_1 \geq 0$ such that

$$\frac{\alpha_n}{\beta_n}\|x_n-x_{n-1}\| \leq \ell_1, \quad \text{for each } n \geq 0. \quad (35)$$

Hence, using Equations (34) and (35) in Equation (33) we obtain

$$\|z_n-p\| \leq \|w_n-p\| \leq \|x_n-p\| + \beta_n\ell_1. \quad (36)$$

Using (23) and the condition that f is a contraction mapping, we have

$$\begin{aligned}\|x_{n+1}-p\| &= \|f(h_n)-p\| \\ &= \|f(h_n)-f(p)+f(p)-p\| \\ &\leq \|f(h_n)-f(p)\| + \|f(p)-p\| \\ &\leq k\|h_n-p\| + \|f(p)-p\|.\end{aligned} \quad (37)$$

By Equation (22), we have

$$
\begin{aligned}
\|h_n - p\| &= \|(1-\beta_n)z_n + \beta_n f(z_n) - p\| \\
&\leq (1-\beta_n)\|z_n - p\| + \beta_n\|f(z_n) - p\| \\
&\leq (1-\beta_n)\|z_n - p\| + \beta_n\|f(z_n) - f(p)\| + \beta_n\|f(p) - p\| \\
&\leq (1-\beta_n)\|z_n - p\| + \beta_n k\|z_n - p\| + \beta_n\|f(p) - p\| \\
&= (1-\beta_n(1-k))\|z_n - p\| + \beta_n\|f(p) - p\|.
\end{aligned} \qquad (38)
$$

Using Equation (38) in Equation (37), we obtain:

$$
\begin{aligned}
\|x_{n+1} - p\| &\leq k[(1-\beta_n(1-k))\|z_n - p\| + \beta_n\|f(p) - p\|] + \|f(p) - p\| \\
&= k(1-\beta_n(1-k))\|z_n - p\| + k\beta_n\|f(p) - p\| + \|f(p) - p\| \\
&\leq k(1-\beta_n(1-k))\|z_n - p\| + k\|f(p) - p\| + \|f(p) - p\| \\
&= k(1-\beta_n(1-k))\|z_n - p\| + (1+k)\|f(p) - p\|.
\end{aligned} \qquad (39)
$$

Using Equation (36) in Equation (39), we have

$$
\begin{aligned}
\|x_{n+1} - p\| &\leq k(1-\beta_n(1-k))\|x_n - p\| + k\beta_n\ell_1 + (1+k)\|f(p) - p\| \\
&\leq k(1-\beta_n(1-k))\|x_n - p\| + k\ell_1 + (1+k)\|f(p) - p\| \\
&\leq \max\{\|x_n - p\|, \ell_1 + 2\|f(p) - p\|\} \\
&\vdots \\
&\leq \max\{\|x_0 - p\|, \ell_1 + 2\|f(p) - p\|\}.
\end{aligned} \qquad (40)
$$

This means that $\{x_n\}$ is bounded. Hence, it follows that $\{z_n\}$, $\{f(z_n)\}$, $\{h_n\}$, $\{f(h_n)\}$ and $\{w_n\}$ are bounded.
Claim II

$$(1-\tau L)\|y_n - w_n\|^2 + (1-\tau L)\|y_n - x_{n+1}\|^2 \leq \|x_n - p\|^2 - \|x_{n+1} - p\|^2 + \beta_n\ell_5, \qquad (41)$$

for some $\ell_5 > 0$. By Equation (23), we have

$$
\begin{aligned}
\|x_{n+1} - p\|^2 &= \|f(h_n) - p\|^2 \\
&= \|f(h_n) - f(p) + f(p) - p\|^2 \\
&\leq (\|f(h_n) - f(p)\| + \|f(p) - p\|)^2 \\
&\leq (k\|h_n - p\| + \|f(p) - p\|)^2 \\
&\leq \|h_n - p\|^2 + 2\|h_n - p\|\|f(p) - p\| + \|f(p) - p\|^2 \\
&\leq \|h_n - p\|^2 + \ell_2,
\end{aligned} \qquad (42)
$$

for some $\ell_2 > 0$. From Equation (22), we have

$$
\begin{aligned}
\|h_n - p\|^2 &= \|(1-\beta_n)z_n + \beta_n f(z_n) - p\|^2 \\
&\leq (1-\beta_n)\|z_n - p\|^2 + \beta_n\|f(z_n) - p\|^2 \\
&\leq (1-\beta_n)\|z_n - p\|^2 + \beta_n(\|f(z_n) - f(p)\| + \|f(p) - p\|)^2 \\
&\leq (1-\beta_n)\|z_n - p\|^2 + \beta_n(k\|z_n - p\| + \|f(p) - p\|)^2 \\
&\leq (1-\beta_n)\|z_n - p\|^2 + 2\beta_n\|z_n - p\|\|f(p) - p\| + \beta_n\|z_n - p\|^2 + \beta_n\|f(p) - p\|^2 \\
&= \|z_n - p\|^2 + \beta_n(2\|z_n - p\|\|f(p) - p\| + \|f(p) - p\|^2) \\
&\leq \|z_n - p\|^2 + \beta_n\ell_3,
\end{aligned} \qquad (43)
$$

for some $\ell_3 > 0$. Using Equation (32) in Equation (43), we obtain

$$\|h_n - p\|^2 \leq \|w_n - p\|^2 - (1-\tau L)\|y_n - x_{n+1}\|^2 - (1-\tau L)\|y_n - w_n\|^2 + \beta_n\ell_3. \qquad (44)$$

From Equation (36), we have
$$\|w_n - p\| \le \|x_n - p\| + \beta_n \ell_1. \tag{45}$$

This implies that
$$\begin{aligned}\|w_n - p\|^2 &\le (\|x_n - p\| + \beta_n \ell_1)^2 \\ &= \|x_n - p\|^2 + \beta_n(2\ell_1 \|x_n - p\| + \beta_n \ell_1^2) \\ &\le \|x_n - p\|^2 + \beta_n \ell_4,\end{aligned} \tag{46}$$

for some $\ell_4 > 0$. Combining Equations (44) and (46), we have

$$\|h_n - p\|^2 \le \|x_n - p\|^2 + \beta_n \ell_4 - (1 - \tau L)\|y_n - x_{n+1}\|^2 - (1 - \tau L)\|y_n - w_n\|^2 + \beta_n \ell_3. \tag{47}$$

Using Equation (47) in Equation (42), we have

$$\|x_{n+1} - p\|^2 \le \|x_n - p\|^2 + \beta_n \ell_4 - (1 - \tau L)\|y_n - x_{n+1}\| - (1 - \tau L)\|y_n - w_n\|^2 + \beta_n \ell_3 + \beta_n \ell_2. \tag{48}$$

This implies that

$$(1 - \tau L)\|y_n - w_n\|^2 + (1 - \tau L)\|y_n - x_{n+1}\|^2 \le \|x_n - p\|^2 - \|x_{n+1} - p\|^2 + \beta_n \ell_5, \tag{49}$$

where $\ell_5 := \ell_2 + \ell_3 + \ell_4$.

Claim III

$$\begin{aligned}\|x_{n+1} - p\|^2 &\le 2k(1 - \beta_n(1-k))\|x_n - p\|^2 + \\ & \quad 2\beta_n(1-k)\left[\tfrac{2k}{1-k}\langle f(p) - p, x_{n+1} - p\rangle + \tfrac{3D}{1-k}\cdot\tfrac{\alpha_n}{\beta_n}\|x_n - x_{n-1}\| + \tfrac{1}{\beta_n(1-k)}\|f(p) - p\|^2\right],\end{aligned} \tag{50}$$

for some $D > 0$. Using Equations (10) and (18), we have

$$\begin{aligned}\|w_n - p\|^2 &= \|x_n + \alpha_n(x_n - x_{n-1}) - p\|^2 \\ &= \|x_n - p\|^2 + 2\alpha_n\langle x_n - p, x_n - x_{n-1}\rangle + \alpha_n^2\|x_n - x_{n-1}\|^2 \\ &\le \|x_n - p\|^2 + 2\alpha_n\|x_n - p\|\|x_n - x_{n-1}\| + \alpha_n^2\|x_n - x_{n-1}\|.\end{aligned} \tag{51}$$

By Equations (10) and (23), we have

$$\begin{aligned}\|x_{n+1} - p\|^2 &= \|f(h_n) - p\|^2 \\ &= \|f(h_n) - f(p) + f(p) - p\|^2 \\ &= \|f(h_n) - f(p)\|^2 + \|f(p) - p\|^2 + 2\langle f(h_n) - f(p), f(p) - p\rangle \\ &\le k^2\|h_n - p\|^2 + \|f(p) - p\|^2 + 2\|f(h_n) - f(p)\|\|f(p) - p\| \\ &\le k^2\|h_n - p\|^2 + \|f(p) - p\|^2 + k^2\|h_n - p\|^2 + \|f(p) - p\|^2 \\ &\le 2k\|h_n - p\|^2 + 2\|f(p) - p\|^2.\end{aligned} \tag{52}$$

Using Equations (9) and (22), we have

$$\begin{aligned}
\|h_n - p\|^2 &= \|\beta_n f(z_n) + (1 - \beta_n)z_n - p\|^2 \\
&= \|\beta_n(f(z_n) - f(p)) + (1 - \beta_n)(z_n - p) + \beta_n(f(p) - p)\|^2 \\
&\leq \|\beta_n(f(z_n) - f(p)) + (1 - \beta_n)(z_n - p)\|^2 + 2\beta_n\langle f(p) - p, x_{n+1} - p\rangle \\
&\leq \beta_n\|f(z_n) - f(p)\|^2 + (1 - \beta_n)\|z_n - p\|^2 + 2\beta_n\langle f(p) - p, x_{n+1} - p\rangle \\
&\leq \beta_n k^2\|z_n - p\|^2 + (1 - \beta_n)\|z_n - p\|^2 + 2\beta_n\langle f(p) - p, x_{n+1} - p\rangle \\
&\leq \beta_n k\|z_n - p\|^2 + (1 - \beta_n)\|z_n - p\|^2 + 2\beta_n\langle f(p) - p, x_{n+1} - p\rangle \\
&= (1 - \beta_n(1 - k))\|z_n - p\|^2 + 2\beta_n\langle f(p) - p, x_{n+1} - p\rangle \\
&\leq (1 - \beta_n(1 - k))\|w_n - p\|^2 + 2\beta_n\langle f(p) - p, x_{n+1} - p\rangle.
\end{aligned} \quad (53)$$

Using Equation (51) in Equation (53), we have

$$\|h_n - p\|^2 \leq (1 - \beta_n(1 - k))\|x_n - p\|^2 + 2\alpha_n\|x_n - p\|\|x_n - x_{n-1}\| + \alpha_n^2\|x_n - x_{n-1}\|^2 + 2\beta_n\langle f(p) - p, x_{n+1} - p\rangle. \quad (54)$$

Using Equation (54) in Equation (52), we have:

$$\begin{aligned}
\|x_{n+1} - p\|^2 &\leq 2k(1 - \beta_n(1 - k))\|x_n - p\|^2 + 4k\alpha_n\|x_n - p\|\|x_n - x_{n-1}\| + 2k\alpha_n^2\|x_n - x_{n-1}\|^2 \\
&\quad + 4k\beta_n\langle f(p) - p, x_{n+1} - p\rangle + 2\|f(p) - p\|^2 \\
&= 2k(1 - \beta_n(1 - k))\|x_n - p\|^2 + 2k\alpha_n\|x_n - x_{n-1}\|(2\|x_n - p\| + \alpha_n\|x_n - x_{n-1}\|) + \\
&\quad 2(1 - k)\left[\frac{2k\beta_n}{1-k}\langle f(p) - p, x_{n+1} - p\rangle + \frac{1}{1-k}\|f(p) - p\|^2\right] \\
&\leq 2k(1 - \beta_n(1 - k))\|x_n - p\|^2 + 2k\alpha_n\|x_n - x_{n-1}\|(2\|x_n - p\| + \alpha\|x_n - x_{n-1}\|) + \\
&\quad 2(1 - k)\left[\frac{2k\beta_n}{1-k}\langle f(p) - p, x_{n+1} - p\rangle + \frac{1}{1-k}\|f(p) - p\|^2\right] \\
&\leq 2k(1 - \beta_n(1 - k))\|x_n - p\|^2 + 6D\alpha_n\|x_n - x_{n-1}\| + \\
&\quad 2(1 - k)\left[\frac{2k\beta_n}{1-k}\langle f(p) - p, x_{n+1} - p\rangle + \frac{1}{1-k}\|f(p) - p\|^2\right] \\
&\leq 2k(1 - \beta_n(1 - k))\|x_n - p\|^2 + \\
&\quad 2\beta_n(1 - k)\left[\frac{2k}{1-k}\langle f(p) - p, x_{n+1} - p\rangle + \frac{3D}{1-k}\cdot\frac{\alpha_n}{\beta_n}\|x_n - x_{n-1}\| + \frac{1}{\beta_n(1-k)}\|f(p) - p\|^2\right],
\end{aligned} \quad (55)$$

where $D := \sup_{n \in \mathbb{N}}\{\|x_n - p\|, \alpha\|x_n - x_{n-1}\|\} > 0$.

Claim IV

We need to prove that the sequence $\{\|x_n - p\|^2\}$ converges to zero by considering two possible cases.

Case I

There exists a number $N \in \mathbb{N}$ such that $\|x_{n+1} - p\|^2 \leq \|x_n - p\|^2$ for each $n \geq N$. This implies that $\lim_{n \to \infty} \|x_n - p\|$ exists and by Claim II, we have

$$\lim_{n \to \infty} \|y_n - w_n\| = 0, \quad \lim_{n \to \infty} \|y_n - x_{n+1}\| = 0. \quad (56)$$

The fact that the sequence $\{x_n\}$ is bounded implies that there exists a subsequence $\{x_{n_k}\}$ of $\{x_n\}$ that converges weakly to some $z \in H$ such that

$$\limsup_{n \to \infty}\langle f(p) - p, x_n - p\rangle = \lim_{k \to \infty}\langle f(p) - p, x_{n_k} - p\rangle = \langle f(p) - p, z - p\rangle. \quad (57)$$

Using Equation (56) and Lemma 3, we get $z \in \Gamma$. From Equation (57) and the fact that $p = P_\Gamma \circ f(p)$, we get

$$\limsup_{n \to \infty}\langle f(p) - p, x_n - p\rangle = \langle f(p) - p, z - p\rangle \leq 0. \quad (58)$$

Next, we prove that
$$\lim_{n\to\infty} \|x_{n+1} - x_n\| = 0. \tag{59}$$

Clearly,
$$\|w_n - x_n\| = \alpha_n \|x_n - x_{n-1}\| = \frac{\alpha_n}{\beta_n} \cdot \beta_n \|x_n - x_{n-1}\| \longrightarrow 0 \text{ as } n \to \infty. \tag{60}$$

Combining Equations (56) and (60) we have
$$\|x_{n+1} - x_n\| \le \|x_{n+1} - y_n\| + \|y_n - w_n\| + \|w_n - x_n\| \longrightarrow 0 \text{ as } n \to \infty. \tag{61}$$

Using Equations (58) and (59) we have
$$\limsup_{n\to\infty} \langle f(p) - p, x_{n+1} - p \rangle \le \limsup_{n\to\infty} \langle f(p) - p, x_n - p \rangle = \langle f(p) - p, z - p \rangle \le 0. \tag{62}$$

Hence by Lemma 3 and Claim III we have $\lim_{n\to\infty} \|x_n - p\| = 0$.

Case II

We can find a subsequence $\{\|x_{n_j} - p\|^2\}$ of $\{\|x_n - p\|^2\}$ satisfying $\|x_{n_j} - p\|^2 < \|x_{n_j+1} - p\|^2$ for each $j \in \mathbb{N}$. Hence, by Lemma 2 it follows that we can find a nondecreasing real sequence $\{m_k\}$ of $\mathbb{N}$ satisfying $\lim_{k\to\infty} m_k = \infty$ and we get the following inequalities for every $k \in \mathbb{N}$:

$$\|x_{m_k} - p\|^2 \le \|x_{m_k+1} - p\|^2, \quad \|x_k - p\|^2 \le \|x_{m_k} - p\|^2. \tag{63}$$

By Claim II we get
$$(1 - \tau L)\|y_{m_k} - w_{m_k}\|^2 + (1 - \tau L)\|y_{m_k} - x_{m_k+1}\|^2 \le \|x_{m_k} - p\|^2 - \|x_{m_k+1} - p\|^2 + \beta_{m_k} \ell_5 \le \beta_{m_k} \ell_5. \tag{64}$$

Hence, we have
$$\lim_{k\to\infty} \|y_{m_k} - w_{m_k}\| = 0, \quad \lim_{k\to\infty} \|y_{m_k} - x_{m_k+1}\| = 0. \tag{65}$$

By similar arguments as in the proof of Case I, we have
$$\|x_{m_k+1} - x_{m_k}\| \longrightarrow 0 \text{ as } k \to \infty, \tag{66}$$

and
$$\limsup_{k\to\infty} \langle f(p) - p, x_{m_k+1} - p \rangle \le 0. \tag{67}$$

By Claim III we obtain
$$\begin{aligned}\|x_{m_k+1} - p\|^2 \le\ & 2k(1 - \beta_{m_k}(1-k))\|x_{m_k} - p\|^2 + \\ & 2\beta_{m_k}(1-k)[\tfrac{2k}{1-k}\langle f(p) - p, x_{m_k+1} - p\rangle + \tfrac{3D}{1-k} \cdot \tfrac{\alpha_{m_k}}{\beta_{m_k}} \|x_{m_k} - x_{m_k-1}\| + \\ & \tfrac{1}{\beta_{m_k}(1-k)} \|f(p) - p\|^2].\end{aligned} \tag{68}$$

By Equations (63) and (68) we have:
$$\begin{aligned}\|x_{m_k+1} - p\|^2 \le\ & 2k(1 - \beta_{m_k}(1-k))\|x_{m_k} - p\|^2 + \\ & 2\beta_{m_k}(1-k)[\tfrac{2k}{1-k}\langle f(p) - p, x_{m_k+1} - p\rangle + \tfrac{3D}{1-k} \cdot \tfrac{\alpha_{m_k}}{\beta_{m_k}} \|x_{m_k} - x_{m_k-1}\| + \\ & \tfrac{1}{\beta_{m_k}(1-k)} \|f(p) - p\|^2].\end{aligned} \tag{69}$$

Hence, we have

$$\|x_{m_k+1} - p\|^2 \leq \frac{2k}{1-k}\langle f(p) - p, x_{m_k+1} - p\rangle + \frac{3D}{1-k} \cdot \frac{\alpha_{m_k}}{\beta_{m_k}}\|x_{m_k} - x_{m_k-1}\| + \frac{1}{\beta_{m_k}(1-k)}\|f(p) - p\|^2. \quad (70)$$

Therefore we obtain:

$$\limsup_{k \to \infty} \|x_{m_k+1} - p\| \leq 0. \quad (71)$$

Combining Equations (63) and (71) we obtain $\limsup_{k\to\infty} \|x_k - p\| \leq 0$, this means that $x_k \longrightarrow p$. The proof of Theorem 1 is completed. □

Remark 1. *Suantai et al. [39] observed that condition (31) can be easily implemented in numerical results since the value of $\|x_n - x_{n-1}\|$ is given before choosing α_n. We can choose α_n as follows:*

$$\alpha_n = \begin{cases} \min\{\alpha, \frac{\varepsilon_n}{\|x_n - x_{n-1}\|}\}, & \text{if } x_n \neq x_{n-1}, \\ \alpha & \text{otherwise,} \end{cases}$$

where $\alpha \geq 0$ and $\{\varepsilon_n\}$ is a positive sequence such that $\varepsilon_n = o(\beta_n)$.

3.2. Picard–Mann Hybrid Type Inertial Subgradient Extragradient Algorithm

We propose the following algorithm

Algorithm 3.2
Step 0: Given $\tau \in (0, \frac{1}{L})$. $\{\alpha_n\} \subset [0, \alpha)$ for some $\alpha > 0$, $\{\lambda_n\} \subset (a, b) \subset (0, 1 - \beta_n)$ and $\{\beta_n\} \subset (0, 1)$ satisfying the following conditions:

$$\lim_{n \to \infty} \beta_n = 0, \quad \sum_{n=1}^{\infty} \beta_n = \infty. \quad (72)$$

Choose initial $x_0, x_1 \in C$ and set $n := 1$.
Step 1: Compute

$$w_n = x_n + \alpha_n(x_n - x_{n-1}), \quad (73)$$

$$y_n = P_C(w_n - \tau A w_n). \quad (74)$$

If $y_n = w_n$, then stop, y_n is a solution of the (**VIP**) problem. Otherwise, go to **Step 2**.
Step 2: Construct the half-space

$$T_n := \{z \in H : \langle w_n - \tau A w_n - y_n, z - y_n\rangle \leq 0\} \quad (75)$$

and compute

$$z_n = P_{T_n}(w_n - \tau A y_n). \quad (76)$$

Step 3: Calculate

$$h_n = (1 - \lambda_n - \beta_n)x_n + \lambda_n z_n, \quad (77)$$

and compute

$$x_{n+1} = f(h_n). \quad (78)$$

Let $n := n + 1$ and return to **Step 1**.

Next, we prove the following important result for Algorithm 3.2.

Theorem 2. *Suppose that $\{\alpha_n\}$ is a real sequence such that the following condition holds:*

$$\lim_{n\to\infty} \frac{\alpha_n}{\beta_n}\|x_n - x_{n-1}\| = 0. \tag{79}$$

Then the sequence $\{x_n\}$ generated by Algorithm 3.2 converges strongly to an element $p \in \Gamma$, where $\|p\| = \min\{\|z\| : z \in \Gamma\}$.

Proof. We now examine the following claims:

Claim I

We claim that the sequence $\{x_n\}$ is bounded. Using similar arguments as in the proof of Theorem 1, we get

$$\|z_n - p\|^2 \leq \|w_n - p\|^2 - (1-\tau L)\|y_n - x_{n+1}\|^2 - (1-\tau L)\|y_n - w_n\|^2. \tag{80}$$

This implies that

$$\|z_n - p\| \leq \|w_n - p\|. \tag{81}$$

Moreover, we have

$$\|z_n - p\| \leq \|w_n - p\| \leq \|x_n - p\| + \beta_n \ell_1, \tag{82}$$

for some $\ell_1 > 0$.

$$\|x_{n+1} - p\| \leq k\|h_n - p\| + \|f(p) - p\|. \tag{83}$$

Using Equation (77) we have

$$\begin{aligned}
\|h_n - p\| &= \|(1 - \lambda_n - \beta_n)x_n + \lambda_n z_n - p\| \\
&= \|(1 - \lambda_n - \beta_n)(x_n - p) + \lambda_n(z_n - p) - \beta_n p\| \\
&\leq \|(1 - \lambda_n - \beta_n)(x_n - p) + \lambda_n(z_n - p)\| + \beta_n\|p\|.
\end{aligned} \tag{84}$$

Using Equations (10) and (82), we have the following estimate:

$$\begin{aligned}
\|(1-\lambda_n-\beta_n)(x_n-p) + \lambda_n(z_n-p)\|^2 &= (1-\lambda_n-\beta_n)^2\|x_n-p\|^2 + \\
&\quad 2(1-\lambda_n-\beta_n)\lambda_n\langle x_n-p, z_n-p\rangle + \lambda_n^2\|z_n-p\|^2 \\
&\leq (1-\lambda_n-\beta_n)^2\|x_n-p\|^2 + \\
&\quad 2(1-\lambda_n-\beta_n)\lambda_n\|x_n-p\|\|z_n-p\| + \lambda_n^2\|z_n-p\|^2 \\
&\leq (1-\lambda_n-\beta_n)^2\|x_n-p\|^2 + (1-\lambda_n-\beta_n)\lambda_n\|x_n-p\|^2 + \\
&\quad (1-\lambda_n-\beta_n)\lambda_n\|z_n-p\|^2 + \lambda_n^2\|z_n-p\|^2 \\
&\leq (1-\lambda_n-\beta_n)(1-\beta_n)\|x_n-p\|^2 + (1-\beta_n)\lambda_n\|z_n-p\|^2
\end{aligned} \tag{85}$$

This implies that

$$\begin{aligned}
\|(1-\lambda_n-\beta_n)(x_n-p) + \lambda_n(z_n-p)\|^2 &\leq (1-\lambda_n-\beta_n)(1-\beta_n)\|x_n-p\|^2 + \\
&\quad (1-\beta_n)\lambda_n(\|x_n-p\| + \beta_n\ell_1)^2 \\
&\leq (1-\lambda_n-\beta_n)(1-\beta_n)\|x_n-p\|^2 + (1-\beta_n)\lambda_n\|x_n-p\|^2 \\
&\quad +2(1-\beta_n)\lambda_n\beta_n\|x_n-p\|\ell_1 + \beta_n^2\ell_1^2 \\
&\leq (1-\beta_n)^2\|x_n-p\|^2 + 2(1-\beta_n)\beta_n\|x_n-p\|\ell_1 + \beta_n^2\ell_1^2 \\
&= \{(1-\beta_n)\|x_n-p\| + \beta_n\ell_1\}^2.
\end{aligned} \tag{86}$$

This implies that

$$\|(1-\lambda_n-\beta_n)(x_n-p) + \lambda_n(z_n-p)\| \leq (1-\beta_n)\|x_n-p\| + \beta_n\ell_1. \tag{87}$$

Using Equation (87) in Equation (84), we get

$$\begin{aligned} \|h_n - p\| &\leq (1-\beta_n)\|x_n - p\| + \beta_n \ell_1 + \beta_n \|p\| \\ &= (1-\beta_n)\|x_n - p\| + \beta_n(\ell_1 + \|p\|). \end{aligned} \qquad (88)$$

Using Equation (88) in Equation (83), we have

$$\begin{aligned} \|x_{n+1} - p\| &\leq (1-\beta_n)\|x_n - p\| + \beta_n(\ell_1 + \|p\|) + \|f(p) - p\| \\ &\leq \max\{\|x_n - p\|, \ell_1 + \|p\| + \|f(p) - p\|\} \\ &\vdots \\ &\leq \max\{\|x_0 - p\|, \ell_1 + \|p\| + \|f(p) - p\|\}. \end{aligned} \qquad (89)$$

Therefore, the sequence $\{x_n\}$ is bounded. It follows that $\{z_n\}$, $\{w_n\}$ and $\{h_n\}$ are all bounded.

Claim II

We want to show that

$$(1-\beta_n)\lambda_n(1-\tau L)\|y_n - x_{n+1}\|^2 + (1-\beta_n)\lambda_n(1-\tau L)\|y_n - w_n\|^2 \leq \|x_n - p\|^2 - \|x_{n+1} - p\|^2 + \beta_n \ell_6, \qquad (90)$$

for some $\ell_6 > 0$. From Equation (42), we have

$$\|x_{n+1} - p\|^2 \leq \|h_n - p\|^2 + \ell_2, \qquad (91)$$

for some $\ell_2 > 0$. Using (10) and (77) we get

$$\begin{aligned} \|h_n - p\|^2 &= \|(1-\lambda_n - \beta_n)x_n + \lambda_n z_n - p\|^2 \\ &= \|(1-\lambda_n - \beta_n)(x_n - p) + \lambda_n(z_n - p) - \beta_n p\|^2 \\ &= \|(1-\lambda_n - \beta_n)(x_n - p) + \lambda_n(z_n - p)\|^2 - \\ &\quad 2\beta_n \langle (1-\lambda_n - \beta_n)(x_n - p) + \lambda_n(z_n - p), p \rangle + \beta_n^2 \|p\|^2 \\ &\leq \|(1-\lambda_n - \beta_n)(x_n - p) + \lambda_n(z_n - p)\|^2 + \beta_n \ell_3, \end{aligned} \qquad (92)$$

for some $\ell_3 > 0$. Using Equation (85) in Equation (92), we get

$$\|h_n - p\|^2 \leq (1-\lambda_n - \beta_n)(1-\beta_n)\|x_n - p\|^2 + (1-\beta_n)\lambda_n \|z_n - p\|^2 + \beta_n \ell_3. \qquad (93)$$

Using Equation (80) in Equation (93), we get

$$\begin{aligned} \|h_n - p\|^2 &\leq (1-\lambda_n - \beta_n)(1-\beta_n)\|x_n - p\|^2 + (1-\beta_n)\lambda_n[\|w_n - p\|^2 - \\ &\quad (1-\tau L)\|y_n - x_{n+1}\|^2 - (1-\tau L)\|y_n - w_n\|^2] + \beta_n \ell_3 \\ &= (1-\lambda_n - \beta_n)(1-\beta_n)\|x_n - p\|^2 + (1-\beta_n)\lambda_n \|w_n - p\|^2 - \\ &\quad (1-\beta_n)\lambda_n(1-\tau L)\|y_n - x_{n+1}\|^2 - (1-\beta_n)\lambda_n(1-\tau L)\|y_n - w_n\|^2 + \beta_n \ell_3. \end{aligned} \qquad (94)$$

From Equation (36), we get

$$\|z_n - p\| \leq \|w_n - p\| \leq \|x_n - p\| + \beta_n \ell_1. \qquad (95)$$

This implies that

$$\|w_n - p\|^2 \leq \|x_n - p\|^2 + \beta_n \ell_4, \qquad (96)$$

for some $\ell_4 > 0$. Using Equation (96) in Equation (94), we get

$$\begin{aligned}
\|h_n - p\|^2 &\leq (1 - \lambda_n - \beta_n)(1 - \beta_n)\|x_n - p\|^2 + (1 - \beta_n)\lambda_n[\|x_n - p\|^2 + \beta_n\ell_4] - \\
&\quad (1 - \beta_n)\lambda_n(1 - \tau L)\|y_n - x_{n+1}\|^2 - (1 - \beta_n)\lambda_n(1 - \tau L)\|y_n - w_n\|^2 + \beta_n\ell_3 \\
&= (1 - \beta_n)^2\|x_n - p\|^2 + \beta_n(1 - \beta_n)\lambda_n\ell_4 - (1 - \beta_n)\lambda_n(1 - \tau L)\|y_n - x_{n+1}\|^2 - \\
&\quad (1 - \beta_n)\lambda_n(1 - \tau L)\|y_n - w_n\|^2 + \beta_n\ell_3 \\
&\leq \|x_n - p\|^2 - (1 - \beta_n)\lambda_n(1 - \tau L)\|y_n - x_{n+1}\|^2 - \\
&\quad (1 - \beta_n)\lambda_n(1 - \tau L)\|y_n - w_n\|^2 + \beta_n\ell_5,
\end{aligned} \quad (97)$$

for some $\ell_5 > 0$. Using Equation (97) in Equation (91), we have

$$\begin{aligned}
\|x_{n+1} - p\|^2 &\leq \|x_n - p\|^2 - (1 - \beta_n)\lambda_n(1 - \tau L)\|y_n - x_{n+1}\|^2 - (1 - \beta_n)\lambda_n(1 - \tau L)\|y_n - w_n\|^2 \\
&\quad + \beta_n\ell_5 + \ell_2 \\
&\leq \|x_n - p\|^2 - (1 - \beta_n)\lambda_n(1 - \tau L)\|y_n - x_{n+1}\|^2 - (1 - \beta_n)\lambda_n(1 - \tau L)\|y_n - w_n\|^2 \\
&\quad + \beta_n\ell_6,
\end{aligned} \quad (98)$$

for some $\ell_6 > 0$. This implies that

$$(1 - \beta_n)\lambda_n(1 - \tau L)\|y_n - x_{n+1}\|^2 + (1 - \beta_n)\lambda_n(1 - \tau L)\|y_n - w_n\|^2 \leq \|x_n - p\|^2 - \|x_{n+1} - p\|^2 + \beta_n\ell_6. \quad (99)$$

Claim III
We want to show that

$$\|x_{n+1} - p\|^2 \leq 2k\|x_n - p\|^2 + 2k\beta_n[2(1 - \beta_n)\|x_n - p\|\ell_9 + \\
\lambda_n \cdot \frac{\alpha_n}{\beta_n}\|x_n - x_{n-1}\|\ell_9 + \frac{1}{k\beta_n}\|f(p) - p\|^2 - 2\langle(1 - \lambda_n - \beta_n)(x_n - p), p\rangle + 3\ell_9], \quad (100)$$

for some $\ell_9 > 0$.

Using Equations (10) and (78), we have

$$\begin{aligned}
\|x_{n+1} - p\|^2 &= \|f(h_n) - f(p) + f(p) - p\|^2 \\
&= \|f(h_n) - f(p)\|^2 + \|f(p) - p\|^2 + 2\langle f(h_n) - f(p), f(p) - p\rangle \\
&\leq k^2\|h_n - p\|^2 + \|f(p) - p\|^2 + 2\langle f(h_n) - f(p), f(p) - p\rangle \\
&\leq k\|h_n - p\|^2 + \|f(p) - p\|^2 + 2\langle f(h_n) - f(p), f(p) - p\rangle \\
&\leq k\|h_n - p\|^2 + \|f(p) - p\|^2 + 2\|f(h_n) - f(p)\|\|f(p) - p\| \\
&\leq k\|h_n - p\|^2 + \|f(p) - p\|^2 + \|f(h_n) - f(p)\|^2 + \|f(p) - p\|^2 \\
&\leq 2k\|h_n - p\|^2 + 2\|f(p) - p\|^2.
\end{aligned} \quad (101)$$

Next, we have the following estimate, using Equations (10) and (77)

$$\begin{aligned}
\|h_n - p\|^2 &= \|(1 - \lambda_n - \beta_n)x_n + \lambda_n z_n - p\|^2 \\
&= \|(1 - \lambda_n - \beta_n)(x_n - p) + \lambda_n(z_n - p) - \beta_n p\|^2 \\
&= \|(1 - \lambda_n - \beta_n)(x_n - p) + \lambda_n(z_n - p)\|^2 - \\
&\quad 2\beta_n\langle(1 - \lambda_n - \beta_n)(x_n - p) + \lambda_n(z_n - p), p\rangle + \beta_n^2\|p\|^2.
\end{aligned} \quad (102)$$

Using Equation (86) in Equation (102), we have

$$
\begin{aligned}
\|h_n - p\|^2 &\leq \{(1-\beta_n)\|x_n - p\| + \beta_n \ell_1\}^2 - 2\beta_n \langle (1-\lambda_n-\beta_n)(x_n-p) + \lambda_n(z_n-p), p \rangle + \beta_n^2 \|p\|^2 \\
&\leq (1-\beta_n)\|x_n - p\|^2 + 2(1-\beta_n)\beta_n \|x_n - p\|\ell_7 + \beta_n \ell_7^2 - \\
&\quad 2\beta_n \langle (1-\lambda_n-\beta_n)(x_n-p) + \lambda_n(z_n-p), p \rangle + \beta_n^2 \|p\|^2 \\
&= (1-\beta_n)\|x_n - p\|^2 + 2(1-\beta_n)\beta_n \|x_n - p\|\ell_7 + \beta_n \ell_7^2 - 2\beta_n \langle (1-\lambda_n-\beta_n)(x_n-p), p \rangle \\
&\quad -2\lambda_n \beta_n \langle z_n - p, p \rangle + \beta_n^2 \|p\|^2 \\
&= (1-\beta_n)\|x_n - p\|^2 + 2(1-\beta_n)\beta_n \|x_n - p\|\ell_7 + \beta_n \ell_7^2 - 2\beta_n \langle (1-\lambda_n-\beta_n)(x_n-p), p \rangle \\
&\quad + 2\lambda_n \beta_n \langle p - z_n, p \rangle + \beta_n^2 \|p\|^2 \\
&\leq (1-\beta_n)\|x_n - p\|^2 + 2(1-\beta_n)\beta_n \|x_n - p\|\ell_7 + \beta_n \ell_7^2 - 2\beta_n \langle (1-\lambda_n-\beta_n)(x_n-p), p \rangle \\
&\quad + 2\lambda_n \beta_n \|p - z_n\|\|p\| + \beta_n^2 \|p\|^2 \\
&\leq (1-\beta_n)\|x_n - p\|^2 + 2(1-\beta_n)\beta_n \|x_n - p\|\ell_7 + \beta_n \ell_7^2 - 2\beta_n \langle (1-\lambda_n-\beta_n)(x_n-p), p \rangle \\
&\quad + \lambda_n \beta_n \|z_n - p\|^2 + \lambda_n \beta_n \|p\|^2 + \beta_n^2 \|p\|^2 \\
&\leq (1-\beta_n)\|x_n - p\|^2 + 2(1-\beta_n)\beta_n \|x_n - p\|\ell_7 + \beta_n \ell_7^2 - 2\beta_n \langle (1-\lambda_n-\beta_n)(x_n-p), p \rangle \\
&\quad + \lambda_n \beta_n \|w_n - p\|^2 + \lambda_n \beta_n \|p\|^2 + \beta_n^2 \|p\|^2.
\end{aligned} \tag{103}
$$

Next, we have

$$
\begin{aligned}
\|w_n - p\|^2 &= \|x_n + \alpha_n(x_n - x_{n-1}) - p\|^2 \\
&= \|(x_n - p) + \alpha_n(x_n - x_{n-1})\|^2 \\
&= \|x_n - p\|^2 + 2\alpha_n \langle x_n - p, x_n - x_{n-1}\rangle + \alpha_n^2 \|x_n - x_{n-1}\|^2 \\
&\leq \|x_n - p\|^2 + 2\alpha_n \|x_n - p\|\|x_n - x_{n-1}\| + \alpha_n^2 \|x_n - x_{n-1}\|^2 \\
&\leq \|x_n - p\|^2 + \alpha_n \|x_n - x_{n-1}\|\{2\|x_n - p\| + \alpha_n \|x_n - x_{n-1}\|\} \\
&\leq \|x_n - p\|^2 + \alpha_n \|x_n - x_{n-1}\|\ell_8,
\end{aligned} \tag{104}
$$

for some $\ell_8 > 0$. Using Equation (104) in Equation (103) we have

$$
\begin{aligned}
\|h_n - p\|^2 &\leq (1-\beta_n)\|x_n - p\|^2 + 2(1-\beta_n)\beta_n \|x_n - p\|\ell_7 + \beta_n \ell_7^2 - 2\beta_n \langle (1-\lambda_n-\beta_n)(x_n-p), p \rangle \\
&\quad + \lambda_n \beta_n \|x_n - p\|^2 + \lambda_n \beta_n \alpha_n \|x_n - x_{n-1}\|\ell_8 + \lambda_n \beta_n \|p\|^2 + \beta_n^2 \|p\|^2 \\
&\leq \|x_n - p\|^2 + 2(1-\beta_n)\beta_n \|x_n - p\|\ell_7 + \beta_n \ell_7^2 - 2\beta_n \langle (1-\lambda_n-\beta_n)(x_n-p), p \rangle \\
&\quad + \lambda_n \beta_n \alpha_n \|x_n - x_{n-1}\|\ell_8 + \lambda_n \beta_n \|p\|^2 + \beta_n^2 \|p\|^2.
\end{aligned} \tag{105}
$$

Using Equation (105) in Equation (101), we have

$$
\begin{aligned}
\|x_{n+1} - p\|^2 &\leq 2k\|x_n - p\|^2 + 4k(1-\beta_n)\beta_n \|x_n - p\|\ell_7 + 2k\beta_n \ell_7^2 - \\
&\quad 4k\beta_n \langle (1-\lambda_n-\beta_n)(x_n-p), p \rangle + 2k\lambda_n \beta_n \alpha_n \|x_n - x_{n-1}\|\ell_8 + 2k\lambda_n \beta_n \|p\|^2 \\
&\quad + 2k\beta_n^2 \|p\|^2 + 2\|f(p) - p\|^2 \\
&\leq 2k\|x_n - p\|^2 + 4k(1-\beta_n)\beta_n \|x_n - p\|\ell_7 + 2k\beta_n \ell_7^2 - \\
&\quad 4k\beta_n \langle (1-\lambda_n-\beta_n)(x_n-p), p \rangle + 2k\lambda_n \alpha_n \|x_n - x_{n-1}\|\ell_8 + 2k\lambda_n \beta_n \|p\|^2 \\
&\quad + 2k\beta_n^2 \|p\|^2 + 2\|f(p) - p\|^2 \\
&\leq 2k\|x_n - p\|^2 + 2k\beta_n [2(1-\beta_n)\|x_n - p\|\ell_7 + \ell_7^2 - \\
&\quad 2\langle (1-\lambda_n-\beta_n)(x_n-p), p \rangle + \lambda_n \cdot \tfrac{\alpha_n}{\beta_n}\|x_n - x_{n-1}\|\ell_8 + \lambda_n \|p\|^2 + \\
&\quad \beta_n \|p\|^2 + \tfrac{1}{k\beta_n}\|f(p) - p\|^2] \\
&\leq 2k\|x_n - p\|^2 + 2k\beta_n [2(1-\beta_n)\|x_n - p\|\ell_9 + \\
&\quad \lambda_n \cdot \tfrac{\alpha_n}{\beta_n}\|x_n - x_{n-1}\|\ell_9 + \tfrac{1}{k\beta_n}\|f(p) - p\|^2 - 2\langle (1-\lambda_n-\beta_n)(x_n-p), p \rangle + 3\ell_9],
\end{aligned} \tag{106}
$$

for some $\ell_9 > 0$.

Claim IV

We need to prove that the real sequence $\{\|x_n - p\|^2\}$ converges to 0 by considering the following two cases:

Case I

There exists a number $N \in \mathbb{N}$ such that for every $n \geq N$, we have $\|x_{n+1} - p\|^2 \leq \|x_n - p\|^2$. Hence, we have that $\lim_{n \to \infty} \|x_n - p\|$ exists so that by Claim II, we have

$$\lim_{n \to \infty} \|y_n - w_n\| = 0, \quad \lim_{n \to \infty} \|y_n - x_{n+1}\| = 0. \tag{107}$$

Since the sequence $\{x_n\}$ is bounded, it follows that there exists a subsequence $\{x_{n_k}\}$ of $\{x_n\}$ such that $\{x_{n_k}\}$ converges weakly to some $z \in H$ such that

$$\limsup_{n \to \infty} \langle f(p) - p, x_n - p \rangle = \lim_{k \to \infty} \langle f(p) - p, x_{n_k} - p \rangle = \langle f(p) - p, z - p \rangle. \tag{108}$$

Using Equation (107) and Lemma 3, we get $z \in \Gamma$. From Equation (108) and the fact that $p = P_\Gamma \circ f(p)$, we get

$$\limsup_{n \to \infty} \langle f(p) - p, x_n - p \rangle = \langle f(p) - p, z - p \rangle \leq 0. \tag{109}$$

Next, we prove that

$$\lim_{n \to \infty} \|x_{n+1} - x_n\| = 0. \tag{110}$$

Clearly,

$$\|w_n - x_n\| = \alpha_n \|x_n - x_{n-1}\| = \frac{\alpha_n}{\beta_n} \cdot \beta_n \|x_n - x_{n-1}\| \longrightarrow 0 \text{ as } n \to \infty. \tag{111}$$

Combining Equations (107) and (111) we have

$$\|x_{n+1} - x_n\| \leq \|x_{n+1} - y_n\| + \|y_n - w_n\| + \|w_n - x_n\| \longrightarrow 0 \text{ as } n \to \infty. \tag{112}$$

Using Equations (109) and (110) we have

$$\limsup_{n \to \infty} \langle f(p) - p, x_{n+1} - p \rangle \leq \limsup_{n \to \infty} \langle f(p) - p, x_n - p \rangle = \langle f(p) - p, z - p \rangle \leq 0. \tag{113}$$

Hence by Lemma 3 and Claim III we have $\lim_{n \to \infty} \|x_n - p\| = 0$.

Case II

We can find a subsequence $\{\|x_{n_j} - p\|^2\}$ of $\{\|x_n - p\|^2\}$ satisfying $\|x_{n_j} - p\|^2 < \|x_{n_j+1} - p\|^2$ for each $j \in \mathbb{N}$. Hence, by Lemma 2 it follows that there is a nondecreasing real sequence $\{m_k\}$ of $\mathbb{N}$ satisfying $\lim_{k \to \infty} m_k = \infty$ so that we get the following inequalities for every $k \in \mathbb{N}$:

$$\|x_{m_k} - p\|^2 \leq \|x_{m_k+1} - p\|^2, \quad \|x_k - p\|^2 \leq \|x_{m_k} - p\|^2. \tag{114}$$

By Claim II we get

$$(1 - \beta_{m_k})\lambda_{m_k}(1 - \tau L)\|y_{m_k} - x_{m_k+1}\|^2 + (1 - \beta_{m_k})\lambda_{m_k}(1 - \tau L)\|y_{m_k} - w_{m_k}\|^2 \leq \|x_{m_k} - p\|^2 - \|x_{m_k+1} - p\|^2 + \beta_{m_k}\ell_6 \leq \beta_{m_k}\ell_6 \tag{115}$$

Hence, we have

$$\lim_{k \to \infty} \|y_{m_k} - w_{m_k}\| = 0, \quad \lim_{k \to \infty} \|y_{m_k} - x_{m_k+1}\| = 0. \tag{116}$$

By similar arguments as in the proof of Case I, we have

$$\|x_{m_k+1} - x_{m_k}\| \longrightarrow 0 \text{ as } k \to \infty, \tag{117}$$

and
$$\limsup_{k \to \infty} \langle f(p) - p, x_{m_k+1} - p \rangle \leq 0. \tag{118}$$

By Claim III we obtain

$$\begin{aligned}\|x_{m_k+1} - p\|^2 &\leq 2k\|x_{m_k} - p\|^2 + 2k\beta_{m_k}[2(1-\beta_{m_k})\|x_{m_k} - p\|\ell_9 + \\ &\quad \lambda_{m_k} \cdot \tfrac{\alpha_{m_k}}{\beta_{m_k}}\|x_{m_k} - x_{m_k-1}\|\ell_9 + \tfrac{1}{k\beta_{m_k}}\|f(p) - p\|^2 - \\ &\quad 2\langle (1 - \lambda_{m_k} - \beta_{m_k})(x_{m_k} - p), p\rangle + 3\ell_9],\end{aligned} \tag{119}$$

for some $\ell_9 > 0$.

By Equations (114) and (119) we have:

$$\begin{aligned}\|x_{m_k+1} - p\|^2 &\leq 2k\|x_{m_k} - p\|^2 + 2k\beta_{m_k}[2(1-\beta_{m_k})\|x_{m_k} - p\|\ell_9 + \\ &\quad \lambda_{m_k} \cdot \tfrac{\alpha_{m_k}}{\beta_{m_k}}\|x_{m_k} - x_{m_k-1}\|\ell_9 + \tfrac{1}{k\beta_{m_k}}\|f(p) - p\|^2 - \\ &\quad 2\langle (1 - \lambda_{m_k} - \beta_{m_k})(x_{m_k} - p), p\rangle + 3\ell_9],\end{aligned} \tag{120}$$

Hence, we have

$$\begin{aligned}\|x_{m_k+1} - p\|^2 &\leq 2(1-\beta_{m_k})\|x_{m_k} - p\|\ell_9 + \lambda_{m_k} \cdot \tfrac{\alpha_{m_k}}{\beta_{m_k}}\|x_{m_k} - x_{m_k-1}\|\ell_9 + \tfrac{1}{k\beta_{m_k}}\|f(p) - p\|^2 - \\ &\quad 2\langle (1 - \lambda_{m_k} - \beta_{m_k})(x_{m_k} - p), p\rangle + 3\ell_9.\end{aligned} \tag{121}$$

Therefore we obtain:

$$\limsup_{k \to \infty} \|x_{m_k+1} - p\| \leq 0. \tag{122}$$

Combining Equations (114) and (122) we obtain $\limsup_{k \to \infty} \|x_k - p\| \leq 0$, this means that $x_k \longrightarrow p$. The proof of Theorem 2 is completed. □

4. Numerical Illustrations

In this section, we consider two numerical examples to illustrate the convergence of Algorithms 3.1, Algorithms 3.2 and compare them with three well-known algorithms. All our numerical illustrations were executed on a HP laptop with the following specifications: Intel(R) Core(TM)i5-6200U CPU 2.3GHz with 4 GB RAM. All our codes were written in MATLAB 2015a. In reporting our numerical results, the following tables, 'Iter.', 'Sec.' and Error denote the number of iterations, the CPU time in seconds and $\|x_{Iter} - x^*\|$, respectively. We choose $\beta_n = \frac{1}{(n+1)}$

$$\alpha_n = \begin{cases} \min\{\alpha_0, \frac{\beta_n^2}{\|x_n - x_{n-1}\|}\}, & \text{if } x_n \neq x_{n-1} \\ \alpha_0, & \text{otherwise.} \end{cases}$$

$f(x) = 0.5x$ for Algorithm 3.1, Algorithm 3.2, $\lambda_n = 1 - \frac{1}{n}$ for Algorithm 3.2.

Example 1. *Suppose that $H = L^2([0,1])$ with the inner product*

$$\langle x, y \rangle := \int_0^1 x(t)y(t)dt, \forall x, y \in H$$

and the included norm

$$\|x\| := \left(\int_0^1 |x(t)|^2 dt\right)^{\frac{1}{2}}, \forall x \in H$$

Let $C := \{x \in H : \|x\| \leq 1\}$ be the unit ball and define an operator $A : C \to H$ by

$$Ax(t) = \max\{0, x(t)\}.$$

and $Q := \{x \in H, \langle a, x \rangle \leq b\}$ where $0 \neq a \in H$ and $b \in \mathbb{R}$,

we can easily see that A is 1-Lipschitz continuous and monotone on C. Considering the condition on C and A, the set of solutions to the variational inequality problem (VIP) is given by

$$T = \{0\} \neq \emptyset.$$

It is known that

$$P_C(x) = \begin{cases} \frac{x}{\|x\|_{L^2}}, & \text{if } \|x\|_{L^2} > 1, \\ x, & \text{if } \|x\|_{L^2} \leq 1. \end{cases}$$

and

$$P_Q(x) = \begin{cases} \frac{b - \langle a, x \rangle}{\|a\|^2} a + x, & \text{if } \langle a, x \rangle > b, \\ x, & \text{if } \langle a, x \rangle \leq b. \end{cases}$$

Now, we apply Algorithm 3.1, Algorithm 3.2, Mainge's algorithm [37] and Kraikaew and Saejung's algorithm [40] to solve the variational inequality problem (VIP). We choose $\alpha_n = \frac{1}{n+1}$ for Mainge's algorithm and Kraikaew and Saejung's algorithm and $\tau = 0.5$ for all algorithms. We use stopping rule $\|x_n - 0\| < 10^{-4}$ or Iter $<= 3000$ for all algorithms. The numerical results of all algorithms with different x_0 are reported in Table 1 below:

Table 1. Numerical results obtained by other algorithms.

Methods	$x_0 = \frac{\sin(-3*t)}{100}$			$x_0 = \frac{(\sin(-3*t) + \cos(-10*t))}{300}$		
	Sec.	Iter.	Error.	Sec.	Iter.	Error.
Algorithm 3.1	0.0022	10	1.1891×10^{-5}	0.0018	9	4.8894×10^{-5}
Algorithm 3.2	0.0019	8	4.7288×10^{-5}	0.0014	7	5.3503×10^{-5}
Algorithm of Kraikaew et al.	0.4063	2287	9.9981×10^{-5}	0.1719	1065	9.9924×10^{-5}
Algorithm of Mainge	0.1250	2287	9.9981×10^{-5}	0.0469	1065	9.9924×10^{-5}

The convergence behaviour of algorithms with different starting point is given in Figures 1 and 2. In these figures, we represent the value of errors $\|x_n - 0\|$ for all algorithms by the y-axis and the number of iterations by the x-axis.

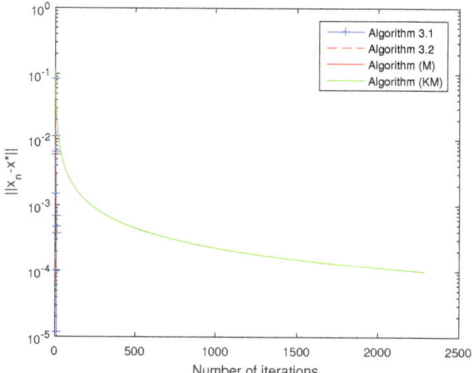

Figure 1. Comparison of all algorithms with $x_0 = \frac{sin(-3t)}{100}$.

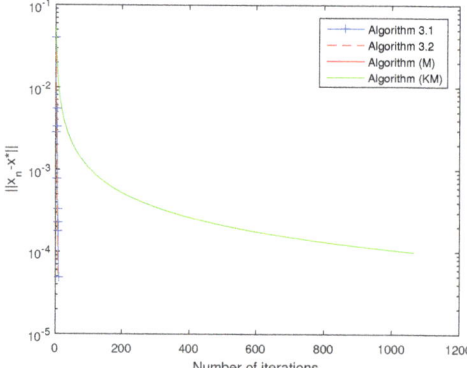

Figure 2. Comparison of all algorithms with $x_0 = \frac{(sin(-3*t)+cos(-10*t))}{300}$.

Example 2. *Assume that $A : \mathbb{R}^m \to \mathbb{R}^m$ is defined by $A(x) = Mx + q$ with $M = BB^T + S + D$, where S is an $m \times m$ skew-symmetric matrix, B is an $m \times m$ matrix, D is an $m \times m$ diagonal matrix, whose diagonal entries are positive (so M is positive definite), q is a vector in $\mathbb{R}^m$ and*

$$C := \{x \in \mathbb{R}^m : -5 \leq x_i \leq 5, i = 1, \cdots, m\}.$$

Clearly, we can see that the operator A is monotone and Lipschitz continuous with a Lipschitz constant $L = \|M\|$. Given that $q = 0$, the unique solution of the corresponding (VIP) is $\{0\}$.

We will compare Algorithm 3.1, Algorithm 3.2 with Tseng's extragradient method (TEGM) [41], Inertial Tseng extragradient algorithm (ITEGM) of Thong and Hieu [33], subgradient extragradient method (SEGM) of Censor et al. [5]. We choose $\tau = \frac{0.9}{L}$ for all algorithm, $\alpha_n = \alpha = 0.99 \frac{\sqrt{1 + 8\epsilon} - 1 - 2\epsilon}{2(1 - \epsilon)}$ where $\epsilon = \frac{1 - \lambda L}{1 + \lambda L}$ for inertial Tseng extragradient algorithm. The starting points are $x_0 = (1, 1, ..., 1)^T \in \mathbb{R}^m$.

For experiment, all entries of B, S and D are generated randomly from a normal distribution with mean zero and unit variance. We use stopping rule $\|x_n - 0\| < 10^{-4}$ or Iter $<= 1000$ for all algorithms. The results are described in Table 2 and Figures 3 and 4.

Table 2. Numerical results obtained by other algorithms.

Methods	m = 50			m = 100		
	Sec.	Iter.	Error.	Sec.	Iter.	Error.
Algorithm 3.1	0.08	10	6.9882×10^{-5}	0.14063	10	6.6947×10^{-5}
Algorithm 3.2	0.078	8	9.0032×10^{-5}	0.1	9	9.9385×10^{-5}
TEGM	4.2438	1000	0.0849	9.4531	1000	0.2646
ITEGM	4.5188	1000	0.0790	9.6875	1000	0.2594
SEGM	4.3969	1000	0.0850	9.5156	1000	0.2647

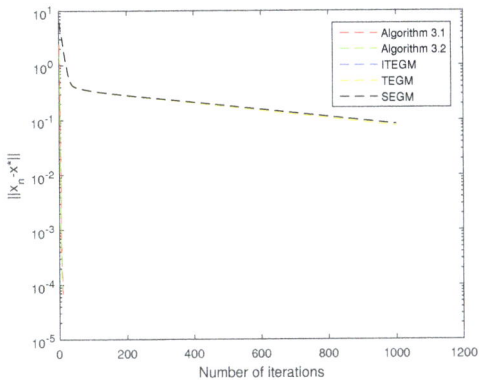

Figure 3. Comparison of all algorithms with $m = 50$.

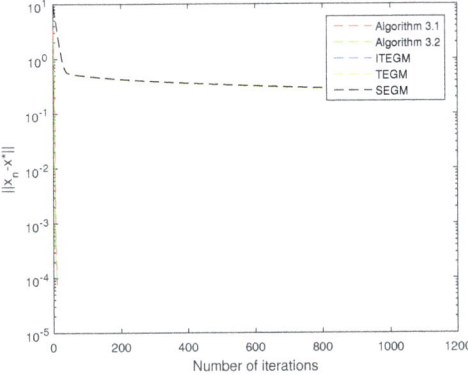

Figure 4. Comparison of all algorithms with $m = 100$.

Tables 1 and 2 and Figures 1–4, give the errors of the Mainge's algorithm [37] and Kraikaew and Saejung's algorithm [40], Tseng's extragradient method (TEGM) [41], Inertial Tseng extragradient algorithm

(ITEGM) [33], subgradient extragradient method (SEGM) of Censor et al. [5] and Algorithms 3.1, 3.2 as well as their execution times. They show that Algorithms 3.1 and 3.2 are less time consuming and more accurate than those of Mainge [37], Kraikaew and Saejung [40], Tseng [41], Thong and Hieu [33] and Censor et al. [5].

5. Conclusions

In this study, we developed two new iterative algorithms for solving K-pseudomonotone variational inequality problems in the framework of real Hilbert spaces. We established some strong convergence theorems for our proposed algorithms under certain conditions. We proved via several numerical experiments that our proposed algorithms performs better in comparison than those of Mainge [37], Kraikaew and Saejung [40], Tseng [41], Thong and Hieu [33] and Censor et al. [5].

Author Contributions: All authors contributed equally to the writing of this paper. All authors have read and agreed to the published version of the manuscript.

Funding: This research was funded by Basque Government: IT1207-19.

Acknowledgments: This paper was completed while the first author was visiting the Abdus Salam School of Mathematical Sciences (ASSMS), Government College University Lahore, Pakistan as a postdoctoral fellow. The authors wish to thank the referees for their comments and suggestions.

Conflicts of Interest: The authors declare no conflict of interests.

Data Availability: The authors declare that all data relating to our results in this paper are available within the paper.

References

1. Stampacchia, G. Formes bilineaires coercitives sur les ensembles convexes. *C. R. Acad. Sci.* **1964**, *258*, 4413–4416.
2. Browder, F.E. The fixed point theory of multivalued mapping in topological vector spaces. *Math. Ann.* **1968**, *177*, 283–301. [CrossRef]
3. Censor, Y.; Gibali, A.; Reich, S. The subgradient extragradient method for solving variational inequalities in Hilbert space. *J. Optim. Theory Appl.* **2011**, *148*, 318–335. [CrossRef] [PubMed]
4. Censor, Y.; Gibali, A.; Reich, S. Strong convergence of subgradient extragradient methods for the variational inequality problem in Hilbert space. *Optim. Methods Softw.* **2011**, *26*, 827–845. [CrossRef]
5. Censor, Y.; Gibali, A.; Reich, S. Algorithms for the split variational inequality problem. *Numer. Algorithm* **2012**, *59*, 301–323. [CrossRef]
6. Censor, Y.; Gibali, A.; Reich, S. Extensions of Korpelevich's extragradient method for the variational inequality problem in Euclidean space. *Optimization* **2012**, *61*, 1119–1132. [CrossRef]
7. Gibali, A.; Shehu, Y. An efficient iterative method for finding common fixed point and variational inequalities in Hilbert spaces. *Optimization* **2019**, *68*, 13–32. [CrossRef]
8. Gibali, A.; Ha, N.H.; Thuong, N.T.; Trang, T.H.; Vinh, N.T. Polyak's gradient method for solving the split convex feasibility problem and its applications. *J. Appl. Numer. Optim.* **2019**, *1*, 145–156.
9. Khan, A.R.; Ugwunnadi, G.C.; Makukula, Z.G.; Abbas, M. Strong convergence of inertial subgradient extragradient method for solving variational inequality in Banach space. *Carpathian J. Math.* **2019**, *35*, 327–338.
10. Maingé, P.E. Projected subgradient techniques and viscosity for optimization with variational inequality constraints. *Eur. J. Oper. Res.* **2010**, *205*, 501–506. [CrossRef]
11. Thong, D.V.; Vinh, N.T.; Cho, Y.J. Accelerated subgradient extragradient methods for variational inequality problems. *J. Sci. Comput.* **2019**, *80*, 1438–1462. [CrossRef]
12. Wang, F.; Pham, H. On a new algorithm for solving variational inequality and fixed point problems. *J. Nonlinear Var. Anal.* **2019**, *3*, 225–233.

13. Wang, L.; Yu, L.; Li, T. Parallel extragradient algorithms for a family of pseudomonotone equilibrium problems and fixed point problems of nonself-nonexpansive mappings in Hilbert space. *J. Nonlinear Funct. Anal.* **2020**, *2020*, 13.
14. Alber, Y.I.; Iusem, A.N. Extension of subgradient techniques for nonsmooth optimization in Banach spaces. *Set Valued Anal.* **2001**, *9*, 315–335. [CrossRef]
15. Xiu, N.H.; Zhang, J.Z. Some recent advances in projection-type methods for variational inequalities. *J. Comput. Appl. Math.* **2003**, *152*, 559–587. [CrossRef]
16. Korpelevich, G.M. The extragradient method for finding saddle points and other problems. *Ekon. Mat. Metod.* **1976**, *12*, 747–756.
17. Farouq, N.E. Pseudomonotone variational inequalities: Convergence of proximal methods. *J. Optim. Theory Appl.* **2001**, *109*, 311–326. [CrossRef]
18. Hadjisavvas, N.; Schaible, S.; Wong, N.-C. Pseudomonotone operators: A survey of the theory and its applications. *J. Optim. Theory Appl.* **2012**, *152*, 1–20. [CrossRef]
19. Kien, B.T.; Lee, G.M. An existence theorem for generalized variational inequalities with discontinuous and pseudomonotone operators. *Nonlinear Anal.* **2011**, *74*, 1495–1500. [CrossRef]
20. Yao, J.-C. Variational inequalities with generalized monotone operators. *Math. Oper. Res.* **1994**, *19*, 691–705. [CrossRef]
21. Karamardian, S. Complementarity problems over cones with monotone and pseudomonotone maps. *J. Optim. Theory Appl.* **1976**, *18*, 445–454. [CrossRef]
22. Attouch, H.; Goudon, X.; Redont, P. The heavy ball with friction. I. The continuous dynamical system. *Commun. Contemp. Math.* **2000**, *2*, 1–34. [CrossRef]
23. Attouch, H.; Czamecki, M.O. Asymptotic control and stabilization of nonlinear oscillators with non-isolated equilibria. *J. Differ. Equ.* **2002**, *179*, 278–310. [CrossRef]
24. Alvarez, F.; Attouch, H. An inertial proximal method for maximal monotone operators via discretization of a nonlinear oscillator with damping. *Set-Valued Anal.* **2001**, *9*, 3–11. [CrossRef]
25. Maingé, P.E. Inertial iterative process for fixed points of certain quasi-nonexpansive mappings. *Set Valued Anal.* **2007**, *15*, 67–69. [CrossRef]
26. Maingé, P.E. Convergence theorems for inertial KM-type algorithms. *J. Comput. Appl. Math.* **2008**, *219*, 223–236. [CrossRef]
27. Bot, R.I.; Csetnek, E.R.; Hendrich, C. Inertial Douglas-Rachford splitting for monotone inclusion problems. *Appl. Math. Comput.* **2015**, *256*, 472–487.
28. Attouch, H.; Peypouquet, J.; Redont, P. A dynamical approach to an inertial forward-backward algorithm for convex minimization. *SIAM J. Optim.* **2014**, *24*, 232–256. [CrossRef]
29. Chen, C.; Chan, R.H.; Ma, S.; Yang, J. Inertial proximal ADMM for linearly constrained separable convex optimization. *SIAM J. Imaging Sci.* **2015**, *8*, 2239–2267. [CrossRef]
30. Bot, R.I.; Csetnek, E.R. A hybrid proximal-extragradient algorithm with inertial effects. *Numer. Funct. Anal. Optim.* **2015**, *36*, 951–963. [CrossRef]
31. Bot, R.I.; Csetnek, E.R. An inertial forward-backward-forward primal-dual splitting algorithm for solving monotone inclusion problems. *Numer. Algor.* **2016**, *71*, 519–540. [CrossRef]
32. Dong, L.Q.; Cho, Y.J.; Zhong, L.L.; Rassias, T.M. Inertial projection and contraction algorithms for variational inequalities. *J. Glob. Optim.* **2018**, *70*, 687–704. [CrossRef]
33. Thong, D.V.; Hieu, D.V. Modified Tseng's extragradient algorithms for variational inequality problems. *J. Fixed Point Theory Appl.* **2018**, *20*, 152. [CrossRef]
34. Moudafi, A. Viscosity approximations methods for fixed point problems. *J. Math. Anal. Appl.* **2000**, *241*, 46–55. [CrossRef]
35. Khan, S.H. A Picard-Mann hybrid iterative process. *Fixed Point Theory Appl.* **2013**, *2013*, 69. [CrossRef]
36. Goebel, K.; Reich, S. *Uniform Convexity, Hyperbolic Geometry and Nonexpansive Mappings*; Marcel Dekker: New York, NY, USA; Basel, Switzerland, 1984.

37. Maingé, P.E. A hybrid extragradient-viscosity method for monotone operators and fixed point problems. *SIAM J. Control Optim.* **2008**, *47*, 1499–1515. [CrossRef]
38. Liu, L.S. Ishikawa and Mann iteration process with errors for nonlinear strongly accretive mappings in Banach spaces. *J. Math. Anal. Appl.* **1995**, *194*, 114–125. [CrossRef]
39. Suantai, S.; Pholasa, N.; Cholamjiak, P. The modified inertial relaxed CQ algorithm for solving the split feasibility problems. *J. Ind. Manag. Optim.* **2018**, *14*, 1595–1615. [CrossRef]
40. Kraikaew, R.; Saejung, S. Strong convergence of the Halpern subgradient extragradient method for solving variational inequalities in Hilbert spaces. *J. Optim. Theory Appl.* **2014**, *163*, 399–412. [CrossRef]
41. Tseng, P. A modified forward-backward splitting method for maximal monotone mappings. *SIAM J. Control Optim.* **2000**, *38*, 431–446. [CrossRef]

© 2020 by the authors. Licensee MDPI, Basel, Switzerland. This article is an open access article distributed under the terms and conditions of the Creative Commons Attribution (CC BY) license (http://creativecommons.org/licenses/by/4.0/).

Article

Fuzzy b-Metric Spaces: Fixed Point Results for ψ-Contraction Correspondences and Their Application

Mujahid Abbas [1,2,†], Fatemeh Lael [3,†] and Naeem Saleem [4,*,†]

1. Department of Mathematics, Government College University, Lahore 54770, Pakistan; abbas.mujahid@gmail.com
2. Department of Mathematics and Applied Mathematics, University of Pretoria, Lynnwood road, Pretoria 0002, South Africa
3. Department of Mathematics, Buein Zahra Technical University, Buein Zahra, Qazvin 3451745346, Iran; f_lael@dena.kntu.ac.ir
4. Department of Mathematics, University of Management and Technology, Lahore 54400, Pakistan
* Correspondence: naeem.saleem2@gmail.com; Tel.: +92-321-426-2145
† These authors contributed equally to this work.

Received: 24 February 2020; Accepted: 26 March 2020; Published: 31 March 2020

Abstract: In this paper we introduce the concepts of ψ-contraction and monotone ψ-contraction correspondence in "fuzzy b-metric spaces" and obtain fixed point results for these contractive mappings. The obtained results generalize some existing ones in fuzzy metric spaces and "fuzzy b-metric spaces". Further we address an open problem in b-metric and "fuzzy b-metric spaces". To elaborate the results obtained herein we provide an example that shows the usability of the obtained results.

Keywords: fixed point; correspondence; fuzzy b-metric space; b-metric space

MSC: 47H09; 47H10; 54H25

---

1. Introduction

Several kinds of nonlinear problems arising in various branches of the sciences can be formulated as a "fixed point problem" mathematically $\mathfrak{f}\mathfrak{x} = \mathfrak{x}$ (an operator equation) where $\mathfrak{f}$ is some nonlinear operator defined on some topological structure. The Banach [1] contraction principle is a significant tool for solving fixed point problems. The simple and constructive nature of its proof has attracted the attention of several researchers around the globe to generalize this famous tool. There are several generalizations; among them, one is to modify the underlying space. In this regard, a framework of "probabilistic metric spaces" is a matter of great interest for scientists and mathematicians (for details, see [2–4]). Kramosil and Michalek [5] defined the "fuzzy metric space". In [6], George and Veeramani modified the concept of "fuzzy metric spaces" using the continuous t-norm. This modification is the generalization of the "probabilistic metric space" to the fuzzy situation. Afterwards the "fuzzy b-metric space" was defined in [7] which generalizes the "fuzzy metric space" and "b-metric space".

The fixed point results in "fuzzy metric space" have deep roots (for details, see [8]). This work has been appreciated by researchers (see [9,10]). This work was extended by several researchers in various ways (compare with [11–21]). Among one of them, in 1969, Nadler proposed Banach's contraction principle for correspondence in Hausdorff metric spaces (see [22]). Various extensions of this work were subsequently proposed by several authors (for details, see [23]). In 1993, Czerwik [24,25] proposed the first Banach fixed point theorem for both single and multivalued mappings in "b-metric spaces", introduced by Bourbaki and Bakhtin [26,27]. Afterwards, this concept was extended for some particular types of contractions in the context of "b-metric spaces" (see [28]). In this direction, many researchers studied and extended various well known fixed point results for several types of contractive mappings in the framework of "b-metric spaces" [29–31].

In general, fixed point theory remained successful in challenging and solving various problems and has contributed significantly to many real-world problems. However, various strong fixed point theorems are proven under strong assumptions. Particularly, in "fuzzy metric spaces", some of these assumptions can lead to some induced norms. Some assumptions do not hold in general or can lead to reformulations as a particular problem in normed vector spaces. The recent trend of research has been dedicated to studying the fundamentals of fixed point theorems and relaxing their conditions by replacing these strong assumptions with weaker ones.

The aim of the work presented in this paper is to provide some fixed point results in "fuzzy b-metric spaces" and to improve their conditions and assumptions by addressing the open questions and challenges outlined in the literature by identifying the ties between "fuzzy b-metric spaces" and "pseudo b-metric spaces".

This paper starts with a brief introduction to "b-metric spaces" and "fuzzy b-metric spaces" along with the required concepts. Afterwards we describe the relation between these two particular spaces. Along with these details, some basic techniques and ways of improving some current fixed point result are also discussed. Finally, an application of Banach's contraction to linear equations is provided.

2. Background and Relevant Literature

This section will serve as an introduction to some fundamental concepts related to "b-metric spaces" and "fuzzy b-metric spaces". Further, some basic definitions and known results are discussed which will be needed in the sequel.

Definition 1. [32] *Let $\mathcal{X}$ be a nonempty set, define a real valued function $\mathfrak{d} : \mathcal{X} \times \mathcal{X} \to [0, \infty)$ such that for a given real number $\mathfrak{s} \geq 1$ satisfies the conditions:*

1. $\mathfrak{d}(\mathfrak{x}, \mathfrak{y}) = 0$ *if and only if* $\mathfrak{x} = \mathfrak{y}$,
2. $\mathfrak{d}(\mathfrak{x}, \mathfrak{y}) = \mathfrak{d}(\mathfrak{y}, \mathfrak{x})$,
3. $\mathfrak{d}(\mathfrak{x}, \mathfrak{z}) \leq \mathfrak{s}[\mathfrak{d}(\mathfrak{x}, \mathfrak{y}) + \mathfrak{d}(\mathfrak{y}, \mathfrak{z})]$, *for all* $\mathfrak{x}, \mathfrak{y}, \mathfrak{z} \in \mathcal{X}$,

the pair $(\mathcal{X}, \mathfrak{d})$ is called a "b-metric space".

It is important to discuss that every "b-metric space" is not necessarily a "metric space" [32]. With $\mathfrak{s} = 1$, every "b-metric space" is a "metric space". If we replace Condition 1 with the following:
- If $\mathfrak{x} = \mathfrak{y}$ implies $\mathfrak{d}(\mathfrak{x}, \mathfrak{y}) = 0$,

then $(\mathcal{X}, \mathfrak{d})$ is called a "pseudo b-metric space". Moreover, it has been shown that several metric fixed point theorems can be extended to "b-metric spaces" (see [33]). It is important to mention that the "b-metric" is not continuous (see [34]). The notion of the "b-metric space" was introduced for the generalization of the fixed point theorem for single valued mappings and correspondences (see [24,25]).

Definition 2. *Let $\mathcal{X}$ be a nonempty set and $\mathfrak{s} \geq 1$ be a real number. A fuzzy subset $\mathcal{M}$ of $\mathcal{M} : \mathcal{X} \times \mathcal{X} \times [0, +\infty) \to [0, 1]$ is called a "fuzzy b-metric" on $\mathcal{X}$ if the following conditions are satisfied for all $\mathfrak{x}, \mathfrak{y}, \mathfrak{z} \in \mathcal{X}$ and $c \in \mathbb{R}$.*

$\mathcal{M}(\mathfrak{x}, \mathfrak{y}, t) = 0$, *for all non-positive real numbers* t,
$\mathcal{M}(\mathfrak{x}, \mathfrak{y}, t) = 1$, *for all* $t \in \mathbb{R}^+$ *if and only if* $\mathfrak{x} = \mathfrak{y}$,
$\mathcal{M}(\mathfrak{x}, \mathfrak{y}, t) = \mathcal{M}(\mathfrak{y}, \mathfrak{x}, t)$,
$\mathcal{M}(\mathfrak{x}, \mathfrak{z}, \mathfrak{s}(t + h)) \geq \min\{\mathcal{M}(\mathfrak{x}, \mathfrak{y}, t), \mathcal{M}(\mathfrak{y}, \mathfrak{z}, h)\}$, *for all* $\mathfrak{s} \geq 1$,

$\mathcal{M}(c\mathfrak{x}, \mathfrak{y}, t) = \mathcal{M}(\mathfrak{x}, \mathfrak{y}, \frac{t}{|c|})$, for $c \neq 0$,

$\mathcal{M}(\mathfrak{x}, \mathfrak{y}, \cdot)$ is a non-decreasing function on $\mathbb{R}$ and $\sup_{t}\{\mathcal{M}(\mathfrak{x}, \mathfrak{y}, t)\} = 1$.

The pair $(\mathcal{X}, \mathcal{M})$ is said to be a "fuzzy b-metric space".

It is important to discuss that for $\mathfrak{s} = 1$, every "fuzzy b-metric space" will reduced to a "fuzzy metric space". The following example explains the concept of the "fuzzy b-metric space".

Example 1. Suppose that $(\mathcal{X}, \partial)$ is a "b-metric space". Define

$$\mathcal{M}(\mathfrak{x}, \mathfrak{y}, t) = \begin{cases} \frac{t}{t + \partial^{\mathfrak{r}}(\mathfrak{x}, \mathfrak{y})} & t > 0, \\ 0 & t \leq 0. \end{cases}$$

Then, $\mathcal{M}(\mathfrak{x}, \mathfrak{y}, t)$ is a "fuzzy b-metric space" for all $\mathfrak{r} \in \mathbb{R}^+$.

Definition 3. Let $(\mathcal{X}, \mathcal{M})$ be a "fuzzy b-metric space". We define the following subset of $\mathcal{X}$, as:

$$B_{\mathfrak{r}}(\mathfrak{x}_0, t_0) = \{\mathfrak{x} \in \mathcal{X} : \mathcal{M}(\mathfrak{x}, \mathfrak{x}_0, t_0) > \mathfrak{r}\},$$

where $\mathfrak{x}_0 \in \mathcal{X}$, $\mathfrak{r} \in (0, 1)$ and $t_0 > 0$.

Let $(\mathcal{X}, \mathcal{M})$ be a "fuzzy b-metric space" and define an open set $O \subseteq \mathcal{X}$ as follows. An element $\mathfrak{x} \in O$ if and only if there exist $\mathfrak{r} \in (0, 1)$ and $t_0 > 0$ such that $B_{\mathfrak{r}}(\mathfrak{x}, t_0) \subseteq O$. Let $\tau_{\mathcal{M}}$ be a topology induced by $\mathcal{M}$ on $\mathcal{X}$ which contains all open sets (for details, see [35,36]). Therefore, with $\tau_{\mathcal{M}}$, some topological notions such that the convergent sequence, Cauchy sequence, closed set, complete set and closure of a set are meaningful. Let $\mathcal{X}$ be a "fuzzy metric space" and suppose that $\partial_{\mathfrak{r}} : \mathcal{X} \times \mathcal{X} \to \mathbb{R}$ for each $\mathfrak{r} \in (0, 1)$ is defined as:

$$\partial_{\mathfrak{r}}(\mathfrak{x}, \mathfrak{y}) = \sup\{t : \mathcal{M}(\mathfrak{x}, \mathfrak{y}, t) \leq \mathfrak{r}\}.$$

Then, $\partial_{\mathfrak{r}}$ is known as a pseudo metric. One can verify that if $\mathcal{X}$ is a fuzzy b-metric then $\partial_{\mathfrak{r}}$ is a pseudo b-metric (for details, see [37]). The family of the pseudo b-metric $\partial_{\mathfrak{r}}(\mathfrak{x}, \mathfrak{y})$ generates a topology on $\mathcal{X}$ which is the same as the topology generated by $\tau_{\mathcal{M}}$. Therefore, $(\mathcal{X}, \mathcal{M})$ is complete if and only if $(\mathcal{X}, \partial_{\mathfrak{r}})$ is complete. It is easy to show that for $\mathfrak{x}, \mathfrak{y}, \mathfrak{z}, \mathfrak{w} \in \mathcal{X}$ and $q \in [0, \infty)$, $\mathcal{M}(\mathfrak{z}, \mathfrak{w}, qt) \geq \mathcal{M}(\mathfrak{x}, \mathfrak{y}, t)$ give that $\partial_{\mathfrak{r}}(\mathfrak{z}, \mathfrak{w}) \leq q\partial_{\mathfrak{r}}(\mathfrak{x}, \mathfrak{y})$, for each $\mathfrak{r} \in (0, 1)$. If we define such pseudo metrics in a "fuzzy b-metric space" then it can lead to a smooth proof for many fixed point theorems in "fuzzy b-metric spaces".

In the following lemma, some equivalences are provided as:

Lemma 1. Let $(\mathcal{X}, \mathcal{M})$ be a "fuzzy b-metric space".

- A sequence $\{\mathfrak{x}_n\} \in \mathcal{X}$ is convergent and converges to $\mathfrak{x} \in \mathcal{X}$ if $\lim_{n} \mathcal{M}(\mathfrak{x}_n, \mathfrak{x}, t) = 1$ for all $t > 0$ and denoted as $\mathfrak{x}_n \to \mathfrak{x}$.
- If $\lim_{n,m} \mathcal{M}(\mathfrak{x}_n, \mathfrak{x}_m, t) = 1$ for all sufficiently large m, n and for any $t > 0$ then $\mathfrak{x}_n$ is called a Cauchy sequence in $\mathcal{X}$.
- If every Cauchy sequence is convergent in $\mathcal{X}$ then $\mathcal{X}$ is called a "complete fuzzy b-metric space".
- A subset $\mathfrak{C}$ of $\mathcal{X}$ is a complete space if and only if it is complete with induced pseudo b-metric $\partial_{\mathfrak{r}}$ for every $\mathfrak{r} \in (0, 1)$.
- A subset $\mathfrak{C} \subset \mathcal{X}$ is open if for every $\mathfrak{x} \in \mathfrak{C}$ there exist $t, \mathfrak{r} > 0$ such that $\mathcal{M}(\mathfrak{x}, \mathfrak{y}, t) > \mathfrak{r}$ implies $\mathfrak{y} \in \mathfrak{C}$.
- A subset $\mathfrak{C} \subset \mathcal{X}$ is closed if it contains all of its limit points.
- The closure of $\mathfrak{C}$ denoted by $\overline{\mathfrak{C}}$ is defined as the set of all points of $\mathcal{X}$ that are the limit points of some sequence in $\mathfrak{C}$.

The following theorem is an equivalent to the "Banach fixed point theorem" in "fuzzy metric space".

Theorem 1. *[8] Let $(\mathcal{X}, \mathcal{M})$ be a "complete fuzzy metric space" and $\mathfrak{T}: \mathcal{X} \to \mathcal{X}$. If*

$$\mathcal{M}(\mathfrak{T}\mathfrak{x}, \mathfrak{T}\mathfrak{y}, t) \geq \mathcal{M}(\mathfrak{x}, \mathfrak{y}, \frac{t}{\mathfrak{k}}),$$

for $\mathfrak{x}, \mathfrak{y} \in \mathcal{X}$, $\mathfrak{k} \in (0,1)$ and $t \in \mathbb{R}$. Then $\mathfrak{T}$ has a fixed point.

Theorem 2. *[37] Suppose $\mathcal{X}$ is a "complete fuzzy metric space", $\mathfrak{T}: \mathcal{X} \to \mathcal{X}$ is a single valued mapping and for every $\mathfrak{d}_\mathfrak{r}$ there exists a constant $\mathfrak{k}_\mathfrak{r}$ with $0 < \mathfrak{k}_\mathfrak{r} < 1$ such that $\mathfrak{d}_\mathfrak{r}(\mathfrak{T}(\mathfrak{x}), \mathfrak{T}(\mathfrak{y})) \leq \mathfrak{k}_\mathfrak{r} \mathfrak{d}_\mathfrak{r}(\mathfrak{x}, \mathfrak{y})$ for all $\mathfrak{x}, \mathfrak{y} \in \mathcal{X}$. Then there exists a unique point $\mathfrak{z} \in \mathcal{X}$ such that $\mathfrak{T}(\mathfrak{z}) = \mathfrak{z}$.*

By a correspondence $\mathfrak{f}$ on a set $\mathcal{X}$ we mean a relation that assigns to each $\mathfrak{x}$ in $\mathcal{X}$ a nonempty subset of $\mathcal{X}$. For a correspondence $\mathfrak{f}$ an element $\mathfrak{x} \in \mathcal{X}$ is said to be a fixed point if $\mathfrak{x} \in \mathfrak{f}(\mathfrak{x})$. It is worthwhile mentioning that it is not necessary for the fixed point of a correspondence to be unique (see Example 3). Define

$$\mathcal{M}(\mathfrak{a}, \mathfrak{f}(\mathfrak{b}), t_0) = \sup_{t < t_0} \sup_{\mathfrak{y} \in \mathfrak{f}(\mathfrak{b})} \mathcal{M}(\mathfrak{a}, \mathfrak{y}, t).$$

In this paper we define the ψ-contractive and monotone ψ-contractive correspondence and prove some results for the existence of fixed points for these contractive conditions in "fuzzy $\mathfrak{b}$-metric spaces", where $\psi \in \Psi$ and Ψ consists of all the functions $\psi : \mathbb{R}^+ \cup \{0\} \to \mathbb{R}^+ \cup \{0\}$ being continuous, nondecreasing and $\psi(1) = 1$. It is important to mention that several researchers have obtained fixed points of correspondence satisfying the contractive conditions via the Hausdorff distance [38–43]. We improve Theorem 1 in a short and comprehensive way and obtain the result without using the Hausdorff distance. Further we answer an open problem related to the "Banach fixed point theorem" in "$\mathfrak{b}$-metric space".

3. Main Results

In the sequel, it is assumed that $(\mathcal{X}, \mathcal{M})$ is a "complete fuzzy $\mathfrak{b}$-metric space" with some $\mathfrak{s} > 1$ and $\mathfrak{f}$ is a closed correspondence i.e. for every $\mathfrak{y}_n \in \mathcal{X}$ such that $\mathfrak{y}_n \in \mathfrak{f}(\mathfrak{x}_n)$, for all $\mathfrak{x}_n \in \mathcal{X}$ then the following implication holds:

$$\mathfrak{x}_n \to \mathfrak{x}, \ \mathfrak{y}_n \to \mathfrak{y} \ \text{implies} \ \mathfrak{y} \in \mathfrak{f}(\mathfrak{x}).$$

The following lemma is a handy tool that will be used in the sequel.

Lemma 2. *[38] A sequence $\{\mathfrak{x}_n\}$ in a "$\mathfrak{b}$-metric space" $(\mathcal{X}, \mathfrak{d})$ is a $\mathfrak{b}$-Cauchy sequence if there exists $\mathfrak{k} \in [0,1)$ such that:*

$$\mathfrak{d}(\mathfrak{x}_n, \mathfrak{x}_{n+1}) \leq \mathfrak{k}\mathfrak{d}(\mathfrak{x}_{n-1}, \mathfrak{x}_n),$$

for every $n \in \mathbb{N}$.

It is also verified that Lemma 2 holds for "pseudo $\mathfrak{b}$-metric spaces" as well.

Definition 4. *Let $\mathfrak{C}$ be a nonempty subset of $(\mathcal{X}, \mathcal{M})$. A correspondence $\mathfrak{f} : \mathfrak{C} \leadsto \mathcal{X}$ is said to be an $\epsilon - \psi$-contraction ($\epsilon.\psi$. $\mathfrak{C}$) if $\psi \in \Psi$, $\epsilon \geq 1$, $\mathfrak{L} > 0$ and for every $\mathfrak{y} \in \mathfrak{C}$ there exists $\mathfrak{w} \in \mathfrak{f}(\mathfrak{y})$ such that:*

$$\psi(\mathcal{M}(\mathfrak{z}, \mathfrak{w}, \frac{t}{\mathfrak{s}^\epsilon})) \geq \min\{\psi(S(\mathfrak{x}, \mathfrak{y}, t)), \psi(I(\mathfrak{x}, \mathfrak{y}, \frac{t}{\mathfrak{L}}))\}, \tag{1}$$

for every $\mathfrak{x} \in \mathfrak{C}$, $\mathfrak{z} \in \mathfrak{f}(\mathfrak{x})$ where

$$S(\mathfrak{x},\mathfrak{y},t) = \min\{\mathcal{M}(\mathfrak{x},\mathfrak{y},t), \mathcal{M}(\mathfrak{x},\mathfrak{f}(\mathfrak{x}),t), \frac{\mathcal{M}(\mathfrak{y},\mathfrak{f}(\mathfrak{y}),t)\mathcal{M}(\mathfrak{x},\mathfrak{f}(\mathfrak{x}),t)}{\mathcal{M}(\mathfrak{x},\mathfrak{y},t)},$$
$$\mathcal{M}(\mathfrak{x},\mathfrak{f}(\mathfrak{y}),2\mathfrak{s}t), \mathcal{M}(\mathfrak{y},\mathfrak{f}(\mathfrak{x}),2\mathfrak{s}t)\},$$

and

$$I(\mathfrak{x},\mathfrak{y},t) = \max\{\min\{\mathcal{M}(\mathfrak{x},\mathfrak{f}(\mathfrak{x}),t), \mathcal{M}(\mathfrak{y},\mathfrak{f}(\mathfrak{y}),t)\}, \mathcal{M}(\mathfrak{x},\mathfrak{f}(\mathfrak{y}),t), \mathcal{M}(\mathfrak{y},\mathfrak{f}(\mathfrak{x}),t)\}.$$

For $\epsilon = 1$, $(\epsilon.\psi. \mathfrak{C})$ is called ψ-contractive or $(\psi. \mathfrak{C})$. The following theorem is a generalization of theorem 1 for $(\psi.\mathfrak{C})$ correspondences in "fuzzy $\mathfrak{b}$-metric spaces".

Theorem 3. *Every $(\psi.\mathfrak{C})$ correspondence $\mathfrak{f}$ has a fixed point.*

Proof. Let $\mathfrak{x}_0$ be any element in the domain of $\mathfrak{f}$. If $\mathfrak{x}_0 \in \mathfrak{f}(\mathfrak{x}_0)$ then $\mathfrak{x}_0$ is the fixed point of $\mathfrak{f}$ and we have obtained the required result. However, if $\mathfrak{x}_0 \notin \mathfrak{f}(\mathfrak{x}_0)$ then choose an arbitrary element $\mathfrak{x}_1 \in \mathfrak{f}(\mathfrak{x}_0)$. By the definition of the $(\psi.\mathfrak{C})$ correspondence there exists an $\mathfrak{x}_2 \in \mathfrak{f}(\mathfrak{x}_1)$ such that

$$\psi(\mathcal{M}(\mathfrak{x}_2,\mathfrak{x}_1,\frac{t}{\mathfrak{s}})) \geq \min\{\psi(S(\mathfrak{x}_1,\mathfrak{x}_0,t)), \psi(I(\mathfrak{x}_1,\mathfrak{x}_0,\frac{t}{\mathfrak{s}}))\}, \tag{2}$$

Now we have to compute $S(\mathfrak{x}_1,\mathfrak{x}_0,t)$ and $I(\mathfrak{x}_1,\mathfrak{x}_0,\frac{t}{\mathfrak{s}})$ where

$$\begin{aligned}S(\mathfrak{x}_1,\mathfrak{x}_0,t) &= \min\{\mathcal{M}(\mathfrak{x}_1,\mathfrak{x}_0,t), \mathcal{M}(\mathfrak{x}_1,\mathfrak{f}(\mathfrak{x}_1),t), \frac{\mathcal{M}(\mathfrak{x}_0,\mathfrak{f}(\mathfrak{x}_0),t)\mathcal{M}(\mathfrak{x}_1,\mathfrak{f}(\mathfrak{x}_1),t)}{\mathcal{M}(\mathfrak{x}_1,\mathfrak{x}_0,t)}, \mathcal{M}(\mathfrak{x}_1,\mathfrak{f}(\mathfrak{x}_0),2\mathfrak{s}t),\\ &\quad \mathcal{M}(\mathfrak{x}_0,\mathfrak{f}(\mathfrak{x}_1),2\mathfrak{s}t)\},\\ &\geq \min\{\mathcal{M}(\mathfrak{x}_1,\mathfrak{x}_0,t), \mathcal{M}(\mathfrak{x}_1,\mathfrak{x}_2,t), \frac{\mathcal{M}(\mathfrak{x}_0,\mathfrak{x}_1,t)\mathcal{M}(\mathfrak{x}_1,\mathfrak{x}_2,t)}{\mathcal{M}(\mathfrak{x}_1,\mathfrak{x}_0,t)}, 1, \mathcal{M}(\mathfrak{x}_0,\mathfrak{x}_2,2\mathfrak{s}t)\},\\ &= \min\{\mathcal{M}(\mathfrak{x}_1,\mathfrak{x}_0,t), \mathcal{M}(\mathfrak{x}_1,\mathfrak{x}_2,t), \mathcal{M}(\mathfrak{x}_0,\mathfrak{x}_2,2\mathfrak{s}t)\},\\ &\geq \min\{\mathcal{M}(\mathfrak{x}_1,\mathfrak{x}_0,t), \mathcal{M}(\mathfrak{x}_1,\mathfrak{x}_2,t), 1, \min\{\mathcal{M}(\mathfrak{x}_0,\mathfrak{x}_1,t), \mathcal{M}(\mathfrak{x}_1,\mathfrak{x}_2,t)\},\end{aligned} \tag{3}$$

and

$$\begin{aligned}I(\mathfrak{x}_1,\mathfrak{x}_0,\frac{t}{\mathfrak{s}}) &= \max\{\min\{\mathcal{M}(\mathfrak{x}_1,\mathfrak{f}(\mathfrak{x}_1),\frac{t}{\mathfrak{s}}), \mathcal{M}(\mathfrak{x}_0,\mathfrak{f}(\mathfrak{x}_0),\frac{t}{\mathfrak{s}})\}, \mathcal{M}(\mathfrak{x}_1,\mathfrak{f}(\mathfrak{x}_0),\frac{t}{\mathfrak{s}}),\\ &\quad \mathcal{M}(\mathfrak{x}_0,\mathfrak{f}(\mathfrak{x}_1),\frac{t}{\mathfrak{s}})\}\\ &\geq \max\{\min\{\mathcal{M}(\mathfrak{x}_1,\mathfrak{x}_2,\frac{t}{\mathfrak{s}}), \mathcal{M}(\mathfrak{x}_0,\mathfrak{x}_1,\frac{t}{\mathfrak{s}})\}, 1, \mathcal{M}(\mathfrak{x}_0,\mathfrak{x}_2,\frac{t}{\mathfrak{s}})\} = 1,\end{aligned}$$

if $\min\{\mathcal{M}(\mathfrak{x}_0,\mathfrak{x}_1,t), \mathcal{M}(\mathfrak{x}_1,\mathfrak{x}_2,t)\} = \mathcal{M}(\mathfrak{x}_1,\mathfrak{x}_2,t)$ then from Inequality 3 we have $S(\mathfrak{x}_1,\mathfrak{x}_0,t) \geq \mathcal{M}(\mathfrak{x}_1,\mathfrak{x}_2,t)$. Then, Inequality 2 becomes

$$\begin{aligned}\psi(\mathcal{M}(\mathfrak{x}_2,\mathfrak{x}_1,\frac{t}{\mathfrak{s}})) &\geq \min\{\psi(\mathcal{M}(\mathfrak{x}_1,\mathfrak{x}_2,t)), \psi(1)\}\\ &\geq \min\{\psi(\mathcal{M}(\mathfrak{x}_1,\mathfrak{x}_2,t)), 1\}\\ &\geq \psi(\mathcal{M}(\mathfrak{x}_1,\mathfrak{x}_2,t)).\end{aligned}$$

From the above inequality we have

$$\psi(\mathcal{M}(\mathfrak{x}_2,\mathfrak{x}_1,\frac{t}{\mathfrak{s}})) \geq \psi(\mathcal{M}(\mathfrak{x}_1,\mathfrak{x}_2,t)).$$

Since ψ is an increasing function, hence we have

$$\mathcal{M}(\mathfrak{x}_2,\mathfrak{x}_1,\frac{t}{\mathfrak{s}}) \geq \mathcal{M}(\mathfrak{x}_1,\mathfrak{x}_2,t).$$

As discussed in the previous section, the related correspondence implies

$$\eth_\mathfrak{r}(\mathfrak{r}_1,\mathfrak{r}_2) \leq \frac{1}{\mathfrak{s}}\eth_\mathfrak{r}(\mathfrak{r}_1,\mathfrak{r}_2),$$

where $\eth_\mathfrak{r}$ is a pseudo b-metric induced by a b-fuzzy metric $\mathcal{M}$. Then the above inequality is true if $\mathfrak{r}_1 = \mathfrak{r}_2$. In this case, $\mathfrak{r}_1$ is a fixed point of $\mathfrak{f}$ and the proof is complete. If not then $\mathfrak{r}_1 \neq \mathfrak{r}_2$. In this case $\min\{\mathcal{M}(\mathfrak{r}_0,\mathfrak{r}_1,t), \mathcal{M}(\mathfrak{r}_1,\mathfrak{r}_2,t)\} = \mathcal{M}(\mathfrak{r}_0,\mathfrak{r}_1,t)$. From Inequality 3 we have $S(\mathfrak{r}_1,\mathfrak{r}_0,t) \geq \mathcal{M}(\mathfrak{r}_0,\mathfrak{r}_1,t)$ and $I(\mathfrak{r}_1,\mathfrak{r}_0,\frac{t}{\mathfrak{s}}) \geq 1$.

$$\begin{aligned}
\psi(\mathcal{M}(\mathfrak{r}_2,\mathfrak{r}_1,\frac{t}{\mathfrak{s}})) &\geq \min\{\psi(\mathcal{M}(\mathfrak{r}_1,\mathfrak{r}_0,t)), \psi(1)\} \\
&\geq \min\{\psi(\mathcal{M}(\mathfrak{r}_1,\mathfrak{r}_0,t)), 1\} \\
&\geq \psi(\mathcal{M}(\mathfrak{r}_1,\mathfrak{r}_0,t)).
\end{aligned}$$

Hence, $\psi(\mathcal{M}(\mathfrak{r}_2,\mathfrak{r}_1,\frac{t}{\mathfrak{s}})) \geq \psi(\mathcal{M}(\mathfrak{r}_1,\mathfrak{r}_2,t))$. Continuing in this way we obtain a sequence $\{\mathfrak{r}_n\}$ for each $n \geq 1$ such that $\mathfrak{r}_{n+1} \in \mathfrak{f}(\mathfrak{r}_n)$ and it satisfies:

$$\psi(\mathcal{M}(\mathfrak{r}_{n+1},\mathfrak{r}_n,\frac{t}{\mathfrak{s}})) \geq \min\{\psi(S(\mathfrak{r}_n,\mathfrak{r}_{n-1},t)), \psi(I(\mathfrak{r}_n,\mathfrak{r}_{n-1},\frac{t}{\mathfrak{s}}))\}.$$

If $\mathfrak{r}_{n+1} = \mathfrak{r}_n$ for some $n \in \mathbb{N}$ then $\mathfrak{f}$ has a fixed point. We assume that $\mathfrak{r}_{n+1} \neq \mathfrak{r}_n$. It is easy to show that $I(\mathfrak{r}_{n-1},\mathfrak{r}_n,t) = 1$. Now we have

$$\begin{aligned}
S(\mathfrak{r}_{n-1},\mathfrak{r}_n,t) &= \min\{\mathcal{M}(\mathfrak{r}_{n-1},\mathfrak{r}_n,t), \mathcal{M}(\mathfrak{r}_{n-1},\mathfrak{f}(\mathfrak{r}_{n-1}),t), \\
&\quad \frac{\mathcal{M}(\mathfrak{r}_n,\mathfrak{f}(\mathfrak{r}_n),t)\mathcal{M}(\mathfrak{r}_{n-1},\mathfrak{f}(\mathfrak{r}_{n-1}),t)}{\mathcal{M}(\mathfrak{r}_{n-1},\mathfrak{r}_n,t)}, \mathcal{M}(\mathfrak{r}_{n-1},\mathfrak{f}(\mathfrak{r}_n),2\mathfrak{s}t), \mathcal{M}(\mathfrak{r}_n,\mathfrak{f}(\mathfrak{r}_{n-1}),2\mathfrak{s}t)\}, \\
&\geq \min\{\mathcal{M}(\mathfrak{r}_{n-1},\mathfrak{r}_n,t), \mathcal{M}(\mathfrak{r}_{n-1},\mathfrak{r}_n,t), \mathcal{M}(\mathfrak{r}_n,\mathfrak{r}_{n+1},t), \\
&\quad \mathcal{M}(\mathfrak{r}_{n-1},\mathfrak{r}_{n+1},2\mathfrak{s}t), \mathcal{M}(\mathfrak{r}_n,\mathfrak{r}_n,2\mathfrak{s}t)\}, \\
&\geq \min\{\mathcal{M}(\mathfrak{r}_{n-1},\mathfrak{r}_n,t), \mathcal{M}(\mathfrak{r}_n,\mathfrak{r}_{n+1},t), \min\{\mathcal{M}(\mathfrak{r}_{n-1},\mathfrak{r}_n,t), \mathcal{M}(\mathfrak{r}_n,\mathfrak{r}_{n+1},t)\}, \\
&= \min\{\mathcal{M}(\mathfrak{r}_{n-1},\mathfrak{r}_n,t), \mathcal{M}(\mathfrak{r}_n,\mathfrak{r}_{n+1},t)\}.
\end{aligned}$$

If $S(\mathfrak{r}_{n-1},\mathfrak{r}_n,t) \geq \mathcal{M}(\mathfrak{r}_n,\mathfrak{r}_{n+1},t)$ then we have

$$\begin{aligned}
\psi(\mathcal{M}(\mathfrak{r}_n,\mathfrak{r}_{n+1},\frac{t}{\mathfrak{s}})) &\geq \min\{\psi(\mathcal{M}(\mathfrak{r}_n,\mathfrak{r}_{n+1},t)), \psi(1)\} \\
&= \psi(\mathcal{M}(\mathfrak{r}_n,\mathfrak{r}_{n+1},t)).
\end{aligned}$$

Since ψ is nondecreasing, so

$$\mathcal{M}(\mathfrak{r}_n,\mathfrak{r}_{n+1},\frac{t}{\mathfrak{s}}) \geq \mathcal{M}(\mathfrak{r}_n,\mathfrak{r}_{n+1},t).$$

This implies that $\eth_\mathfrak{r}(\mathfrak{r}_n,\mathfrak{r}_{n+1}) \leq \frac{1}{\mathfrak{s}}\eth_\mathfrak{r}(\mathfrak{r}_n,\mathfrak{r}_{n+1})$ where $\eth_\mathfrak{r}$ is a "pseudo b-metric" induced by a "b-fuzzy metric" $\mathcal{M}$. Since $\mathfrak{r}_n \neq \mathfrak{r}_{n+1}$ then above inequality generates a contradiction. Hence we have

$$S(\mathfrak{r}_{n-1},\mathfrak{r}_n,t) \geq \mathcal{M}(\mathfrak{r}_{n-1},\mathfrak{r}_n,t).$$

This implies

$$\psi(\mathcal{M}(\mathfrak{r}_n,\mathfrak{r}_{n+1},\frac{t}{\mathfrak{s}})) \geq \min\{\psi(\mathcal{M}(\mathfrak{r}_{n-1},\mathfrak{r}_n,t)), \psi(1)\} = \psi(\mathcal{M}(\mathfrak{r}_{n-1},\mathfrak{r}_n,t)).$$

Thus we have $\mathcal{M}(\mathfrak{r}_n,\mathfrak{r}_{n+1},\frac{t}{\mathfrak{s}}) \geq \mathcal{M}(\mathfrak{r}_{n-1},\mathfrak{r}_n,t)$, for all $n \in \mathbb{N}$. Hence

$$\eth_\mathfrak{r}(\mathfrak{r}_{n+1},\mathfrak{r}_n) \leq \frac{1}{\mathfrak{s}}\eth_\mathfrak{r}(\mathfrak{r}_{n-1},\mathfrak{r}_n).$$

Lemma 2 implies that $\{\mathfrak{x}_n\}$ is a Cauchy sequence by $\partial_{\mathfrak{r}}$ for each $\mathfrak{r} \in (0,1)$. Thus, $\{\mathfrak{x}_n\}$ is a Cauchy sequence in $(\mathcal{X}, \mathcal{M})$. Since $(\mathcal{X}, \mathcal{M})$ is a "complete fuzzy b-metric space" there exists an $\mathfrak{x} \in \mathcal{X}$ such that $\lim_{n\to\infty} \mathcal{M}(\mathfrak{x}_n, \mathfrak{x}, \mathfrak{t}) = 1$. As $\mathfrak{x}_n \in \mathfrak{f}(\mathfrak{x}_{n-1})$, $\mathfrak{x}_n \to \mathfrak{x}$, $\mathfrak{x}_{n-1} \to \mathfrak{x}$ and $\mathfrak{f}$ is closed, this implies $\mathfrak{x} \in \mathfrak{f}(\mathfrak{x})$. □

Clearly, $(\epsilon.\psi.\mathfrak{C})$ is $(\psi.\mathfrak{C})$. Thus, the Theorem 3 also holds for $(\epsilon.\psi.\mathfrak{C})$. The following theorem is equivalent to Nadler's theorem in [22] in the "fuzzy b-metric space".

Theorem 4. *Suppose that* $\mathfrak{f}: \mathcal{X} \rightsquigarrow \mathcal{X}$ *is a correspondence in the "fuzzy b-metric space" such that for* $\mathfrak{x}, \mathfrak{y} \in \mathcal{X}$ *and* $\mathfrak{z} \in \mathfrak{f}(\mathfrak{x})$ *there is* $\mathfrak{w} \in \mathfrak{f}(\mathfrak{y})$ *satisfying the following condition*

$$\mathcal{M}(\mathfrak{z}, \mathfrak{w}, \mathfrak{t}) \geq \mathcal{M}(\mathfrak{x}, \mathfrak{y}, \frac{\mathfrak{t}}{\mathfrak{k}}),$$

where $\mathfrak{k} \in (0,1)$ *and* $\mathfrak{t} \in \mathbb{R}$. *Then,* $\mathfrak{f}$ *has a fixed point.*

Proof. The proof follows using similar arguments as in Theorem 3. □

The following example supports Theorems 3 and 4.

Example 2. *Let* $\mathcal{X} = [0,1]$ *and* $(\mathcal{X}, \mathcal{M})$ *be a "fuzzy b-metric space" where*

$$\mathcal{M}(\mathfrak{x}, \mathfrak{y}, \mathfrak{t}) = \begin{cases} \frac{\mathfrak{t}}{\mathfrak{t}+(\mathfrak{x}-\mathfrak{y})^2} & \mathfrak{t} > 0, \\ 0 & \mathfrak{t} \leq 0. \end{cases}$$

As in Example 1, $(\mathcal{X}, \mathcal{M})$ is a "complete fuzzy b-metric space" with $\mathfrak{s} = 2$. Let $\mathfrak{f}: \mathcal{X} \rightsquigarrow \mathcal{X}$ be defined as $\mathfrak{f}(\mathfrak{x}) = \{\frac{\mathfrak{x}}{2}\}$. It is straightforward to see that for each $\mathfrak{x}, \mathfrak{y} \in \mathcal{X}$,

$$\mathcal{M}(\frac{\mathfrak{x}}{2}, \frac{\mathfrak{y}}{2}, \frac{\mathfrak{t}}{4}) \geq \min\{S(\mathfrak{x}, \mathfrak{y}, \mathfrak{t}), I(\mathfrak{x}, \mathfrak{y}, \mathfrak{t})\}.$$

Thus, for $\mathfrak{L} = 1$, $\epsilon = 2$ and $\psi(\mathfrak{t}) = \mathfrak{t}$, Theorem 3 is satisfied and $\mathfrak{f}$ has a fixed point.

Example 3. *Let* $\mathcal{X} = [0, \frac{1}{2}] \cup \{1\}$ *and* $(\mathcal{X}, \mathcal{M})$ *be a "fuzzy b-metric space" where*

$$\mathcal{M}(\mathfrak{x}, \mathfrak{y}, \mathfrak{t}) = \begin{cases} 0 & |\mathfrak{x} - \mathfrak{y}| \geq \mathfrak{t}, \\ 1 & |\mathfrak{x} - \mathfrak{y}| < \mathfrak{t}. \end{cases}$$

Define a correspondence $\mathfrak{f}$ on $\mathcal{X}$ as

$$\mathfrak{f}(\mathfrak{x}) = \begin{cases} \frac{1}{4} & \mathfrak{x} = 1, \\ \{\frac{1}{4}, \frac{1}{2}\} & \mathfrak{x} \neq 1. \end{cases}$$

We claim that for some $\mathfrak{x}, \mathfrak{y} \in \mathcal{X}$ and $\mathfrak{z} \in \mathfrak{f}(\mathfrak{x})$ there exists a $\mathfrak{w} \in \mathfrak{f}(\mathfrak{y})$ such that $\mathcal{M}(\mathfrak{z}, \mathfrak{w}, \mathfrak{t}) \geq \mathcal{M}(\mathfrak{x}, \mathfrak{y}, \frac{\mathfrak{t}}{\mathfrak{k}})$ where $\mathfrak{k} \in (0,1)$. Suppose that for $\mathfrak{k} = \frac{1}{2}$ and

$$\mathcal{M}(\mathfrak{z}, \mathfrak{w}, \mathfrak{t}) = 1 \text{ or } \mathcal{M}(\mathfrak{x}, \mathfrak{y}, \frac{\mathfrak{t}}{\mathfrak{k}}) = 0,$$

then the claim holds. However,

$$\mathcal{M}(\mathfrak{z}, \mathfrak{w}, \mathfrak{t}) = 0 \text{ and } \mathcal{M}(\mathfrak{x}, \mathfrak{y}, \frac{\mathfrak{t}}{\mathfrak{k}}) = 1,$$

is impossible. Without loss of generality we can suppose that $\mathfrak{y} = 1$ and $\mathfrak{x} \neq 1$. Indeed, if $\mathfrak{x} \neq 1$ and $\mathfrak{y} \neq 1$ we can choose $\mathfrak{z} = \mathfrak{w}$ and therefore, $\mathcal{M}(\mathfrak{z}, \mathfrak{w}, \mathfrak{t}) = 1$. Take $\mathfrak{z} = \frac{1}{4}$ and $\mathfrak{w} \in \{\frac{1}{4}, \frac{1}{2}\}$. If $\mathfrak{z} = \mathfrak{w}$ then $\mathcal{M}(\mathfrak{z}, \mathfrak{w}, \mathfrak{t}) = 1$ is a contradiction. If $\mathfrak{z} = \frac{1}{4}$ and $\mathfrak{w} = \frac{1}{2}$ then

$$\mathcal{M}(\frac{1}{4}, \frac{1}{2}, \mathfrak{t}) = 0 \text{ and } \mathcal{M}(\mathfrak{x}, 1, 2\mathfrak{t}) = 1$$

implies that $\mathfrak{t} \leq \frac{1}{4}$ and $|1 - \mathfrak{x}| < 2\mathfrak{t} \leq \frac{1}{2}$, that is $\frac{1}{2} < \mathfrak{x}$ which is impossible. By Theorem 4, $\mathfrak{x} = \frac{1}{2}$ is a fixed point of the mapping $\mathfrak{f}$.

Corollary 1. *Suppose that* $\mathfrak{T} : \mathcal{X} \to \mathcal{X}$ *is a single valued mapping on a "fuzzy b-metric space" satisfying*

$$\mathcal{M}(\mathfrak{T}(\mathfrak{x}), \mathfrak{T}(\mathfrak{y}), \mathfrak{t}) \geq \mathcal{M}(\mathfrak{x}, \mathfrak{y}, \frac{\mathfrak{t}}{\mathfrak{k}}),$$

for every $\mathfrak{x}, \mathfrak{y} \in \mathcal{X}$, $\mathfrak{k} \in (0, 1)$ *and* $\mathfrak{t} \in \mathbb{R}$. *Then,* $\mathfrak{T}$ *has a unique fixed point.*

Corollary 1 is a "fuzzy b-metric" version of the "Banach fixed point theorem".

Corollary 2. *Suppose that* $\mathfrak{T} : \mathcal{X} \to \mathcal{X}$ *is a single valued mapping on a "complete b-metric space"* $(\mathcal{X}, \mathfrak{d})$ *and*

$$\mathfrak{d}(\mathfrak{T}(\mathfrak{x}), \mathfrak{T}(\mathfrak{y})) \leq \mathfrak{k}\mathfrak{d}(\mathfrak{x}, \mathfrak{y}),$$

holds for every $\mathfrak{x}, \mathfrak{y} \in \mathcal{X}$, $\mathfrak{k} \in (0, 1)$. *Then,* $\mathfrak{T}$ *has a fixed point.*

Proof. The inequality $\mathfrak{d}(\mathfrak{T}(\mathfrak{x}), \mathfrak{T}(\mathfrak{y})) \leq \mathfrak{k}\mathfrak{d}(\mathfrak{x}, \mathfrak{y})$ implies that

$$\frac{\mathfrak{t}}{\mathfrak{t} + \mathfrak{d}(\mathfrak{T}(\mathfrak{x}), \mathfrak{T}(\mathfrak{y}))} \geq \frac{\mathfrak{t}}{\mathfrak{t} + \mathfrak{k}\mathfrak{d}(\mathfrak{x}, \mathfrak{y})}.$$

Therefore

$$\mathcal{M}(\mathfrak{T}(\mathfrak{x}), \mathfrak{T}(\mathfrak{y}), \mathfrak{t}) \geq \mathcal{M}(\mathfrak{x}, \mathfrak{y}, \frac{\mathfrak{t}}{\mathfrak{k}})$$

Note that every "b-metric" is a "fuzzy b-metric", as shown in Example 1. The rest of the proof follows by using Corollary 1. □

Remark 1. *Corollary 2 has been proven for* $\mathfrak{k} \in [\frac{1}{\mathfrak{s}}, 1)$. *It is an open problem whether* $\mathfrak{T}$ *has a fixed point when* $\frac{1}{\mathfrak{s}} \leq \mathfrak{k} < 1$. *Actually we replied to this important question in the Corollary 2.*

Theorem 5. *Let* $\mathfrak{x}_0 \in \mathcal{X}$, $\mathfrak{r} > 0$ *and* $\mathfrak{t}_0 > 0$. *Suppose that* $\mathfrak{f} : \overline{B_{\mathfrak{r}}(\mathfrak{x}_0, \mathfrak{t}_0)} \rightsquigarrow \mathcal{X}$ *is an* $(\epsilon.\psi.\mathfrak{C})$ *correspondence where* $\epsilon > 1$. *Suppose that there exists* $\mathfrak{x}_1 \in \mathfrak{f}(\mathfrak{x}_0)$ *such that*

$$\mathcal{M}(\mathfrak{x}_1, \mathfrak{x}_0, \frac{\mathfrak{s}^{\epsilon-1} - 1}{\mathfrak{s}^{\epsilon}} \mathfrak{t}_0) > \mathfrak{r}.$$

Then, $\mathfrak{f}$ *has a fixed point.*

Proof. Since $\mathcal{M}(\mathfrak{x}_1, \mathfrak{x}_0, \frac{\mathfrak{s}^{\epsilon-1}-1}{\mathfrak{s}^{\epsilon}} \mathfrak{t}_0) > \mathfrak{r}$ we have $\mathcal{M}(\mathfrak{x}_1, \mathfrak{x}_0, \mathfrak{t}_0) \geq \mathcal{M}(\mathfrak{x}_1, \mathfrak{x}_0, \frac{\mathfrak{s}^{\epsilon-1}-1}{\mathfrak{s}^{\epsilon}} \mathfrak{t}_0) > \mathfrak{r}$. This implies that $\mathfrak{x}_1 \in B_{\mathfrak{r}}(\mathfrak{x}_0, \mathfrak{t}_0)$. By the similar arguments as in the proof of Theorem 3 there exists $\mathfrak{x}_2 \in \mathfrak{f}(\mathfrak{x}_1)$ such that

$$\mathcal{M}(\mathfrak{x}_2, \mathfrak{x}_1, \frac{\mathfrak{t}_0}{\mathfrak{s}^{\epsilon}}) \geq \mathcal{M}(\mathfrak{x}_1, \mathfrak{x}_0, \mathfrak{t}_0) > \mathfrak{r}.$$

Therefore, $\mathfrak{d}_{\mathfrak{r}}(\mathfrak{x}_2, \mathfrak{x}_1) \leq \frac{1}{\mathfrak{s}^\epsilon}\mathfrak{d}_{\mathfrak{r}}(\mathfrak{x}_1, \mathfrak{x}_0) \leq \frac{\mathfrak{s}^{\epsilon-1}-1}{\mathfrak{s}^\epsilon} \mathfrak{t}_0$. Following on the same lines we have

$$\mathfrak{d}_{\mathfrak{r}}(\mathfrak{x}_{n+1}, \mathfrak{x}_n) \leq \left(\frac{1}{\mathfrak{s}^\epsilon}\right)^n \mathfrak{d}_{\mathfrak{r}}(\mathfrak{x}_1, \mathfrak{x}_0),$$

for all $n \in \mathbb{N}$. Hence

$$\begin{aligned}
\mathfrak{d}_{\mathfrak{r}}(\mathfrak{x}_n, \mathfrak{x}_0) &\leq \mathfrak{s}\mathfrak{d}_{\mathfrak{r}}(\mathfrak{x}_1, \mathfrak{x}_0) + \mathfrak{s}^2\mathfrak{d}_{\mathfrak{r}}(\mathfrak{x}_1, \mathfrak{x}_2) + \mathfrak{s}^3\mathfrak{d}_{\mathfrak{r}}(\mathfrak{x}_2, \mathfrak{x}_3) + \ldots + \mathfrak{s}^n\mathfrak{d}_{\mathfrak{r}}(\mathfrak{x}_{n-1}, \mathfrak{x}_n), \\
&\leq \mathfrak{s}\mathfrak{d}_{\mathfrak{r}}(\mathfrak{x}_1, \mathfrak{x}_0)[1 + \left(\frac{\mathfrak{s}}{\mathfrak{s}^\epsilon}\right) + \left(\frac{\mathfrak{s}}{\mathfrak{s}^\epsilon}\right)^2 + \ldots + \left(\frac{\mathfrak{s}}{\mathfrak{s}^\epsilon}\right)^{n-2} + \frac{\mathfrak{s}^{n-2}}{\mathfrak{s}^{\epsilon(n-1)}}], \\
&\leq \mathfrak{s}\mathfrak{d}_{\mathfrak{r}}(\mathfrak{x}_1, \mathfrak{x}_0)[1 + \left(\frac{\mathfrak{s}}{\mathfrak{s}^\epsilon}\right) + \left(\frac{\mathfrak{s}}{\mathfrak{s}^\epsilon}\right)^2 + \ldots + \left(\frac{\mathfrak{s}}{\mathfrak{s}^\epsilon}\right)^{n-2} + \left(\frac{\mathfrak{s}}{\mathfrak{s}^\epsilon}\right)^{n-1}], \\
&\leq \frac{\mathfrak{s}}{1 - \frac{1}{\mathfrak{s}^{\epsilon-1}}}\mathfrak{d}_{\mathfrak{r}}(\mathfrak{x}_1, \mathfrak{x}_0) < \frac{1}{\mathfrak{s}^\epsilon}\mathfrak{d}_{\mathfrak{r}}(\mathfrak{x}_1, \mathfrak{x}_0) \\
&< \mathfrak{t}_0,
\end{aligned}$$

for all $n \in \mathbb{N}$. Therefore, the sequence $\mathfrak{x}_n \in B_{\mathfrak{r}}(\mathfrak{x}_0, \mathfrak{t}_0)$, for all $n \in \mathbb{N}$. Following similar arguments to those in the proof of Theorem 3, we deduce that $\{\mathfrak{x}_n\}$ is a Cauchy sequence. By the closeness of $\overline{B_{\mathfrak{r}}(\mathfrak{x}_0, \mathfrak{t}_0)}$ and the completeness of $\mathcal{X}$ there exists an $\mathfrak{x} \in \overline{B_{\mathfrak{r}}(\mathfrak{x}_0, \mathfrak{t}_0)}$ such that $\mathfrak{x}_n \to \mathfrak{x}$. $\mathfrak{f}$ is closed and we have $\mathfrak{x} \in \mathfrak{f}(\mathfrak{x})$. □

A correspondence $\mathfrak{f} : \mathcal{X} \rightsquigarrow \mathcal{X}$ is called monotone if for all $\mathfrak{x} \preceq \mathfrak{y}$, $u \in \mathfrak{f}(\mathfrak{x})$ and $v \in \mathfrak{f}(\mathfrak{y})$ we have $u \preceq v$ (for details, see [44,45]). Suppose that in Definition 4 (defined in [46]), $(\mathcal{X}, \mathcal{M})$ is equipped with a partial order relation $\preceq$ and Inequality 1 holds for $\mathfrak{x}, \mathfrak{y} \in \mathfrak{C}$ where $\mathfrak{x} \preceq \mathfrak{y}$. Then, $\mathfrak{f}$ is said to be monotone ψ-contractive (briefly, monotone $(\psi.\mathfrak{C})$). The following theorem is a generalization of theorem 1 to monotone ψ. $\mathfrak{C}$ correspondences in "ordered fuzzy b-metric space".

Theorem 6. Let $(\mathcal{X}, \mathcal{M})$ be a complete order "fuzzy b-metric space" and $\mathfrak{f}$ be a monotone $(\psi.\mathfrak{C})$ such that $\mathfrak{x}_0 \preceq \mathfrak{f}(\mathfrak{x}_0)$ for some $\mathfrak{x}_0 \in \mathcal{X}$. Then, $\mathfrak{f}$ has a fixed point.

Proof. The proof is closely modeled on Theorem 3. □

4. Application to Linear Equations

In this section we will provide an application of the "Banach fixed point theorem" on "fuzzy b-metric spaces" to linear equations. Now, consider the linear system

$$\mathfrak{a}_{11}\mathfrak{x}_1 + \mathfrak{a}_{12}\mathfrak{x}_2 + \cdots + \mathfrak{a}_{1n}\mathfrak{x}_n = \mathfrak{b}_1,$$

$$\mathfrak{a}_{21}\mathfrak{x}_1 + \mathfrak{a}_{22}\mathfrak{x}_2 + \cdots + \mathfrak{a}_{2n}\mathfrak{x}_n = \mathfrak{b}_2,$$

$$\vdots$$

$$\mathfrak{a}_{n1}\mathfrak{x}_1 + \mathfrak{a}_{n2}\mathfrak{x}_2 + \cdots + \mathfrak{a}_{nn}\mathfrak{x}_n = \mathfrak{b}_n,$$

which has a unique solution. It is equivalent to show that the following linear system has a unique solution.

$$\mathfrak{c}_{11}\mathfrak{x}_1 + \mathfrak{c}_{12}\mathfrak{x}_2 + \cdots + \mathfrak{c}_{1n}\mathfrak{x}_n = \mathfrak{b}'_1$$

$$\mathfrak{c}_{21}\mathfrak{x}_1 + \mathfrak{c}_{22}\mathfrak{x}_2 + \cdots + \mathfrak{c}_{2n}\mathfrak{x}_n = \mathfrak{b}'_2$$

$$\vdots$$

$$\mathfrak{c}_{n1}\mathfrak{x}_1 + \mathfrak{c}_{n2}\mathfrak{x}_2 + \cdots + \mathfrak{c}_{nn}\mathfrak{x}_n = \mathfrak{b}'_n$$

where $c_{ij} = \frac{a_{ij}}{2nM}, i,j \in \{1,\ldots,n\}$, $\mathcal{M} = \sqrt{\max\limits_{i,j} a_{ij}^2}$ and $\mathfrak{b}' = [\frac{b_1}{2nM},\ldots,\frac{b_n}{2nM}]^{\mathfrak{T}}$. For this we consider the "fuzzy b-metric space" generated by

$$\mathcal{M}(\mathfrak{x},\mathfrak{y},t) = \frac{t}{t + \max\limits_{1 \leq j \leq n}|\mathfrak{x}_j - \mathfrak{y}_j|^2},$$

for all $\mathfrak{x},\mathfrak{y} \in \mathbb{R}^n$. Consider the mapping $\mathfrak{T}: \mathbb{R}^n \to \mathbb{R}^n$ defined as

$$\mathfrak{T}(\mathfrak{x}) = \mathfrak{C}\mathfrak{x} + \mathfrak{b}'$$

where $\mathfrak{x} \in \mathbb{R}^n$, $\mathfrak{b}'$ is a column matrix having entries from $\mathbb{R}$ and $\mathfrak{C}$ is an $n \times n$ matrix with c_{ij} arrays. It is essay to show that $\max\limits_{i,j} c_{ij}^2 = \max\limits_{i,j} \frac{a_{ij}^2}{4n^2\mathcal{M}^2}$. Now we have to show that the self-mapping $\mathfrak{T}$ satisfies the "Banach's contraction principle" on "fuzzy b-metric spaces".

$$\begin{aligned}
\max\limits_{1 \leq i \leq n}(\sum_{j=1}^{n} c_{ij}|\mathfrak{x}_j - \mathfrak{y}_j|)^2 &\leq \max\limits_{1 \leq i \leq n}\sum_{j=1}^{n} c_{ij}^2 \sum_{j=1}^{n}|\mathfrak{x}_j - \mathfrak{y}_j|^2, \\
&\leq n^2 \max\limits_{i,j} c_{ij}^2 \max\limits_{j}|\mathfrak{x}_j - \mathfrak{y}_j|^2, \\
&\leq \tfrac{1}{4}\max\limits_{j}|\mathfrak{x}_j - \mathfrak{y}_j|^2.
\end{aligned}$$

This implies that $\mathcal{M}(\mathfrak{T}(\mathfrak{x}),\mathfrak{T}(\mathfrak{y}),t) \geq \mathcal{M}(\mathfrak{x},\mathfrak{y},4t)$. Corollary 1 implies that $\mathfrak{T}$ has a fixed point. Therefore, the linear system has a unique solution.

5. Conclusions

In this article we defined the ψ-contraction and monotone ψ-contraction correspondence and obtained fixed point result in the "fuzzy b-metric space". As a consequence of our main result we obtained the Banach contraction principle in the "fuzzy b-metric space". Further we addressed an open problem in which we generalized the interval of contraction and proved that our results were also valid if contractive constant $\mathfrak{k}$ lied in $[\frac{1}{\mathfrak{s}},1)$, where $\mathfrak{s} \geq 1$. As an application of our result we obtained a solution of the system of n linear equations in the "fuzzy b-metric space". Further we provided examples that further elaborated the useability of our result.

Author Contributions: Supervision and editing, N.S.; Investigation and Writing, F.L.; review, M.A. All authors have read and agreed to the published version of the manuscript.

Funding: This research received no external funding.

Conflicts of Interest: The authors declare no conflict of interest.

References

1. Banach, S. Sur les opérations dans les ensembles abstraits et leur application aux équations intégrales. *Fund. Math.* **1922**, *3*, 133–181. [CrossRef]
2. Schweizer, B.; Sklar, A. Statistical Metric Spaces. *Pac. J. Math.* **1960**, *10*, 313–334. [CrossRef]
3. Schweizer, B.; Sklar, A.; Thorp, E. The Metrization of Statistical Metric Spaces. *Pac. J. Math.* **1960**, *10*, 673–675. [CrossRef]
4. Schweizer, B.; Sklar, A. Triangle Inequalities in a Class of Statistical Metric Spaces. *J. Lond. Math. Soc.* **1963**, *38*, 401–406. [CrossRef]
5. Kramosil, I.; Michalek, J. Fuzzy metric and statistical metric spaces. *Kyber-Netica* **1975**, *11*, 326–334.
6. George, A.; Veeramani, P. On some results in fuzzy metric spaces. *Fuzzy Sets Syst.* **1994**, *64*, 395–399. [CrossRef]

7. Yamaod, O.; Sintunavarat, W. Fixed point theorems for $(\alpha, \beta) - (\psi, \varphi)$- contractive mapping in "b-metric spaces" with some numerical results and applications. *J. Nonlinear Sci. Appl.* **2016**, *9*, 22–34. [CrossRef]
8. Grabiec, M. Fixed points in fuzzy metric spaces. *Fuzzy Sets Syst.* **1983**, *27*, 385–389. [CrossRef]
9. Shena, Y.; Qiub, D.; Chenc, W. Fixed point theorems in fuzzy metric spaces. *Appl. Math. Lett.* **2012**, *25*, 138–141. [CrossRef]
10. Subrahmanyam, P.V. A common fixed point theorem in fuzzy metric spaces. *Inf. Sci.* **1995**, *83*, 109–112. [CrossRef]
11. Abbas, M.; Ali, B.; Vetro, C. Some fixed point results for admissible Geraghty contraction type mappings in fuzzy metric spaces. *Iran. J. Fuzzy Syst.* **2017**, *14*, 161–177.
12. Gregori, V.; Sapena, A. On fixed point theorem in fuzzy metric spaces. *Fuzzy Set Syst.* **2002**, *125*, 245–252. [CrossRef]
13. Kaleva, O. On the convergence of fuzzy sets. *Fuzzy Set Syst.* **1985**, *17*, 53–65. [CrossRef]
14. Razani, A. Existence of fixed point for the nonexpansive mapping of intuitionistic fuzzy metric spaces. *Chaos Solitons Fractals* **2006**, *30*, 367–373. [CrossRef]
15. Saleem, N.; Abbas, M.; Raza, Z. Optimal coincidence best approximation solution in non-archimedean fuzzy metric spaces. *Iran. J. Fuzzy Syst.* **2016**, *13*, 113–124.
16. Saleem, N.; Habib, I.;De la Sen, M. Some new results on coincidence points for multivalued Suzuki type mappings in fairly complete spaces. *Computation* **2020**, *8*, 17. [CrossRef]
17. Saleem, N.; Abbas, M.; Ali, V.; Raza, Z. Fixed points of Suzuki type generalized multivalued mappings in fuzzy metric spaces with applications. *Fixed Point Theory Appl.* **2015**, *2015*, 36. [CrossRef]
18. Sen, M.; Abbas, M.; Saleem, N. On optimal fuzzy best proximity coincidence points of proximal contractions involving cyclic mappings in non-Archimedean fuzzy metric spaces. *Mathematics.* **2017**, *5*, 22. [CrossRef]
19. Abbas, M.; Saleem, N.; De la Sen, M. Optimal coincidence point results in partially ordered non-Archimedean fuzzy metric spaces. *Fixed Point Theory Appl.* **2016**, *2016*, 44. [CrossRef]
20. Abbas, M.; Saleem, N.; Sohail, K. Optimal coincidence best approximation solution in b-fuzzy metric spaces. *Commun. Nonlinear Anal.* **2019**, *6*, 1–12.
21. Raza, Z.; Saleem, N.; Abbas, M. Optimal coincidence points of proximal quasi-contraction mappings in non-Archimedean fuzzy metric spaces. *J. Nonlinear Sci. Appl.* **2016**, *9*, 3787–3801. [CrossRef]
22. Nadler, S.B., Jr. Multi-valued contraction mappings. *Pacific J. Math.* **1969**, *30*, 475–488. [CrossRef]
23. Hadžíc, O. Fixed point theorems for multivalued mappings in some classes of fuzzy metric spaces. *Fuzzy Sets Syst.* **1989**, *29*, 115–125. [CrossRef]
24. Czerwik, S. Contraction mappings in *b*-metric spaces. *Acta Math. Inform. Univ. Ostrav.* **1993**, *1*, 5–11.
25. Czerwik, S. Nonlinear set-valued contraction mappings in *b*-metric spaces. *Atti Semin. Mat. Fis. Univ. Modena* **1998**, *46*, 263–276.
26. Bakhtin, I.A. The contraction mapping principle in almost metric spaces. *Funct. Anal. Unianowsk Gos. Ped. Inst.* **1989**, *30*, 26–37.
27. Bourbaki, N. ; Topologie generale. Herman: Paris, France, **1974**.
28. Kir, N.; Kiziltun, H. On Some well known fixed point theorems in *b*-Metric spaces. *Turk. J. Anal. Number Theory* **2013**, *1*, 13–16. [CrossRef]
29. Saleem, N.; Ali, B.; Abbas, M.; Raza, Z. Fixed points of Suzuki type generalized multivalued mappings in fuzzy metric spaces with applications. *Fixed Point Theory Appl.* **2015**, *1*,1–18. [CrossRef]
30. Abdeljawad, T.; Mlaiki, N.L.; Aydi, H.; Souayah, N. Double controlled metric type spaces and some fixed point results. *Mathematics* **2018**, *6*, 320. [CrossRef]
31. Abdeljawad, T.; Abodayeh, K.; Mlaiki, N. On fixed point generalizations to partial *b*-metric spaces. *J. Comput. Anal. Appl.* **2015**, *19*, 883–891.
32. Singh, S.L.; Prasad, B. Some coincidence theorems and stability of iterative procedures. *Comput. Math. Appl.* **2008**, *55*, 2512–2520. [CrossRef]
33. Van An, T.; Tuyen, L.Q.; Van Dung, N. Stone-type theorem on *b*-metric spaces and applications. *Topol. Its Appl.* **2015**, *185*, 50–64. [CrossRef]
34. Aghajani, A.; Abbas, M.; Roshan, J.R. Common fixed point of generalized weak contractive mappings in partially ordered *b*-metric spaces. *Math. Slovaca* **2014**, *64*, 941–960. [CrossRef]
35. Fallahi, K.; Rad, G.S. Fixed point results in cone metric spaces endowed with a graph. *SCMA* **2017**, *6*, 39–47.

36. Faraji, H.; Nourouzi, K. Fixed and common fixed points for (ψ,φ)-weakly contractive mappings in b-metric spaces. *SCMA* **2017**, *1*, 49–62.
37. Cain, G.L.; Kasriel, R.H. Fixed and periodic points of local contraction mappings on probabilistic metric spaces. *Math. System. Theory* **1976**, *9*, 289–297. [CrossRef]
38. Miculescu, R.; Mihail, A. New fixed point theorems for set-valued contractions in b-metric spaces *arXiv* **2015**, arXiv:1512.03967v1.
39. Sintunavarat, W.; Kumam, P. Common fixed point theorem for cyclic generalized multi-valued contraction mappings. *Appl. Math. Lett.* **2012**, *25*, 1849–1855. [CrossRef]
40. Dinevari, T.; Frigon, M. Fixed point results for multivalued contractions on a metric space with a graph. *J. Math. Anal. Appl.* **2013**, *405*, 507–517. [CrossRef]
41. Pathak, H.K.; Agarwal, R.P.; Cho, Y.J. Coincidence and fixed points for multi-valued mappings and its application to nonconvex integral inclusions. *J. Comput. Appl. Math.* **2015**, *283*, 201–217. [CrossRef]
42. Mehmood, N.; Azam, A.; Kočinac, L.D.R. Multivalued fixed point results in cone metric spaces. *Topology Its Appl.* **2015**, *179*, 156–170. [CrossRef]
43. Tiammee, J.; Charoensawan, P.; Suantai, S. Fixed Point Theorems for Multivalued Nonself G-Almost Contractions in Banach Spaces Endowed with Graphs. *Hindawi J. Funct. Spaces* **2017**, *2017*, 7053849.
44. Lael, F.; Heidarpour, Z. Fixed point theorems for a class of generalized nonexpansive mappings. *Fixed Point Theory Appl.* **2016**, *2016*, 1–7. [CrossRef]
45. Petrusel, G. Fixed point results for multivalued contractions on ordered gauge spaces. *Central Eur. J. Math.* **2009**, *7*, 520–528. [CrossRef]
46. Kumam, P.; Sintunavarat, W. A new contractive condition approach to φ-fixed point results in metric spaces and its applications. *J. Comput. Appl. Math.* **2017**, *311*, 194–204.

© 2020 by the authors. Licensee MDPI, Basel, Switzerland. This article is an open access article distributed under the terms and conditions of the Creative Commons Attribution (CC BY) license (http://creativecommons.org/licenses/by/4.0/).

Article

On a Common Jungck Type Fixed Point Result in Extended Rectangular b-Metric Spaces

Hassen Aydi [1,2], **Zoran D. Mitrović** [3], **Stojan Radenović** [4] **and Manuel de la Sen** [5,*]

1. Institut Supérieur d'Informatique et des Techniques de Communication, H. Sousse, Université de Sousse, Sousse 4000, Tunisia; hassen.aydi@isima.rnu.tn
2. China Medical University Hospital, China Medical University, Taichung 40402, Taiwan
3. Faculty of Electrical Engineering, University of Banja Luka, Patre 5, 78000 Banja Luka, Bosnia and Herzegovina; zoran.mitrovic@etf.unibl.org
4. Faculty of Mechanical Engineering, University of Belgrade, Kraljice Marije 16, 11120 Beograd, Serbia; radens@beotel.rs
5. Institute of Research and Development of Processes IIDP, University of the Basque Country, 48940 Bizkaia, Spain
* Correspondence: manuel.delasen@ehu.eus

Received: 10 December 2019; Accepted: 24 December 2019; Published: 27 December 2019

Abstract: In this paper, we present a Jungck type common fixed point result in extended rectangular b-metric spaces. We also give some examples and a known common fixed point theorem in extended b-metric spaces.

Keywords: fixed points; common fixed points; extended rectangular b-metric space

1. Introduction

The notion of b-metric spaces was first introduced by Bakhtin [1] and Czerwik [2]. This metric type space has been generalized in several directions. Among of them, we may cite, extended b-metric spaces [3], controlled metric spaces [4] and double controlled metric spaces [5]. Within another vision, Branciari [6] initiated rectangular metric spaces. In same direction, Asim et al. [7] included a control function to initiate the concept of extended rectangular b-metric spaces, as a generalization of rectangular b-metric spaces [8].

Definition 1 ([7])**.** *Let X be a nonempty set and $e : X \times X \to [1, \infty)$ be a function. If $d_e : X \times X \to [0, \infty)$ is such that*

(ERbM1) $d_e(\omega, \Omega) = 0$ iff $\omega = \Omega$;
(ERbM2) $d_e(\omega, \Omega) = d_e(\Omega, \omega)$;
(ERbM3) $d_e(\omega, \Omega) \leq e(\omega, \Omega)[d_e(\omega, \zeta) + d_e(\zeta, \sigma) + d_e(\sigma, \Omega)]$;

for all $\omega, \Omega \in X$ and all distinct elements $\zeta, \sigma \in X \setminus \{\omega, \Omega\}$, then d_e is an extended rectangular b-metric on X with mapping e.

Definition 2 ([7])**.** *Let (X, d_e) be an extended rectangular b-metric space, $\{\Omega_n\}$ be a sequence in X and $\Omega \in X$.*

(a) $\{\Omega_n\}$ *converges to* Ω, *if for each* $\tau > 0$ *there is* $n_0 \in \mathbb{N}$ *so that* $d_e(\Omega_n, \Omega) < \tau$ *for any* $n > n_0$. *We write it as* $\lim_{n \to \infty} \Omega_n = \Omega$ *or* $\Omega_n \to \Omega$ *as* $n \to \infty$.
(b) $\{\Omega_n\}$ *is Cauchy if for each* $\tau > 0$ *there is* $n_0 \in \mathbb{N}$ *so that* $d_e(\Omega_n, \Omega_{n+p}) < \tau$ *for any* $n > n_0$ *and* $p > 0$.
(c) (X, d) *is complete if each Cauchy sequence is convergent.*

Note that the topology of rectangular metric spaces need not be Hausdorff. For more examples, see the papers of Sarma et al. [9] and Samet [10]. The topological structure of rectangular metric spaces is not compatible with the topology of classic metric spaces, see Example 7 in the paper of Suzuki [11]. Going in same direction, extended rectangular b-metric spaces can not be Hausdorff. The following example (a variant of Example 1.7 of George et al. [8]) explains this fact.

Example 1. *Let $X = \Gamma_1 \cup \Gamma_2$, where $\Gamma_1 = \{\frac{1}{n}, n \in \mathbb{N}\}$ and Γ_2 is the set of all positive integers. Define $d_e : X \times X \to [0, \infty)$ so that d_e is symmetric and for all $\Omega, \omega \in X$,*

$$d_e(\Omega, \omega) = \begin{cases} 0, & \text{if } \Omega = \omega, \\ 8, & \text{if } \Omega, \omega \in \Gamma_1, \\ \frac{2}{n}, & \text{if } \Omega \in \Gamma_1 \text{ and } \omega \in \{2,3\}, \\ 4 & \text{otherwise.} \end{cases}$$

Here, (X, d_e) is an extended rectangular b-metric space with $e(\Omega, \omega) = 2$. Note that there exist no $\tau_1, \tau_2 > 0$ such that $B_{\tau_1}(2) \cap B_{\tau_2}(3) = \emptyset$ (where $B_x(\tau)$ denotes the ball of center x and radius τ). That is, (X, d_e) is not Hausdorff.

The main result of Jungck [12] is following.

Theorem 1 ([12]). *If f and H are commuting self-maps on a complete metric space (X, d) such that $f(X) \subseteq H(X)$, H is continuous and*

$$d(f\Omega, f\omega) \leq \delta d(H\Omega, H\omega), \tag{1}$$

for all $\Omega, \omega \in X$, where $0 < \delta < 1$, then there is a unique common fixed point of f and H.

Our goal is to get the analogue of Theorem 1 in the setting of extended rectangular b-metric spaces. Some examples are also provided.

2. Main Results

Definition 3. *Let X be a nonempty set and f, H be two commuting self-mappings of X so that $f(X) \subseteq H(X)$. Then (f, H) is called a Jungck pair of mappings on X.*

Example 2. *Let $X = \mathbb{R} \times \mathbb{R}$. Define $f, H : X \to X$ by $f(\omega, \Omega) = (2\omega, (\Omega/2) + 3)$ and $H(\omega, \Omega) = (3\omega, (\Omega/3) + 4)$. Then $f(H(\omega, \Omega)) = (6\omega, (\Omega/6) + 5) = H(f(\omega, \Omega))$, so that (f, H) is a Jungck pair of mappings on X.*

Lemma 1. *Let X be a nonempty set and (f, H) be a Jungck pair of mappings on X. Given $\Omega_0 \in X$. Then there is a sequence $\{\Omega_n\}$ in X so that $H\Omega_{n+1} = f\Omega_n, n \geq 0$.*

Proof. For such $\Omega_0 \in X$, $f\Omega_0$ and $H\Omega_0$ are well defined. Since $f\Omega_0 \in H(X)$, there is $\Omega_1 \in X$ so that $H\Omega_1 = f\Omega_0$. Going in same direction, we arrive to $H\Omega_{n+1} = f\Omega_n$. □

Definition 4. *Let (f, H) be a Jungck pair of mappings on a nonempty set X. Given $e : X \times X \to [1, \infty)$. Let $\{\Omega_n\}$ be a sequence such that $H\Omega_{n+1} = f\Omega_n$, for each $n \geq 0$. Then $\{\Omega_n\}$ is called a (f, H) Jungck sequence in X. We say that $\{\Omega_n\}$ is e-bounded if $\limsup\limits_{n,m \to \infty} e(H\Omega_n, H\Omega_m) < \infty$.*

Remark 1.
1. If $H = id$, $(id(\omega) = \omega, \omega \in X)$ then a (f, id) Jungck sequence is a Picard sequence.
2. Note that each sequence in a rectangular b-metric space with coefficient $s \geq 1$ (see [8]) is e-bounded $(e(\Omega_m, \Omega_n) = s$, for all $m, n \in \mathbb{N})$.

Theorem 2. Let (f, H) be a Jungck pair of mappings on a complete extended rectangular b-metric space (X, d_e) so that
$$d_e(f\Omega, f\omega) \leq \rho d_e(H\Omega, H\omega), \tag{2}$$
for all $\Omega, \omega \in X$, where $0 < \rho < 1$. If H is continuous and there is an e-bounded (f, H) Jungck sequence, then there is a unique common fixed point of f and H.

Proof. Let $\{\Omega_n\}$ be an e-bounded (f, H) Jungck sequence. Then for $\Omega_0 \in X$, $f\Omega_{n+1} = H\Omega_n$, for each $n \geq 0$. We show that $\{f\Omega_n\}$ is Cauchy. From (2), we have

$$\begin{aligned} d_e(H\Omega_{m+k}, H\Omega_{n+k}) &= d_e(f\Omega_{m+k-1}, f\Omega_{n+k-1}) \\ &\leq \rho d_e(H\Omega_{m+k-1}, H\Omega_{n+k-1}). \end{aligned}$$

So,
$$d_e(H\Omega_{m+k}, H\Omega_{n+k}) \leq \rho^k d_e(H\Omega_m, H\Omega_n), \tag{3}$$
for each $k \in \mathbb{N}$.

Case 1:

If $H\Omega_n = H\Omega_{n+1}$ for some n, define $\theta := f\Omega_n = H\Omega_n$. We claim that $f\theta = H\theta = \theta$ and θ is unique. First,
$$f\theta = fH\Omega_n = Hf\Omega_n = H\theta.$$

Let $d_e(\theta, f\theta) > 0$. Here,

$$\begin{aligned} d_e(\theta, f\theta) &= d_e(f\Omega_n, f\theta) \\ &\leq \rho d_e(H\Omega_n, H\theta) \\ &= \rho d_e(\theta, H\theta) \\ &= \rho d_e(\theta, f\theta) \\ &< d_e(\theta, f\theta), \end{aligned}$$

which is a contradiction. Recall that (2) yields that $f\Omega_n = H\Omega_n = \theta$ is the unique common fixed point of f and H.

Case 2:

If $H\Omega_n \neq H\Omega_{n+1}$ for all $n \geq 0$, then $H\Omega_n \neq H\Omega_{n+k}$ for all $n \geq 0$ and $k \geq 1$. Namely, if $H\Omega_n = H\Omega_{n+k}$ for some $n \geq 0$ and $k \geq 1$, we have that

$$\begin{aligned} d_e(H\Omega_{n+1}, H\Omega_{n+k+1}) &= d_e(f\Omega_n, f\Omega_{n+k}) \\ &\leq \rho d_e(H\Omega_n, H\Omega_{n+k}) \\ &= 0. \end{aligned}$$

So, $H\Omega_{n+1} = H\Omega_{n+k+1}$. Then (3) implies that

$$d_e(H\Omega_{n+1}, H\Omega_n) = d_e(H\Omega_{n+k+1}, H\Omega_{n+k}) \leq \rho^k d_e(H\Omega_{n+1}, H\Omega_n) < d_e(H\Omega_{n+1}, H\Omega_n).$$

It is a contradiction. Thus we assume that $H\Omega_n \neq H\Omega_m$ for all integers $n \neq m$. Note that $H\Omega_{m+k} \neq H\Omega_{n+k}$ for any $k \in \mathbb{N}$. Also, $H\Omega_{n+k}, H\Omega_{m+k} \in X \setminus \{H\Omega_n, H\Omega_m\}$. Since (X, d_e) is an extended rectangular b-metric space, by (ERbM3), we get

$$\begin{aligned} d_e(H\Omega_m, H\Omega_n) &\leq e(H\Omega_m, H\Omega_n)[d_e(H\Omega_m, H\Omega_{m+n_0}) + d_e(H\Omega_{m+n_0}, H\Omega_{n+n_0}) \\ &\quad + d_e(H\Omega_{n+n_0}, H\Omega_n)], \end{aligned}$$

where $n_0 \in \mathbb{N}$ so that $\lim\sup_{n,m\to\infty} e(H\Omega_m, H\Omega_n) < \frac{1}{\rho^{n_0}}$. Then

$$\begin{aligned} d_e(H\Omega_m, H\Omega_n) &\leq e(H\Omega_m, H\Omega_n)[\rho^m d_e(H\Omega_0, H\Omega_{n_0}) + \rho^{n_0} d_e(H\Omega_m, H\Omega_n) \\ &\quad + \rho^n d_e(H\Omega_0, H\Omega_{n_0})]. \end{aligned}$$

So,

$$(1 - e(H\Omega_m, H\Omega_n)\rho^{n_0}) d_e(H\Omega_m, H\Omega_n) \leq e(H\Omega_m, H\Omega_n)(\rho^m + \rho^n) d_e(H\Omega_0, H\Omega_{n_0}).$$

From this, we obtain

$$d_e(H\Omega_m, H\Omega_n) \leq \frac{e(H\Omega_m, H\Omega_n)(\rho^m + \rho^n)}{1 - e(H\Omega_m, H\Omega_n)\rho^{n_0}} d_e(H\Omega_0, H\Omega_{n_0}). \tag{4}$$

Thus $\{H\Omega_n\}$ is Cauchy in $H(X)$, which is complete, so there is $u \in X$ so that

$$\lim_{n\to\infty} H\Omega_n = \lim_{n\to\infty} f\Omega_{n-1} = u. \tag{5}$$

The continuity of H together with (2) implies that f is itself continuous. The commutativity of f and H leads to

$$Hu = H(\lim_{n\to\infty} f\Omega_n) = \lim_{n\to\infty} Hf\Omega_n = \lim_{n\to\infty} fH\Omega_n = f(\lim_{n\to\infty} H\Omega_n) = fu. \tag{6}$$

Let $v = Hu = fu$. Then

$$fv = fHu = Hfu = Hv. \tag{7}$$

If $fu \neq fv$, by (2) we find that

$$\begin{aligned} d_e(fu, fv) &\leq \rho d_e(Hu, Hv) \\ &= \rho d_e(fu, fv) \\ &< d_e(fu, fv). \end{aligned}$$

It is a contradiction, hence $fu = fv$. Thus,

$$fv = Hv = v.$$

Condition (2) yields that v is the unique common fixed point. □

Example 3. *If we take in Example 3.1. of [7], $H = id$ and f as*

$$f1 = f2 = f3 = f4 = 2 \quad \text{and} \quad f5 = 1,$$

then all the other conditions of Theorem 2 are satisfied, and so f and H have a unique fixed point, which is, $\theta = 2$. Here, the space (X, d_e) is extended rectangular b-metric space, but it is not extended b-metric space. Hence Theorem 2 generalizes, compliments and improves several known results in existing literature.

A variant of Banach theorem in extended rectangular b-metric spaces is given as follows.

Theorem 3. *Let (X, d_e) be a complete extended rectangular b-metric space and $f : X \to X$ be so that*

$$d_e(f\Omega, f\omega) \leq \rho d_e(\Omega, \omega) \tag{8}$$

for all $\Omega, \omega \in X$, where $\rho \in [0, 1)$. If there is an e-bounded Picard sequence in X, then f has a unique fixed point.

Remark 2. *Theorem 3.1 in [7] is a consequence of Theorem 3. Indeed, instead of condition $\lim_{n,m\to\infty} d_e(\Omega_n, \Omega_m) < \frac{1}{\rho}$ of Theorem 3.1 in [7], we used a weaker condition, that is, $\limsup_{n,m\to\infty} d_e(\Omega_n, \Omega_m) < \infty$.*

3. A Jungck Theorem in Extended b-Metric Spaces

Let (X, d_e) be an extended b-metric space (see Definition 3 in [3]) and $\{\Omega_n\}$ be a (f, H) e-bounded Jungck sequence in X. Then

$$\begin{aligned} d_e(H\Omega_m, H\Omega_n) &\leq e(H\Omega_m, H\Omega_n)[d_e(H\Omega_m, H\Omega_{m+n_0}) + d_e(H\Omega_{m+n_0}, H\Omega_n)] \\ &\leq e(H\Omega_m, H\Omega_n)[d_e(H\Omega_m, H\Omega_{m+n_0}) + \\ &\quad e(H\Omega_{m+n_0}, H\Omega_n)[d_e(H\Omega_{m+n_0}, H\Omega_{n+n_0}) + d_e(H\Omega_{n+n_0}, H\Omega_n)]] \\ &\leq e(H\Omega_m, H\Omega_n)e(H\Omega_{m+n_0}, H\Omega_n)[d_e(H\Omega_m, H\Omega_{m+n_0}) + \\ &\quad d_e(H\Omega_{m+n_0}, H\Omega_{n+n_0}) + d_e(H\Omega_{n+n_0}, H\Omega_n)]. \end{aligned}$$

Since $\{\Omega_n\}$ is a (f, H) e-bounded Jungck sequence, we find that

$$\limsup_{m,n\to\infty} e(H\Omega_m, H\Omega_n)e(H\Omega_{m+n_0}, H\Omega_n) < \infty.$$

By Theorem 2, we obtain the following.

Theorem 4. *Let (f, H) be a Jungck pair of mappings on a complete extended b-metric space (X, d_e) so that*

$$d_e(f\Omega, f\omega) \leq \rho d_e(H\Omega, H\omega), \tag{9}$$

for all $\Omega, \omega \in X$, where $0 < \rho < 1$. If H is continuous and there is an e-bounded (f, H) Jungck sequence, then f and H have a unique common fixed point.

Remark 3. *By Theorem 4, we obtain the Banach contraction principle in extended b-metric spaces. It improves Theorem 2.1 in [13], Theorem 2 in [3] and Theorem 2.1 in [14]. Also Theorem 3 generalizes an open problem raised by George et al. [8].*

Example 4. *Let $X = [0, \infty), e : X \times X \to [1, \infty)$. Consider $d_e : X \times X \to [0, \infty)$ as*

$$d_e(\Omega, \omega) = (\Omega - \omega)^2,$$

where $e(\Omega, \omega) = \Omega + \omega + 2$. Then (X, d_e) is an extended b-metric space. Define $f\Omega = \frac{3\Omega}{4}$. Then (8) holds for $\rho = \frac{9}{16}$. Let $\Omega_0 \in X$ and $\Omega_n = f^n \Omega_0, n \in \mathbb{N}$. Then $\lim_{m,n\to\infty} e(\Omega_m, \Omega_n) = 2$. So, $\lim_{m,n\to\infty} e(\Omega_m, \Omega_n) > \frac{16}{9}$ and Theorem 3.1 in [7] is not applicable. Applying Theorem 3, we conclude that f has a unique fixed point.

Author Contributions: All authors contributed equally and significantly in writing this paper. All authors have read and agreed to the published version of the manuscript.

Funding: This work has been partially supported by Basque Governmnet through Grant IT1207-19.

Conflicts of Interest: The authors declare no conflict of interest.

References

1. Bakhtin, I.A. The contraction mapping principle in quasimetric spaces. *Funct. Anal. Ulianowsk Gos. Ped. Inst.* **1989**, *3*, 26–37.
2. Czerwik, S. Contraction mappings in b-metric spaces. *Acta Math. Inform. Univ. Ostrav.* **1993**, *1*, 5–11.
3. Kamran, T.; Samreen, M.; UL Ain, Q. A Generalization of b-metric space and some fixed point theorems. *Mathematics* **2017**, *5*, 19. [CrossRef]
4. Abdeljawad, T.; Mlaiki, N.; Aydi, H.; Souayah, N. Double controlled metric type spaces and some fixed point results. *Mathematics* **2018**, *6*, 320. [CrossRef]
5. Mlaiki, N.; Aydi, H.; Souayah, N.; Abdeljawad, T. Controlled metric type spaces and the related contraction principle. *Mathematics* **2018**, *6*, 194. [CrossRef]
6. Branciari, A. A fixed point theorem of Banach-Caccippoli type on a class of generalised metric spaces. *Publ. Math. Debr.* **2000**, *57*, 31–37.
7. Asim, M.; Imdad, M.; Radenović, S. Fixed point results in extended rectangular b-metric spaces with an application. *UPB Sci. Bull. Ser. A* **2019**, *20*, 43–50.
8. George, R.; Radenović, S.; Reshma, K.P.; Shukla, S. Rectangular b-metric space and contraction principles. *J. Nonlinear Sci. Appl.* **2015**, *8*, 1005–1013. [CrossRef]
9. Sarma, I.R.; Rao, J.M.; Rao, S.S. Contractions over generalized metric spaces. *J. Nonlinear Sci. Appl.* **2009**, *2*, 180–182. [CrossRef]
10. Samet, B. Discussion on "A fixed point theorem of Banach-Caccioppoli type on a class of generalized metric spaces" by A. Branciari. *Publ. Math. Debr.* **2010**, *76*, 493–494.
11. Suzuki, T. Generalized metric spaces do not have the compatible topology. *Abstr. Appl. Anal.* **2014**, *2014*. [CrossRef]
12. Jungck, G. Commuting mappings and fixed points. *Am. Math. Mon.* **1976**, *83*, 261–263. [CrossRef]
13. Dung, N.V.; Hang, V.T.L. On relaxations of contraction constants and Caristi's theorem in b-metric spaces. *J. Fixed Point Theory Appl.* **2016**, *1*, 267–284. [CrossRef]
14. Mitrović, Z.D.; Radenović, S. A common fixed point theorem of Jungck in rectangular b-metric spaces. *Acta Math. Hungr.* **2017**, *15*, 401–407. [CrossRef]

© 2019 by the authors. Licensee MDPI, Basel, Switzerland. This article is an open access article distributed under the terms and conditions of the Creative Commons Attribution (CC BY) license (http://creativecommons.org/licenses/by/4.0/).

MDPI
St. Alban-Anlage 66
4052 Basel
Switzerland
Tel. +41 61 683 77 34
Fax +41 61 302 89 18
www.mdpi.com

Axioms Editorial Office
E-mail: axioms@mdpi.com
www.mdpi.com/journal/axioms